Societal and Environmental Ramifications of Plastic Pollution

Nisha Gaur
Gautam Buddha University, India

Eti Sharma
Gautam Buddha University, India

Tuan Anh Nguyen
Vietnam Academy of Science and Technology, Vietnam

Muhammad Bilal
Gdansk University of Technology, Poland

Niranjan Prakash Melkania
Gautam Buddha University, India

IGI Global
Publishing Tomorrow's Research Today

Published in the United States of America by
IGI Global
701 E. Chocolate Avenue
Hershey PA, USA 17033
Tel: 717-533-8845
Fax: 717-533-8661
E-mail: cust@igi-global.com
Web site: https://www.igi-global.com

Library of Congress Cataloging-in-Publication Data

CIP PENDING

ISBN13: 9798369391631
Isbn13Softcover: 9798369391648
EISBN13: 9798369391655

Vice President of Editorial: Melissa Wagner
Managing Editor of Acquisitions: Mikaela Felty
Managing Editor of Book Development: Jocelynn Hessler
Production Manager: Mike Brehm
Cover Design: Phillip Shickler

British Cataloguing in Publication Data
A Cataloguing in Publication record for this book is available from the British Library.

Table of Contents

Detailed Table of Contents

The mismanagement of plastic waste has led to the environmental pollution and bioaccumulation of these plastic particles into living organisms through various food chains and food webs in the ecosystems. The breakdown of plastic into microplastic (MP) and nanoplastic (NP) intensifies its adverse impact on soil fertility, food security and living organisms. MP and NP can enter the soil through various pathways such as plastic mulching, compost and sewage sludge and atmospheric deposition. These particles migrate through various ecosystems. Small sizes limit the efficiency of detection of these particles and thus require urgent, sustainable solutions to mitigate MP and NP pollution. These particles can adsorb harmful chemicals like pesticides, PAHs, PCBs and heavy metals on their surface. Therefore, this chapter discusses the sources, impact and mitigation strategies for MP and NP pollution. Moreover, this chapter enhances understanding of the environmental threats by MP and NP pollution and provides sustainable approaches to mitigate their detrimental effects..

Economy & growth of society depend on access to clean, safe water. However, rapid population increase, industrialization, & urbanization have led to generation of wastewater that pollutes (contaminates) water leading to several health issues. Millions die annually after consuming pathogen infected water. Many limitations restrict the use of various methods for treating wastewater which includes among others - chemical application, formation of disinfection by products (DBPs), time taken and cost incurred. Nanotechnology has achieved significant successes in different areas including water treatment. Nanomaterials are characterized by high surface area to volume ratio, high reactivity/sensitivity/adsorption capacity/ease of functionalization, hence suitable for use in treating waste water. This chapter aims to provide an overview of techniques for wastewater treatment & management based on nanotechnology Moreover, the fate of nanomaterials in wastewater treatment has also been reported and the hazardous effects of their use.

This thorough analysis examines how nanotechnology is transforming a number of industries, with a primary focus on packaging materials. Through the improvement of biodegradable biopolymer-based materials functional performance, nanotechnology has completely transformed the packaging sector. These developments support food safety and sustainability in addition to enhancing packaging qualities. The study also addresses the use of nanotechnology in water filtration systems to lessen the use of plastic bottles, addressing issues related to plastic pollution that affect the environment and human health. It also draws attention to the difficulties and possibilities associated with applying nanotechnology and biodegradable polymers to reduce the amount of plastic trash generated and boost the effectiveness of environmental cleanup procedures. The paper also explores the advantages and disadvantages of nanoplastic particles, highlighting the need for more investigation to fully comprehend their effects on the environment.

This research examines the economic implications of micro- and nano plastic pollution. The study assesses the substantial costs associated with environmental cleanup, healthcare, and biodiversity loss. Economic impacts on sectors such as fisheries, tourism, and agriculture are analyzed. The paper highlights the urgent need for effective policies, technological advancements, and international cooperation to mitigate these costs. By quantifying the economic consequences, this research emphasizes the imperative for immediate action to address plastic pollution and its detrimental effects on economies and societies worldwide.

Micro- and nano-plastic pollution poses a significant environmental and health challenge. This chapter explores the potential of public awareness campaigns and behavior change strategies in mitigating this issue. It delves into the critical role of social media and influencers in fostering sustainable practices. The chapter examines effective communication channels, messaging strategies, and audience segmentation to optimize campaign impact. It also analyzes the influence of influencers on shaping public perceptions and driving behavioral shifts. By understanding the dynamics of social media and influencer marketing, policymakers, environmental organizations, and businesses can develop targeted interventions to reduce plastic pollution and promote sustainable lifestyles. This research contributes to the growing body of knowledge on effective communication strategies for environmental challenges.

Preface

Plastic, once heralded as a revolutionary material, has now become one of the most pervasive pollutants in our environment. Its durability, while beneficial for various applications, turns into a liability when plastic waste breaks down into tiny fragments, known as microplastics and nanoplastics. These particles, smaller than 5 millimeters, carry a cocktail of harmful chemicals that threaten both the ecosystem and human health. Microplastic research, still in its infancy, faces numerous challenges in understanding the complexities of how these particles degrade and spread through the environment. Scientists around the world are diligently studying these issues, but much remains to be done in order to develop sustainable, circular economy solutions that can effectively combat the spread of microplastics and nanoplastics.

It can often feel overwhelming to consider how one person's efforts could make a difference in such a large-scale environmental crisis. However, the collective impact of individual actions can be significant, especially when it comes to raising awareness and promoting informed change. With this book, tentatively titled *Societal and Environmental Ramifications of Plastic Pollution*, we, the editors, aim to contribute to this growing body of knowledge. Our hope is that it will serve as a vital resource to the scientific community as well as to a broader audience invested in addressing the issue of plastic pollution.

This book offers readers a deep dive into the growing problem of micro- and nanoplastics pollution, particularly in terrestrial ecosystems. It provides a holistic analysis of micro/nanoplastics characterization, exploring their sources, pathways, and detection methods in the environment. Additionally, the book delves into the response of plants and soil to microplastic exposure, detailing the physiochemical and biological changes that occur as a result. Remediation strategies, control measures, and policies aimed at mitigating this pollution are covered extensively, alongside insightful case studies.

We also explore emerging concerns, such as how the COVID-19 pandemic has exacerbated microplastic pollution and the troubling rise of multidrug-resistant pathogens linked to plastic waste. Additionally, we focus on the impacts of microplastics in aquatic and aerial environments, highlighting the broader scope of this pollution problem across different ecosystems.

By compiling this wide-ranging analysis, we hope to support environmental researchers, students, ecologists, toxicologists, policymakers, and even concerned non-experts in their pursuit of understanding and combating the adverse effects of micro- and nanoplastic pollution.

Organization of the Book

Chapter 1, "Global Status of Micro/Nano-Plastic Pollution and its Effect on Climate Change," offers a comprehensive examination of the environmental havoc caused by micro and nano-plastics (MNPs). The chapter highlights how MNPs, due to their minuscule size, have the ability to permeate and disrupt various ecosystems, including the atmosphere, hydrosphere, and lithosphere. It explores how these particles travel far from their source and resist degradation, embedding themselves in soil and water systems. The detrimental effects of MNPs on both aquatic and terrestrial ecosystems are discussed, with particular emphasis on their impact on the food chain, from plants to humans. The chapter underscores the urgent need to address the pervasive nature of MNP pollution and its contribution to long-term environmental damage.

Chapter 2, "Microplastics: An Invisible Air Pollutant of the New Era," addresses the emergence of microplastics (MPs) as a critical global environmental and health issue. Originating from industrial sources and the degradation of larger plastic debris, MPs are now present in ambient air and contribute to particulate matter (PM10 and PM2.5). This chapter delves into the inhalation and ingestion risks of MPs, their impact on plant growth, photosynthesis, and genetic expression, and their broader influence on climate change through energy and carbon cycle disruption. It also reviews mitigation strategies aimed at reducing MPs' presence in the atmosphere and their adverse effects on health and the environment.

Chapter 3, "Status of Micro/Nano-Plastic (MNP) Pollution in the Aquatic System," provides an in-depth exploration of MNP contamination in aquatic ecosystems. The chapter traces the origins of MNPs, highlighting their persistence in the environment and their role as carriers for pollutants such as organic chemicals, heavy metals, and bacteria. Techniques for detecting and identifying MNPs, such as Raman spectroscopy and scanning electron microscopy, are reviewed. Additionally, the chapter discusses the adverse health effects MNPs impose on aquatic organisms and the ecosystem at large, while offering potential solutions for eliminating MNPs from water systems.

Chapter 4, "Plant Stress Management Mitigating Micro/Nano Plastics Pollution," explores the agricultural implications of micro and nano plastics. This chapter details how MNPs infiltrate soil ecosystems and disrupt nutrient uptake and oxidative processes in plants, causing significant stress. The chapter highlights mitigation techniques such as bioaugmentation, nanomaterials, and phytoremediation. Specific attention is given to biochar and mustard plants as potential solutions for capturing MNPs in agricultural soils. Case studies demonstrate successful applications of these methods, showcasing their potential in reducing MNP pollution and promoting sustainable farming practices.

Chapter 5, "Soil Physio-Biochemical Alterations Due to Micro/Nano-Plastic Pollution," examines how MNPs fundamentally alter the physical and biochemical properties of soil. This chapter details how MNPs affect soil texture, structure, and water retention, leading to soil erosion and disruptions in hydrological cycles. Additionally, the chapter discusses how MNPs can mobilize harmful contaminants like pesticides and heavy metals, exacerbating food chain contamination risks. The impact of MNPs on soil microbiomes and plant health is also explored, highlighting the far-reaching consequences of their presence on ecosystem functions such as nutrient cycling and plant growth.

Chapter 6, "Strategies to Mitigate Microplastic (MP) and Nanoplastic (NP) Pollution in Soil," delves into sustainable solutions for reducing MP and NP contamination in soil ecosystems. The chapter discusses the various pathways through which MPs and NPs enter the soil, such as plastic mulching and compost, and how these particles interact with harmful chemicals, leading to further environmental degradation. It highlights the need for advanced detection technologies and outlines mitigation strategies such as sustainable waste management practices. This chapter offers practical approaches to combat the detrimental effects of MP and NP pollution, emphasizing environmental protection and food security.

Chapter 7, "Remediation of Micro/Nanoplatic in Wastewater Using Nanotechnology: A Sustainable Approach," focuses on the promise of nanotechnology in addressing global water contamination challenges. As population growth and industrialization increase wastewater production, traditional treatment methods often fall short. This chapter explores the application of nanomaterials in wastewater treatment, detailing their properties—high surface area, reactivity, and adsorption capacity—which make them effective in contaminant removal. It also discusses the environmental fate of these nanomaterials and the potential risks associated with their use, positioning nanotechnology as a critical tool for sustainable water management.

Chapter 8, "Application of Nanotechnology in Food Packaging for Environmental Sustainability," investigates the role of nanotechnology in enhancing the sustainability and functionality of food packaging materials. By improving the performance of biodegradable biopolymer-based materials, nanotechnology supports food safety while reducing plastic waste. This chapter explores how nanotechnology is being used in water filtration to reduce reliance on plastic bottles, a key source of environmental pollution. The benefits and limitations of these materials are analyzed, as well as the broader implications for plastic waste reduction and environmental cleanup, marking a critical step toward more sustainable packaging practices.

Chapter 9, "Economic Impacts of Micro and Nano Plastic Pollution," evaluates the significant economic burden posed by MNP pollution across various sectors, including fisheries, tourism, and agriculture. This chapter quantifies the costs of environmental degradation, health impacts, and biodiversity loss, underscoring the urgent need for coordinated international policies and technological advancements to mitigate these economic repercussions. The research highlights the imperative for immediate action to prevent further harm to economies worldwide, offering a sobering look at the financial consequences of MNP pollution.

Chapter 10, "Scaling Legal Framework for Plastic Pollution and Advancing Cutting Edge Water Governance: Reducing and Eliminating Marine Pollution in Alignment with SDG 14 (Life below Water)," discusses the legal and governance mechanisms necessary to address marine plastic pollution. This chapter explores how marine debris, primarily from terrestrial sources, wreaks havoc on marine ecosystems and examines innovative governance models that can curb this issue. It aligns with Sustainable Development Goal 14 (SDG 14), calling for enhanced legal frameworks and cross-border cooperation to mitigate marine pollution and protect aquatic biodiversity.

Chapter 11, "Mounting Legal Driven Solutions for Plastic Pollution Focusing on Environment and Coastal Management: Eradicating Marine Pollution in Alignment with SDG 14 (Life below Water)," continues the discussion on legal frameworks but focuses on coastal management and environmental protection. This chapter examines the specific sources of marine debris, such as plastic bags and fishing gear, and advocates for scalable solutions to eradicate this pollution. In line with SDG 14, it offers interdisciplinary approaches to water governance and policy assessments aimed at reducing the global water pollution crisis.

Chapter 12, "Optimizing Public Awareness Campaigns and Behavior Change Strategies for Reducing Micro- and Nano Plastic Pollution: The Role of Social Media and Influencers in Fostering Sustainable Practices," explores the role of public engagement in addressing plastic pollution. The chapter highlights the potential of social media and influencers in driving behavior change and promoting sustainable practices. It provides insights into the design of effective awareness campaigns, emphasizing the importance of messaging strategies, audience targeting, and influencer marketing to foster public responsibility and mitigate MNP pollution.

In Conclusion

In conclusion, this edited volume represents a critical and comprehensive examination of the multifaceted challenges posed by micro- and nano-plastic pollution on a global scale. Each chapter provides unique insights into the mechanisms, effects, and potential solutions to this rapidly escalating environmental threat. From the pervasive impact on soil, water, and air ecosystems to the economic, legal, and social dimensions, the contributors have illuminated the far-reaching consequences of micro- and nano-plastics. By emphasizing the need for interdisciplinary approaches, this collection encourages innovation in scientific research, governance, and public engagement, all of which are essential to mitigating plastic pollution's detrimental effects on the environment and human health.

As editors, we have sought to bring together diverse perspectives and methodologies, reflecting the complex and interconnected nature of this issue. The research presented here highlights not only the urgency of addressing the problem but also the opportunities for technological advancements, policy reforms, and global cooperation in confronting plastic pollution. Ultimately, this volume aims to inspire further research, advocacy, and action, recognizing that the fight against micro- and nano-plastic pollution is integral to achieving broader environmental sustainability goals.

We hope that this book will serve as a valuable resource for researchers, policymakers, and practitioners alike, fostering a deeper understanding of the problem and motivating concrete steps toward a more sustainable future.

Chapter 1

Global Status of Micro/Nano-Plastic Pollution and Its Effect on Climate Change

Ayan Sheikh
https://orcid.org/0009-0009-8181-069X
Gautam Buddha University, India

Nisha Gaur
https://orcid.org/0000-0002-8699-6659
Gautam Buddha University, India

Vaishanavi Shrivastava
Gautam Buddha University, India

Shubham Kumar
University of Minnesota, Minneapolis, India

Eti Sharma
Gautam Buddha University, India

Shashank Mittal
O.P. Jindal Global University, India

ABSTRACT

MNPs, by the virtue of their extremely small size are very potent in causing huge destruction to our environment and local ecosystems. MNPs impart a sustaining damage on our atmosphere, hydrosphere, lithosphere, and collectively on our biosphere. They can be easily transported and spread to distant locations from their source of origination, thanks to their small size and light weight characteristics. Once out in the wild, MNPs are very hard to get filtered out from our environment and degraded. MNPs cause a lot of negative impact on our aquatic as well as the soil ecosystem. They hinder the activities of natural microflora of these ecosystems as well as they get into our food chains via plants, fishes, etc. Once incorporated into our food chain, MNPs leave a drastic effect on every trophic level, with most

DOI: 10.4018/979-8-3693-9163-1.ch001

acute and severe effect on the higher trophic levels i.e. tertiary trophic level which includes humans.

INTRODUCTION

Plastics, one of the greatest inventions of mankind, which is used in almost all aspects of modern-day life ranging from food containers and tap water pipes to the clothing industry. In today's scenario Micro/Nano Plastics are widely being used in the drug delivery industry, personal care products and bioimaging. Microplastics are artificial polymer particles with dimensions less than 5mm as defined by the US National Oceanic and Atmospheric Administration (NOAA) and European Chemical Agency, whereas Nano plastics are artificial polymer particles with dimensions ranging from 1 nm to 1 μm. Micro/Nano Plastics are mostly generated in the environment by the weathering and degradation of plastic waste by natural phenomena like sunlight, winds, waves which break down the plastic waste into micro/nano particles, notably the major part of these micro/nano particles comes from the human-driven activities in the form of microbeads used in cosmetics and personal care products, laundering of synthetic clothes, abrasion of tires, wide use of plastic in industries including food, textiles, etc. (Lai et al., 2022)

Once out in the wild, these particles are rapidly carried away by the medium of water bodies, wind, and are found to be widespread all across the biosphere, ranging from the highest peak Mount Everest to the deepest trough Mariana Trench. Surprisingly, these Microplastics have taken over our food chain, and are now known to be existing in blood, lungs and even in feces of several living organisms, including humans too. Since its inception, Plastics have been one of the biggest sources of pollution in our biosphere, thanks to its non-biodegradable property, which helps it reign over our globe for thousands of years. In the past couple of decades, a new form of this plastic pollution has been noticed in the form of micro/nano plastics, which has overtaken nearly everything in our biodome, from oceans to mountains, from plants to microorganisms, even animals and humans (Parker et al., 2019).

Micro/nano plastics are easily ingested or inhaled by living organisms, several scientific studies on animals and humans have shown that these microplastics, once ingested, can spread to liver, spleen, heart, lung, thymus, reproductive organs, and even the brain. They are also reported to act as carriers of heavy metals and persistent organic pollutants (POPs) like poly chlorinated biphenyls (PCBs) (Dube & Okuthe, 2023a) and pesticides carrying them into the food chain. Microplastics which are adsorbed with these heavy metals, and unwanted chemicals, once ingested into our body, can prove to be carcinogenic and detrimental to our lives. The recent global pandemic of COVID-19, apart from affecting human life, has also threatened our environment due to the uncontrolled discard of facemasks and other medical equipment, made up of artificial polymers such as polypropylene, which has worsened the current situation of microplastic pollution (Ziani et al., 2023).

Therefore, this chapter aims to investigate and spread awareness among the masses for the threat posed by micro/nano plastics to our environment, climate and even the living organisms, including humans. This study also provides a thorough examination of the issue, covering the environmental, health and regulation policy aspects, emphasizing more upon the climate change caused due to the increased greenhouse gas emissions, impact on the carbon fixation which will help in setting up the stage for exploring solutions and future research fostering a sustainable environment ahead. The chapter is illustrated by various case-studies, research data gathered over the last 5 years, giving the reader a fruitful insight into the topic.

1. SOURCES AND DISTRIBUTION OF MICRO/NANO-PLASTICS

Plastics are known for their versatile nature. Their properties include thermal, electrical resistance, and resistance to microbes and oxygen, being good for the sustainability of the plastics; these properties could be very harmful for the environment when considering the degradation of the plastics. In 2004 the term "microplastics" appeared for the first time (Song et al., 2024), identifying both granular and fibrous plastic fragments with a dimension of around 20 μm.

1.1 Primary vs Secondary Microplastics

Plastics intentionally produced at the microscopic level are termed as primary microplastics, whereas the small plastic fragments that originate from the disintegration of the plastics, occurring in both marine and terrestrial environments are referred to as secondary microplastics. Every source has a different definition of what microplastics are, according to some primary microplastics are intentionally added to products like exfoliating personal care items and arise due to the wear and tear of larger plastics in the production, some categorize microplastic fibers from washing machine as the secondary microplastics, accurately identifying the sources is crucial (Schmid et al., 2021).

Microplastics are not the end product of the plastic waste, they themselves degrade into nano plastics, which can be defined as plastics ranging in the size between 1 nm to 1 μm, which is different from the criteria of classification of engineered nanomaterials. MNPs follow a complex and webbed path.

Figure 1. Entry and occurrence of MNPs in air, soil, and water

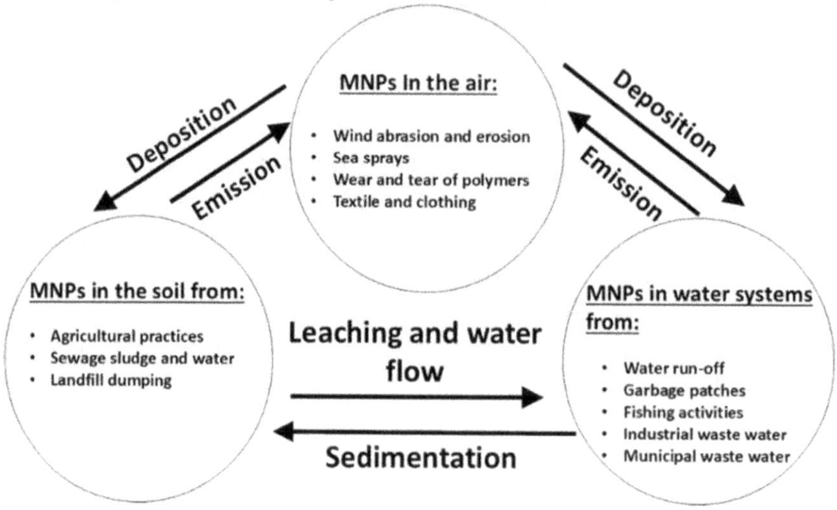

1.1.1 Entry and occurrence of MNPs in Air

Air around the urban areas is most affected by the MPNs, aerosols, dust and atmospheric decomposition all have been reported with MNPs. The source for the entry of the MNPs in air could be plastic wear and tear, tire abrasion, synthetic clothing, farm dust, poorly managed landfills, water incinerations, and more. MNPs because of their small size and lightweight nature are carried over long distances by the wind, their presence in the areas with no such productions confirm their long-distance transportation (Habibi et al., 2022).

MNPs can be released into the air in the following ways: a.) MNPs due to wear of plastic or polymeric materials: Plastics wear due to mechanical friction between surfaces. For example, car tires made of synthetic polymers release wear particles. As a result of mechanical friction, these particles are considered microplastics due to their very small size, physicochemical properties and the presence of synthetic polymers, and these particles can be transported over long distances by wind or carried by water. Using dynamic probabilistic material flow analysis, Sieber et al. concluded that 219 Â± 22 tons of rubber particles had accumulated in the Swiss environment since 1988. b.) MNPs from synthetic fibers and clothing: Artificial hair products such as wigs and hair extensions are made of various synthetic materials (including acrylic, polyester and nylon). These materials release small plastic particles when styled, combed or cut. O'Brien et al. showed when a 660 g blanket was dried in a household dryer for 20 minutes, approximately 1.6-1.8 fibers/m^3

were released into the ambient air, with approximately $1.1 \pm 0.3 \times 10^6$ fibers captured by the built-in filtration device. A study by Chen et al. found fragments of acrylic, rubber and polyurethane (all smaller than 50 micrometers) in nail salon air (Chen et al, 2022). c.) MNPs from sea spray: These particles range in size from 0.3 to 70 micrometers. This finding highlight that the oceans are a major source of plastic pollution in the atmosphere. Strong winds can carry these tiny particles long distances, reaching remote areas over land and sea. A study by Yang et al. found that ocean spray releases a staggering amount of tiny plastic particles, up to 24 billion or 773 tons, each year. d.) MNP due to wind abrasion and erosion: Wind abrasion occurs when coarse particles carried by the wind rub against the surface, destroying the plastic. Agricultural plastic films exposed to ultraviolet light have been shown to readily break down into debris, which eventually form MNP. On the second-hand wind erosion occurs when loose particles including MNPs are moved from one location to another, wind erosion has been shown to facilitate the transport of MNPs from the soil to the atmosphere and aquatic ecosystems (Kiran et al., 2022).

2.1.2. Entry and Occurrence of MNPs in Water

Through a variety of pathways, including runoff, atmospheric deposition, and seepage, MNPs end up in the oceans around the planet, due to sewers and poorly managed landfills), sewage treatment plants, and improper waste disposal. Most low-income countries lack adequate funding. Waste collection and treatment facilities, as well as random dumping of waste, including personal protective equipment for the COVID-19 pandemic. (Wang et al.) found that the middle layer of the mask is susceptible to UV weathering, and a single weathered mask releases more than 1.5 million MNPs. into the water system.

MNP can enter water via the following pathways: a.) MNP from water runoff: road dust, atmospheric degradation, abrasion of road markings, plastics decomposing in open dumps, peeling paint, washed away from agricultural soils that collect in water bodies. The amount of MNP in water poses health problems to marine organisms as well as water consumers, humans and terrestrial plants. Ross et al. estimated that individual urban drains discharge 1.9–9.6 billion MP/kg of precipitation into water bodies, confirming that aquatic runoff is the major pathway for MNP to move from land to aquatic environments. b.) MNP from industrial and municipal waste: Industrial waste from the polymer and textile industries and municipal waste find their fate in water bodies through seepage, discharge into streams or direct discharge into the ocean (Ross et al., 2023). Chan et al. showed that industrial wastewater from a textile factory in China contained fibrous MNPs (361.6 ± 24.5 MNP L^{-1}) (Chan et al. 2021). c.) MNPs from fishing activities: Most fishing gear is made of different types of plastics, including polyethylene (PE), polypropylene (PP), and nylon (PA).

These plastics break down into small plastic particles called microplastics (MPPs) as the gear is used. For example, ropes used in Danish seine nets can be dragged long distances along the seabed, which causes friction and rope wear, releasing microplastics into the water. Syversenet et al. estimated that between 77 and 97 tons of plastic are dumped into Norwegian waters each year due to wear and tear from Danish fishing. The entire Norwegian fishing industry is estimated to release around 208 tons of MNP per year (Kuczenski et al., 2021), with an estimated 45,000 tons of plastic pollution annually from lost fishing gear. Wright et al. reported that lost or abandoned fishing gear on beaches in the south-west of England (a region with high fishing debris density) could release around 1277 $\pm$ 431 MNP m^{-1} fragments. d.) MNP from litter patches: Debris such as plastics, fishing nets and gear accumulate on the sea surface, decompose and contribute to MNP accumulation. These patches are expected to become increasingly larger everywhere, especially in marine environments, due to their deterioration. Zhao et al. demonstrate the presence of MNPs in the subtropical gyre of the eastern North Pacific. The accumulation of coastal debris patches reported in Ashmore Reef Marine Park ("Pulau Pasir"), Australia, amounts to 5,161 g km^{-2} (Mahesh et al., 2023).

2.1.3. Entry and Occurrence of MNPs in Soil

Developing countries like India are heavily dependent on agriculture and with a large portion of the population dependent on agriculture, land and soil quality are critical resources. The harsh truth is that soils are now becoming reservoirs of plastic contamination. MNP can enter the soil through the following pathways: a.) MNP from landfills: Land is considered as landfills globally and is a major reservoir for plastic waste, with 49% of the plastic waste generated globally being disposed of in these landfills. Mahesh et al. reported the presence of MNP (ranging from 180 to 1120 MNP particles per kg soil) in urban open dumps. Afrin et al. MPs from LDPE, HDPE and cellulose acetate (CA) were observed in landfills in Bangladesh. These MNPs can be transported by substances such as air, water or other organisms. Earthworms have been reported to ingest MNPs from the soil surface and excrete them deeper into the soil. b.) MNPs from sewage sludge: Industrial plants discharge treated water along with municipal effluent and domestic wastewater, which contribute to the formation of MNPs in soil. Approximately 90% of MNP remains in the sludge. Di Bella et al. identified the presence of various plastics (fragments and fibers of PE, PET, PP, polybutadiene (PB), and PES) in waste sludge (Di Bella et al., 2022). Harley-Nyang et al. reported that 1.02×10^{10} - 1.61×10^{10} MNP remain in the sludge each month in a wastewater treatment plant in the UK. c.) MNP from agricultural practices: Slow-release fertilizers, pesticides and seed coatings encapsulated in plastic capsules are a major source of MNP in agricultural soils, as they remain in the soil even after extensive irrigation. Irrigation with MNP-contaminated water and wastewater treatment plant sludge also increases MNP in soils. A study by van Schothorst et al. showed that compost samples from some municipal organic waste in the Netherlands contained 2800 ± 616 MPs kg^{-1} (van Schothorst et al., 2021). In China, 2400 ± 358 MPs kg^{-1} (mainly fibers and films) was also reported in compost. Plastic materials used in various agricultural practices, such as plastic mulching films, greenhouse films, plastic drip irrigation, screens, and irrigation pipes, degrade into secondary MNPs in the soil. Li et al. found that after 30 years of plastic mulching, MPs ranging from 7,183 particles/kg to 10,586 particles/kg accumulated in the topsoil (0-10 cm), and plastic mulching accounted for 33-56% of the total MPs (Amobonye et al.., 2021)

Table 1. Distinguishing between microplastics and nanoplastics

	MPs	**NPs**	**References**
Definition	**No legal definition**		
Production	Produced in micro scale for certain purposes	Unintentional	
Shape	Fibers, films, fragments, pellets.	Variety of shapes	(Allouzi et al., 2021)
Size	Medium: 1–5 mm	WG-GES classification <2 mm	
	Small: 25 μm–1 mm	Nano materials general definition <100 nm	
Sources:			
Primary	Plastics produced at micro sized scale for industrial use	Presence of microbeads in packaging of personal care products such as scrubs and shampoos	(Allouzi et al., 2021)
Secondary	Weathering of larger plastic particles due to environmental conditions	Degradation of aged MPs	

3. ENVIRONMENTAL IMPACT OF MICRO/NANO PLASTICS

With the uncontrolled release of plastic waste in our surroundings, the concentration of micro/nano plastics has skyrocketed in our lithosphere, hydrosphere and the atmosphere, which has become a worldwide concern in aspects of environmental sustainability. Alarmingly, these micro/nano plastics have creeped down into our agricultural soil, through multiple pathways. Once in the soil, these micro/nano plastics are assimilated into the crops and plants, reducing the crop biomass and causing adverse effects like increased oxidative stress (Su et al., 2022).

3.1.1 Transport and Fate

The small size and light weight of MNPs are a boosting factor for their transport and spread all across the world, by the means of atmosphere. MNPs can stay suspended in our atmosphere for longer periods of time, and therefore can be transported over long distances from their point of origin by strong air currents. MNPs from the terrestrial landscape and from the ocean and seas are injected into the atmosphere by various ways, for example the formation of sea spray aerosols results in ejection of seawater containing MNPs into the atmosphere. Once carried away by the atmosphere, these MNPs can be deposited over large terrestrial/aquatic ecosystems by wet/dry deposition. These deposition processes depend upon the meteorological conditions (like wind direction and speed, air pressure, precipitation and temperature) and topographic conditions (like elevation, plateau, surface features of landforms).

Figure 2. Schematic depicting various source/sink and migration patterns of atmospheric MNPs

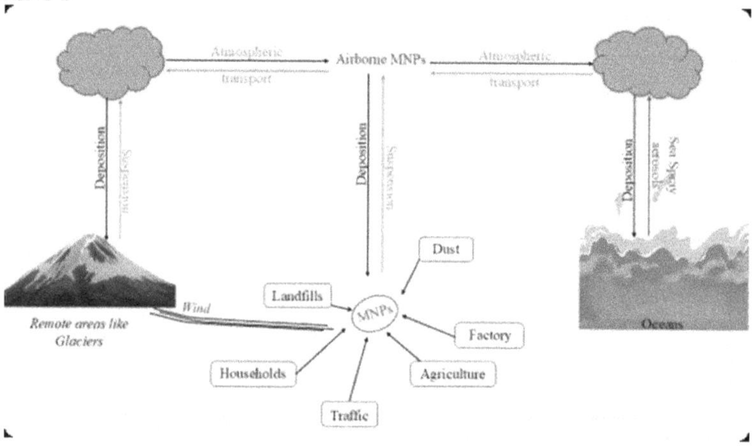

3.1.2 Impact on Water

The aquatic ecosystem, bustling with life, faces a serious threat from MNPs. These particles upset the natural balance of ecosystems, causing big problems such as increasing the potential toxicity of the marine habitat making it unfit for the aquatic flora and fauna, impacting the metabolic activities of marine life and interfering with the important life processes like growth, reproduction, immune system. MNPs are widely transported all across our aquatic ecosystem, from various sources like sewage treatment plants, agricultural activities, and surface water runoff, eventually reaching groundwater and surface water bodies. Soil acts as a large source and sink of MNPs, with agricultural practices and compost contributing to their accumulation. MNPs are transported through surface water, driven by hydrological dynamics and wind patterns, with ocean currents dispersing them across vast marine expanses. Biological transport, including ingestion by zooplankton and attachment to floating debris, plays a crucial role in MNPs' dissemination and potential bioaccumulation in aquatic organisms. Additionally, factors such as temperature, salinity, and water chemistry influence MNPs interactions and transport pathways within aquatic ecosystems. Biofouling of debris enhances MNPs transportability, particularly along coastal regions, where they can accumulate and impact marine ecosystems. A complete and thorough understanding of how these MNPs move through our aquatic ecosystem, will open the doors for future investigations and yield valuable insights for us (Habumugisha et al., 2024).

Figure 3. Presence and interactions of MNPs with aquatic species and the toxicological effect of MNPs in aquatic organisms

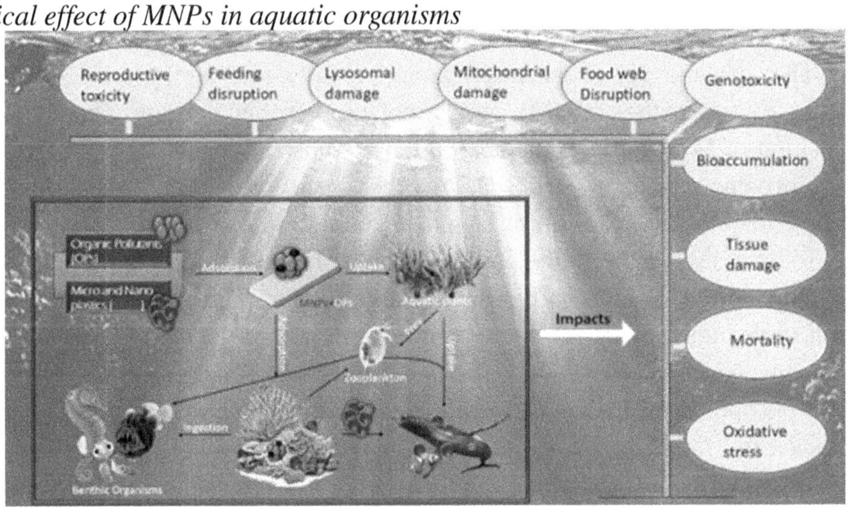

The toxicological effects of MNPs on aquatic organisms vary with particle size (MNPs become increasingly toxic with decreasing dimensions of the particles), influencing interactions with pollutants and organisms by the virtue of their minute dimensions and high surface to volume ratio. MNPs can disrupt algal growth and photosynthesis, increase oxidative stress, and damage cellular structures. MNPs can alter biofilm formation and bind to biomolecules, forming an eco-corona that affects toxicity. Filter feeders and consumers face risks from MNPs' ingestion and bioaccumulation, leading to oxidative stress and disruptions in growth and reproduction. MNPs can also bio magnify in the food web, causing various negative effects on organisms, from inflammation to liver damage. MNPs can accumulate in vital organs, affecting gene expression and enzyme function. Even embryos can accumulate MNPs, impacting growth and organ development. Therefore, understanding these effects is crucial for mitigating MNP pollution in aquatic ecosystems (Russo et al., 2023).

3.1.3 Impact on Land

With the exponentially increasing use of agricultural equipment (e.g. mulching nets, irrigation systems, polystyrene containers, and greenhouse cover films), synthetic fertilizers and chemicals, and reuse of sewage sludge, brings in thousands of tons of plastics into our cultivated soil leading to wide scale soil contamination. Once these MNPs inhabit our soil, they terribly impact the fertility of the soil, they are taken up by the plants and crops from where these MNPs enter into our food chain, causing toxic effects to the wildlife.

Soil contamination by MNPs is a growing concern as they are known to mix with soil aggregates, persist in the soil, and potentially contaminate groundwater and plants. They can alter soil properties (texture, porosity, moisture, and pH), increase dissolved organic matter (DOM), and stimulate microbial activity. The chemical composition of MNPs affects the adsorption of substances such as heavy metals, thereby influencing soil properties and microbial activity. MNPs act as a substrate for microbial colonization, forming a unique community known as the plastisphere, and in some cases, the presence of MNPs can completely suppress the microbial community. BNPs can affect soil respiration and negatively affect enzymes involved in cellulose degradation such as 'β-D-glucosidases' and 'β-cellobiosidases', as well as enzymes involved in the chitin-phosphorus cycle such as β-N-acetyl-β-glucosaminidase and phosphatases. Soil biota, including microorganisms such as '*Paenibacillus pseudomonas*', '*Bacillus, Brevibacillus*', 'cellulosimicrobium', 'Lysinibacillus', and '*Aspergillus flavus*', play an important role in the transformation and degradation of BNP. MNP can affect crop growth, and the effect depends on the concentration of MNP and the diversity of the microbial community. Recent studies show MNP penetration through root systems, affecting plant tissues. MNPs can enter via cracks, root damages, and incomplete Casparian strips. Higher transpiration rates promote absorption, with leaves transferring particles to roots. The effects caused by these MNP particles depend upon polymer type, size, dose, and shape. MNPs can hinder seed germination, root elongation, and nutrient absorption, inducing oxidative stress and cytotoxicity. They are also known to affect photosynthetic organisms by altering photosynthetic processes and pollutant interactions. Plastic pollution in the form of MNPs disrupts soil-plant relationships, affecting water use efficiency and ecosystem health. MNPs can enter the animal body through ingestion and interact with molecules present in the gastrointestinal tract. They can permeate the placenta and internalize into cells, leading to inflammation, oxidative stress, and organ damage. MNPs may carry harmful chemicals, posing risks to animal health.

Figure 4. The interactions between micro and nano plastics with terrestrial photosynthetic organisms

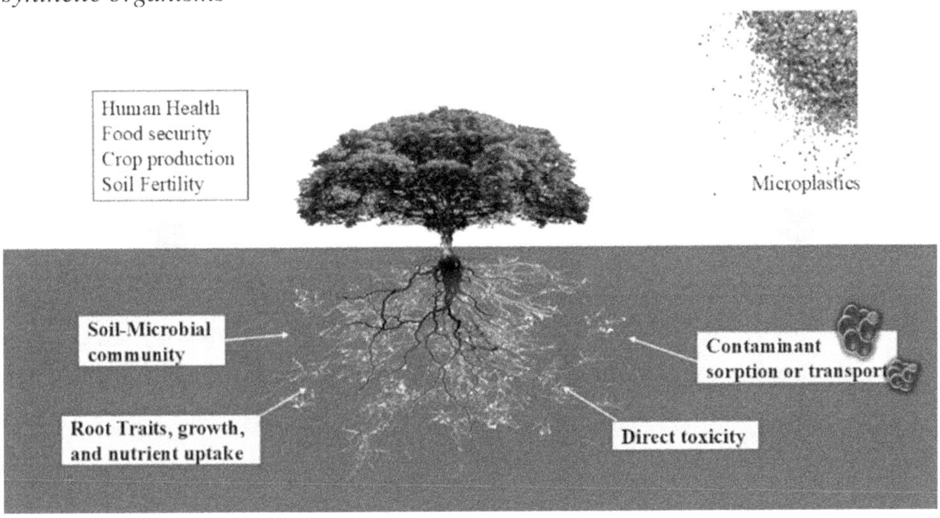

4. HUMAN HEALTH IMPLICATIONS OF MICRO/NANOPLASTICS

Plastic use is accelerating every year, with current figures indicating that plastic production in 2019 exceeded 368 million tons (Yee et al., 2021a). While there are many studies reporting the estimated environmental impacts of MNPs, studies on their effects on humans at the subcellular and molecular levels are limited. In particular, the potential for nano plastics to migrate through the gut, lungs, and skin epithelium, leading to systemic exposure, have not been thoroughly studied. This section will answer many questions regarding how MNPs enter the human body, the possible risks to human health, and gaps in research and knowledge, as well as provide suggestions for additional future research (Dube & Okuthe, 2023b). Humans are susceptible to exposure to MNP through three main routes: ingestion (due to the presence of MNP in food sources, plastic beverage bottles, etc.), inhalation (due to the presence of MNP in the air), and dermal exposure (due to the presence of microbeads used in personal care products). Uncontrolled release of plastic waste into the environment has significantly increased the bioavailability and ubiquity of

MNP in both terrestrial and aquatic environments. Recent studies have shown that these MNP enter our food chain through various routes such as aquatic sources (MNP is present in tap water and fresh water), aquatic foods (fish, shrimp, mussels, and many other edible aquatic organisms are known to contain MNP). MNP is known to be present in fruits and vegetables (MNP is mainly found in crops such as apples, carrots, rice, wheat, corn, and soybeans), processed foods (MNP is widely found in processed seafood, canned foods, salt, honey, milk, beer, and other beverages), food containers and packaging, plastic containers, tea bags, baby bottles, and nipples.

Table 2. Quantity of MNPs found in food items

Type of foods	Detection rate (%)	Abundance (items/kg)	Type of plastic	Size (μm)	Reference
Fresh food					
Cucumbers	14.3	/	PS	0.7	(Z. Li et al., 2021)
Broccoli	100	126150 ± 80715	/	2.1	(Oliveri Conti et al., 2020)
Lettuce	100	50550 ± 25011	/	2.52	
Carrot	100	101950 ± 44368	/	1.51	
Apples	100	195500 ± 128687	/	2.17	
Pears	100	189550 ± 105558	/	1.99	
Processed food					
Sugars	100	343.7 ± 32.08	HDPE	<300	(Afrin et al., 2022)
Store-bought rice	100	52 ± 5.0 (μg/kg)	PE, PP, others	/	(Dessì et al., 2021)
Commercial seaweed nori	95.8	1800 ± 700	PET, PP, PA, others	0.11–4.94 mm	(Q. Li et al., 2020)
Beers	88.5	6.61 ± 2.22	/	<1 mm	(Shruti et al., 2020)
Cold tea	100	3.67 ± 1.75	/	<1 mm	
Beers	93.3	37.14 ± 23.67	HDPE, PPA, PP	13.45–202.29	(Diaz-Basantes et al., 2020)
Refreshing beverage	100	63.71 ± 32.45	HDPE, PPA, PP	15.64–247.54	
Honey	100	104.67 ± 41.74	HDPE, PPA, PP	5.63–226.01	

4.1. Uptake and Bioaccumulation of MNPs in the Human Body

MNPs are up taken and bio accumulated into the human body by three major exposure pathways as following: a) ***Gastric Exposure:*** Experimental studies combined with specific research models have confirmed us with the result indicating that MNPs are being regularly consumed by humans. These MNPs enter into our Gastrointestinal tract through the lymphatic tissues, mainly by the process of phagocytosis or endocytosis. Nano plastics pose a more significant risk of getting absorbed into the GI tract. Once absorbed, these Nano plastics undergo a plethora of transformations, interacting with several biomolecules, especially proteins. Most of the studies were based upon the polystyrene nano plastics and their interaction with proteins. It has been significantly observed that nano plastics are enclosed by a collection or 'corona' of proteins, where polystyrene nano plastics have been found to develop into varying forms of complex coronas, according to the conditions they are in. b) ***Pulmonary Exposure:*** Inhalation of airborne MNPs is the second pathway for human exposure. These airborne MNPs are present both in our indoor and outdoor environments. These MNPs in our atmosphere are unintentionally inhaled by the human population on a regularly. The human lungs have a very intricate and delicate structure, with the total alveolar surface spanning up to 150m2 and the tissue barrier width being as thin as up to 1um. The large alveolar surface promotes larger absorption exposure, even for MNPs, and the thin tissue barrier allows nanoparticles to get transferred into our capillary system, which in turn disperses them all over our human body. The absorption of MNPs through our alveolar surface poses a fatal threat to the human body, in the form of particle and chemical toxicity, and introduction of pathogens and parasites which use these MNPs as a vector. These particles can also get permanently embedded in our lungs, causing damage and failure, or they may also migrate from our lungs to other body parts, having adverse effects on other organs. There are a number of factors such as particle size, hydrophobicity, surface charge, surface functionalization and surrounding protein coronas that determine the rate and intensity of absorption of these MNPs.

c) ***Dermal Exposure:*** Some recent studies have revealed that MNPs are very likely to permeate through the skin barrier and enter our human body. The increased use of nanoparticles especially in the body and facial scrubs used topically over the skin, nanocarriers for drug delivery through dermal application are all major exposure routes for MNPs to enter through our skin. The stratum corneum, outermost layer of our skin, play the role of a barrier against injuries, chemicals, microbial agents and therefore it is very unlikely it easily allows the absorption of nanoparticles through dermal exposure as predicted by latest researches, but these plastic particles can enter into our body via sweat glands, skin wounds and hair follicles. Urea, glycerol and alpha hydroxy acids which are widely used as an ingredient in body lotions are known to amplify the permeation of nano plastics through the skin barrier.

Although all the above studies indicate the absorption of MNPs through dermal exposure routes, these studies are solely based upon the polystyrene particle models. Hence further studies incorporating more types of particle models will help us completely understand and validate the theory behind permeation of MNPs through skin barrier (Dube & Okuthe, 2023b).

4.2. Cellular Uptake and Intracellular Fate of MNPs

Once these MNPs penetrate the protective coating of the stratum corneum and are absorbed into the body, they can interact with a wide range of target cells. The amount of nanoparticles that interact with target cells depends on several factors, including size, surface chemistry, and charge of biological components (e.g., proteins, phospholipids, carbohydrates). MNPs are taken up into cells via a variety of pathways, mainly endocytosis, including phagocytosis, micropinocytosis, clathrin caveolae-mediated endocytosis. Inhibition studies performed on human colonic fibroblasts and bovine oviductal epithelial cells have shown that these cells take up nanoparticles via a clathrin-independent mechanism. In tumor cell lines such as HeLa, 1321N1, and A549, nanoparticles are taken up through an active process that involves actin depolymerization or inhibition of clathrin- or caveolae-mediated endocytosis. Furthermore, cells have been shown to select very different endocytic pathways to take up different types of nanoparticles, highlighting the diverse pathways and cellular responses associated with nanoparticle uptake and the complexity of their interactions with biological systems (Dube & Okuthe, 2023b).

4.3. Detrimental Effects of Micro-/Nanoplastic Pollution

Nanoparticles like PSNPs induce genotoxicity in cells, causing genomic instabilities and alterations that manifest in the form of diseases such as cancer in humans. Studies on human lung cells and fish (Carassius auratus) indicate MNPs can enter nuclei, causing DNA alterations. Intake of MNPs causes oxidative stress in cells of various organ systems such as a. respiratory system, causing lung fibrosis and induced pulmonary fibrosis injury, b. digestive system, causing intestinal barrier dysfunction, metabolic disorders, immune response and inflammation, c. urinary system, causing mitochondrial dysfunction, endoplasmic reticulum stress, inflammation, and autophagy in kidney cells.

Figure 5. Potential health risks of micro- and nano-plastics to nine human organ systems

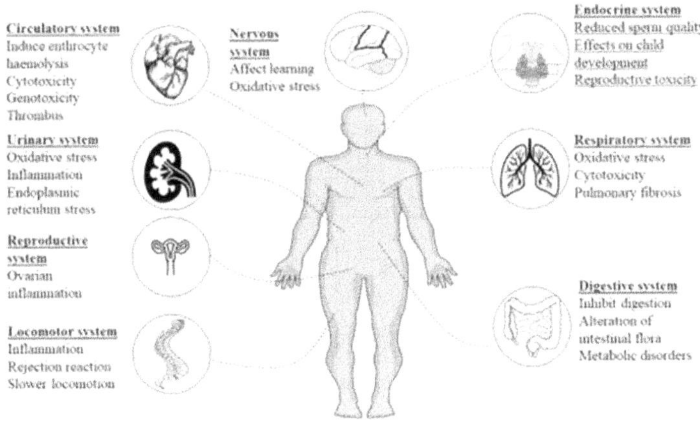

In addition, MNPs can cross the blood-brain barrier and accumulate in the brain, affecting microglia activation and neuronal function. They inhibit acetylcholinesterase activity, alter neurotransmitter levels, and induce neurotoxic effects in animal models, including developmental abnormalities and apoptosis in neurons, disrupting the neurological processes. MNPs' ability to adsorb persistent organic pollutants (POPs) exacerbates environmental health concerns, potentially contributing to rising cancer rates, including childhood cancer, and inducing inflammatory and metabolic disorders in offspring *(Yang et al., 2023).*

Table 3. Toxic effects of MNPs on human health

Toxic Effects	Characteristics of Plastic Particles	Details	References
Inflammation	Polystyrene particles	Upregulation of IL-8 expression. Induced inflammation in human A549 lung cells.	(Yee et al., 2021b)
	Unaltered/carboxylated polystyrene nanoparticles	Upregulation of IL-6 and IL-8 expression. Enhanced inflammation in multiple human malignancies.	
	Carboxylated and amino-modified polystyrene particles	Altered expression of scavenger receptors. M2 cells increased IL-10 production.	
	Unaltered polyethylene particles	Increased the secretion of IL-6, IL-1β, and TNFα in murine macrophages.	
	Polyethylene particles from plastic prosthetic implants	Induced the expression of TNFα, IL-1, and RANKL. Resulted in periprosthetic bone resorption.	
		Induced inflammatory response at the implant area.	
	Polystyrene microplastics particles	Induced inflammation in the liver. Induced adverse effects on neurotransmission.	
Oxidative stress and apoptosis	Amine-modified polystyrene nanoparticles	Strong interaction and aggregation with mucin.	
	Cationic polystyrene nanoparticles	Induced ROS generation and ER stress	
	Unaltered or functionalized polystyrene	Induced apoptosis of several human cell types.	
	polyvinyl chloride (PVC) and poly (methyl methacrylate) (PMMA)	Reduced cell viability with a reduction of ATP and increase of ROS concentrations.	
Metabolic homeostasis	Pristine and fluorescent polystyrene microplastics	Induced gut microbiota dysbiosis and intestinal barrier dysfunction.	
	Anionic carboxylated polystyrene nanoparticles	Altered ion channel function and ionic homeostasis Activated basolateral K^+ channels. Induced Cl^- and HCO^{3-} ion efflux.	
	Polystyrene nanoparticles	Blocked vesicle transport and the distribution of cytokinesis-associated proteins.	

5. REGULATION AND POLICY RESPONSES

Micro/nanoplastic pollution is an urgent problem due to the increased release and persistence of microplastics in the natural environment and the risks to human health and the environment. The global plastic production and the transboundary nature of plastic pollution makes it extremely difficult for individual countries to take mitigation and prevention measures.

5.1. International Policies and Agreements

Over the past decade, the United Nations Environment Assembly (UNEA) came up with a number of UNEA resolutions namely 1/6, 2/11, 3/7, and the latest one being 4/6.The UNEA resolution 4/6 introduced in 2019, came up with a significant positive impact as it enabled the member nations to work closely with the private sectors in order to: a. phase out products containing MNP; b. Innovative product design to reduce secondary MNP emissions; c. Improve MNE waste management; d. Prevent loss of plastic granules, flakes and powder throughout production and supply chain.

A similar kind of effort is being carried out by EUROqCHARM a research project funded by the European Union (EU). There are a handful of other International Agencies such as: a) Science Advice for policy by European Academies (SAPEA), which provides an overview of the current state of knowledge regarding MNP pollution. b) Food and Agriculture Organization (FAO), which has laid down scientific evidence regarding the adverse effects of MNP on the food chain, and the possible policy response to address this issue. c) The World Health Organization (WHO) is currently working on control of MNP pollution via drinking water pathways. d) Organization for Economic Cooperation and Development (OECD) published an extensive collection of policies to mitigate MNP release from tire wear and textiles. e) The International Union for Conservation of Nature (IUCN), the Centre for International Environment Laws (CIEL), Minderoo Foundation have funded and produced relevant policy reports

5.2. National and Local Regulations

National Action Plans against MNP pollution have been seen to neglect/lack at several key measures. Microbeads ban is the most common measure which nearly every nation implemented, but Microbeads form only a small fraction of MNP emission as compared to other sources. Reports have also shown that higher income countries tend to address this issue with more seriousness as compared to middle and lower income countries, which can be attributed to the factors such as capacity to monitor, map, access to relevant information, etc.

Countries like South Korea, China, Malaysia, Norway and the state of California have started their independent marine monitoring and policy action plans to control the MNP pollution. The Plastic Waste Amendment Rules 2021, issued by the Indian Government has permitted single-use plastics to prevent the uncontrolled release of plastics in the environment.

5.3. Effectiveness of Current Policies

Policies aimed at controlling MNP pollution have undergone significant changes over the past five years, moving from knowledge generation to policy implementation in some countries. The positive outcomes are evident in the growing public pressure and political will to accelerate the development of measures to control MNP pollution. As policy development is still in its early stages, there is much work to be done and implemented by international organizations, governments, and the public. Source identification, accurate monitoring and data collection, and knowledge of potential health risks must be carried out on a standardized scale to ensure that all countries can make equal efforts in policy development.

Another important conclusion that can be drawn from the current scenario is that while reducing MNP pollution without systemic changes and sustainability is proving to be more costly and technically unfeasible, systemic changes to reduce MNP emissions instead of using filters and treatment efforts are much more targeted, feasible, and cost-effective (Mohammed et al., 2022). The presence of plastic materials in waterways poses a serious environmental problem due to their low biodegradability. The ability of MP to adsorb other toxic pollutants present in the aquatic environment may cause dangerous diseases to humans and aquatic organisms. Drawing from the overview of current policies and measures, various other potential measures can be listed out as follows: For Plastic Pellets, Flakes and Powder: Measures can be taken to overlook trade restrictions, transport and storage requirement, BAT/BEP for raw materials as before production, manufacturing and recycling. Intentionally added Primary MNPs: The action plans should include measures on product design, essential use, advancements of wastewater systems, bans and restrictions, BAT/BEP for large scale applications. Use-Phase Secondary Microplastics: Similar to the measures mentioned in case of Primary MNPs, policies can be focused on product design, industry standards, bans on products prone to leakage, BAT/BEP. Degradation Based Secondary Microplastics: Measures can be taken to prevent generation of waste, improving design and waste collection systems, ESM disposal, targeted litter cleanups, BAT/BEP for mechanical recycling (Karmakar, 2023).

6. TECHNOLOGICAL AND INNOVATIVE SOLUTIONS

a) **MNP removal from WWTPs:** Wastewater treatment plants remove approximately 95% of microplastics (MPs), primarily through screens (6-150 mm) and secondary clarifiers. Smaller MPs (100-300 µm) are further eliminated via biological processes. Advanced tertiary treatments achieve an additional 5% reduction. MPs, mainly plastic microfibers, accumulate in sludge, with potential environmental release during sludge land application despite high overall removal efficiencies. b) **Coagulation and Flocculation:** Coagulation and flocculation are effective methods for microplastic (MP) removal in wastewater, influenced by coagulant type, pH, and wastewater composition. Electrocoagulation shows promise with high removal efficiency and less sludge. Aluminum-based coagulants excel in MP removal, enhanced by polyacrylamide. Innovative techniques like agglomeration are explored. Inorganic coagulants like ferric chloride achieve over 99% MP removal, suitable for tertiary treatment in wastewater plants. Challenges include sludge disposal and cost. c) **Membrane, Biological & Filtration Technologies:** Various membrane technologies such as microfiltration, ultrafiltration, reverse osmosis, and membrane bioreactors are utilized for microplastic (MP) removal from wastewater. Biological processes involving microorganisms are effective but face challenges like fouling. Membrane technologies, like MBRs and ultrafiltration, achieve high removal efficiencies (up to 99.9%) but may require combination with other treatments for optimal results. d) **Advanced Oxidation Process:** Photocatalytic oxidation breaks down polymeric chains in microplastics using semiconductors like ZnO and TiO_2, activated by sunlight or artificial light. Salinity and surface area affect the efficiency of the oxidation process. Use of carbon nanotubes and N-TiO_2 are promising innovations in this field.. e) **Adsorption Technologies:** Bioinspired silica gels effectively adsorb MPs due to surface characteristics and chemical interactions. Biochar from pine and spruce bark shows strong retention for large MPs but less for smaller particles. Graphene oxide also demonstrates high adsorption capacity for MPs, influenced by various factors. f) **Biodegradation of MNPs:** Microbial enzymes, including hydrolases and oxidases, degrade microplastic particles (MNPs) by breaking down polymers into assimilable monomers. Bacteria like *Alcanivorax borkumensis* and Bacillus species, along with fungi such as *Zalerion maritimum*, utilize enzymatic and biosurfactant mechanisms for effective MNP degradation. Enzyme engineering in Bacillus subtilis enhances PETase activity, boosting MNP breakdown efficiency. g) **Microalgae-MNP Interaction:** Microalgae are utilized for their dual role in sustainable biofuel production and microplastic (MNP) removal from water bodies. They interact with MNPs, accumulating them within cells or on surfaces, facilitating their removal during harvesting and downstream processing. This approach aims to mitigate environmental impacts by using microalgae as bio-scavengers for MNPs.

(Abomohra & Hanelt, 2022). h) **Micro/Nano Robots for removal of MNPs:** Micro/ nanorobots (MNRs) utilize various energy sources like light, electricity, and magnetic fields to efficiently remove micro/nano plastics (MNPs) from water. Photocatalytic MNRs driven by self-electrophoresis and bubble propulsion, magnetic microrobots guided by magnetic fields, and ion-exchange-based MNRs with intelligent assembly capabilities demonstrate effective MNP removal strategies. (W. Li et al., 2023).

7. SUSTAINABLE ALTERNATIVE TO PLASTICS

Finding sustainable alternatives to microplastics is crucial for reducing their impact on the environment and human health. Here are some promising solutions: a. Biodegradable Plastics or Bioplastics (BPLs), biodegradable plastics are plastics that can be decomposed by the action of living organisms, usually microbes, into water, carbon dioxide, and Biomass Researchers are working on biodegradable plastics that break down more efficiently. b. Natural Fiber Textiles, fabrics made from natural fibers like cotton, linen, and hemp do not release microplastics and can be biodegradable under the right conditions. Innovative textile technologies, such as closed-loop recycling systems, also help reduce the fashion-industry's environmental impact (Connected, 2020). c. Edible and Water-Soluble Packaging, some companies are exploring edible and water-soluble packaging materials as alternatives to traditional plastics. These materials break down harmlessly and do not contribute to microplastic pollution. d. Glass or Stainless-Steel Containers, instead of using plastic water bottles, opting for glass or stainless-steel containers can significantly reduce the consumption and disposal of single use plastics. Materials like steel and glass are durable, recyclable, and less likely to release microplastics (Laurel, 2024).

CONCLUSION

The increase of microplastics is an overall global issue, which ultimately contributes to the environmental and human health on a scale hardly known in the history of civilization. Because these little plastic particles are not usually able to be seen by human eyesight, they have invaded every single part of the planet, completing their accretion in oceans, rivers, soil, and even in the air that is breathed. As the chapter discusses, the implications of this contamination are immense, with discussion about such issues as the more obvious ones, like pollution, to less obvious impacts on ecosystems, biodiversity, and human populations.

The environmental concern due to microplastics is alarming. Marine and terrestrial organisms at all levels of ecosystems ingest these particles, mainly because they often mistake such particles for food. This results in physical damage in these organisms and introduces toxic chemicals into the food chain. The paper referred to studies that claimed microplastics are capable of carrying hazardous substances emanating from chemicals known as POPs, for Persistent Organic Pollutants, bioaccumulating in the tissues of wildlife and perhaps entering the human diet by the consumption of seafood. That has raised long-term serious health concerns for humans due to their direct exposure to ingested microplastic particles, including disruption to the endocrine system, among other biological processes.

REFERENCES

Abomohra, A., & Hanelt, D. (2022). Recent Advances in Micro-/Nanoplastic (MNPs) Removal by Microalgae and Possible Integrated Routes of Energy Recovery. *Microorganisms*, 10(12), 12. Advance online publication. DOI: 10.3390/microorganisms10122400

Afrin, S., Rahman, M., Hossain, M. N., Uddin, M. K., & Malafaia, G. (2022). Are there plastic particles in my sugar? A pioneering study on the characterization of microplastics in commercial sugars and risk assessment. *The Science of the Total Environment*, 837, 155849. DOI: 10.1016/j.scitotenv.2022.155849

Allouzi, M. M. A., Tang, D. Y. Y., Chew, K. W., Rinklebe, J., Bolan, N., Allouzi, S. M. A., & Show, P. L. (2021). Micro (nano) plastic pollution: The ecological influence on soil-plant system and human health. *The Science of the Total Environment*, 788, 147815. DOI: 10.1016/j.scitotenv.2021.147815

Amobonye, A., Bhagwat, P., Raveendran, S., Singh, S., & Pillai, S. (2021). Environmental Impacts of Microplastics and Nanoplastics: A Current Overview. *Frontiers in Microbiology*, 12, 768297. Advance online publication. DOI: 10.3389/fmicb.2021.768297

Chan, C. K. M., Park, C., Chan, K. M., Mak, D. C. W., Fang, J. K. H., & Mitrano, D. M. (2021). Microplastic fibre releases from industrial wastewater effluent: A textile wet-processing mill in China. *Environmental Chemistry*, 18(3), 93–100. DOI: 10.1071/EN20143

Chen, E.-Y., Lin, K.-T., Jung, C.-C., Chang, C.-L., & Chen, C.-Y. (2022). Characteristics and influencing factors of airborne microplastics in nail salons. *The Science of the Total Environment*, 806(Pt 4), 151472. DOI: 10.1016/j.scitotenv.2021.151472

Connected, S. (2020, March 25). Will We Ever Find a Microplastics Alternative? *Science Connected Magazine*. https://magazine.scienceconnected.org/2020/03/will-we-ever-find-a-microplastics-alternative/

Dessì, C., Okoffo, E. D., O'Brien, J. W., Gallen, M., Samanipour, S., Kaserzon, S., Rauert, C., Wang, X., & Thomas, K. V. (2021). Plastics contamination of store-bought rice. *Journal of Hazardous Materials*, 416, 125778. DOI: 10.1016/j.jhazmat.2021.125778

Di Bella, G., Corsino, S. F., De Marines, F., Lopresti, F., La Carrubba, V., Torregrossa, M., & Viviani, G. (2022). Occurrence of Microplastics in Waste Sludge of Wastewater Treatment Plants: Comparison between Membrane Bioreactor (MBR) and Conventional Activated Sludge (CAS) Technologies. [PubMed]. *Membranes*, 12(4), 4. Advance online publication. DOI: 10.3390/membranes12040371

Diaz-Basantes, M. F., Conesa, J. A., & Fullana, A. (2020). Microplastics in Honey, Beer, Milk and Refreshments in Ecuador as Emerging Contaminants. *Sustainability (Basel)*, 12(14), 14. Advance online publication. DOI: 10.3390/su12145514

Dube, E., & Okuthe, G. E. (2023). Plastics and Micro/Nano-Plastics (MNPs) in the Environment: Occurrence, Impact, and Toxicity. *International Journal of Environmental Research and Public Health*, 20(17), 17. Advance online publication. DOI: 10.3390/ijerph20176667

Habibi, N., Uddin, S., Fowler, S. W., & Behbehani, M. (2022). Microplastics in the atmosphere: A review. Journal of Environmental Exposure Assessment, 1(1). doi:DOI: 10.20517/jeea.2021.07

Habumugisha, T., Zhang, Z., Uwizewe, C., Yan, C., Ndayishimiye, J. C., Rehman, A., & Zhang, X. (2024). Toxicological review of micro- and nano-plastics in aquatic environments: Risks to ecosystems, food web dynamics and human health. *Ecotoxicology and Environmental Safety*, 278, 116426. DOI: 10.1016/j.ecoenv.2024.116426

Karmakar, M. (2023, July 24). *Managing microplastic pollution is important for meeting sustainable development goals in India*. Down To Earth. https://www.downtoearth.org.in/water/managing-microplastic-pollution-is-important-for-meeting-sustainable-development-goals-in-india-90789

Kiran, B. R., Kopperi, H., & Venkata Mohan, S. (2022). Micro/nano-plastics occurrence, identification, risk analysis and mitigation: Challenges and perspectives. *Reviews in Environmental Science and Biotechnology*, 21(1), 169–203. DOI: 10.1007/s11157-021-09609-6

Lai, H., Liu, X., & Qu, M. (2022). Nanoplastics and Human Health: Hazard Identification and Biointerface. *Nanomaterials (Basel, Switzerland)*, 12(8), 1298. DOI: 10.3390/nano12081298

Li, Q., Feng, Z., Zhang, T., Ma, C., & Shi, H. (2020). Microplastics in the commercial seaweed nori. *Journal of Hazardous Materials*, 388, 122060. DOI: 10.1016/j.jhazmat.2020.122060

Li, W., Wang, J., Xiong, Z., & Li, D. (2023). Micro/nanorobots for efficient removal and degradation of micro/nanoplastics. *Cell Reports. Physical Science*, 4(11), 101639. DOI: 10.1016/j.xcrp.2023.101639

Li, Z., Li, Q., Li, R., Zhou, J., & Wang, G. (2021). The distribution and impact of polystyrene nanoplastics on cucumber plants. *Environmental Science and Pollution Research International*, 28(13), 16042–16053. DOI: 10.1007/s11356-020-11702-2

Mahesh, S., Gowda, N. K., & Mahesh, S. (2023). Identification of microplastics from urban informal solid waste landfill soil; MP associations with COD and chloride. *Water Science and Technology: A Journal of the International Association on Water Pollution Research, 87*(1), 115–129. doi:DOI: 10.2166/wst.2022.412

Oliveri Conti, G., Ferrante, M., Banni, M., Favara, C., Nicolosi, I., Cristaldi, A., Fiore, M., & Zuccarello, P. (2020). Micro- and nano-plastics in edible fruit and vegetables. The first diet risks assessment for the general population. *Environmental Research*, 187, 109677. DOI: 10.1016/j.envres.2020.109677

Ross, M. S., Loutan, A., Groeneveld, T., Molenaar, D., Kroetch, K., Bujaczek, T., Kolter, S., Moon, S., Huynh, A., Khayam, R., Franczak, B. C., Camm, E., Arnold, V. I., & Ruecker, N. J. (2023). Estimated discharge of microplastics via urban stormwater during individual rain events. *Frontiers in Environmental Science*, 11, 1090267. Advance online publication. DOI: 10.3389/fenvs.2023.1090267

Russo, M., Oliva, M., Hussain, M. I., & Muscolo, A. (2023). The hidden impacts of micro/nanoplastics on soil, crop and human health. *Journal of Agriculture and Food Research*, 14, 100870. DOI: 10.1016/j.jafr.2023.100870

Schmid, C., Cozzarini, L., & Zambello, E. (2021). Microplastic's story. *Marine Pollution Bulletin*, 162, 111820. DOI: 10.1016/j.marpolbul.2020.111820

Shruti, V. C., Pérez-Guevara, F., Elizalde-Martínez, I., & Kutralam-Muniasamy, G. (2020). First study of its kind on the microplastic contamination of soft drinks, cold tea and energy drinks—Future research and environmental considerations. *The Science of the Total Environment*, 726, 138580. DOI: 10.1016/j.scitotenv.2020.138580

Simple Kitchen Swaps to Reduce Your Exposure to Microplastics. (2024, February 5). https://www.healthcentral.com/chronic-health/simple-kitchen-swaps-to-reduce-your-exposure-to-microplastics

Song, J., Wang, C., & Li, G. (2024). Defining Primary and Secondary Microplastics: A Connotation Analysis. *ACS ES&T Water*, 4(6), 2330–2332. DOI: 10.1021/acsestwater.4c00316

Su, Y., Hu, X., Tang, H., Lu, K., Li, H., Liu, S., Xing, B., & Ji, R. (2022). Steam disinfection releases micro(nano)plastics from silicone-rubber baby teats as examined by optical photothermal infrared microspectroscopy. [PubMed]. *Nature Nanotechnology*, 17(1), 76–85. Advance online publication. DOI: 10.1038/s41565-021-00998-x

Treatment technologies for the removal of micro plastics from aqueous medium | AIP Conference Proceedings | AIP Publishing. (n.d.). Retrieved August 21, 2024, from https://pubs.aip.org/aip/acp/article-abstract/2463/1/030009/2825284/Treatment-technologies-for-the-removal-of-micro?redirectedFrom=fulltext

van Schothorst, B., Beriot, N., Huerta Lwanga, E., & Geissen, V. (2021). Sources of Light Density Microplastic Related to Two Agricultural Practices: The Use of Compost and Plastic Mulch. *Environments (Basel, Switzerland)*, 8(4), 4. Advance online publication. DOI: 10.3390/environments8040036

Yee, M. S.-L., Hii, L.-W., Looi, C. K., Lim, W.-M., Wong, S.-F., Kok, Y.-Y., Tan, B.-K., Wong, C.-Y., & Leong, C.-O. (2021). Impact of Microplastics and Nanoplastics on Human Health. [PubMed]. *Nanomaterials (Basel, Switzerland)*, 11(2), 2. Advance online publication. DOI: 10.3390/nano11020496

Ziani, K., Ioni ă-Mîndrican, C.-B., Mititelu, M., Neac u, S. M., Negrei, C., Moro an, E., Drăgănescu, D., & Preda, O.-T. (2023). Microplastics: A Real Global Threat for Environment and Food Safety: A State of the Art Review. *Nutrients*, 15(3), 3. Advance online publication. DOI: 10.3390/nu15030617

Chapter 2
Microplastics:
An Invisible Air Pollutant of New Era

Mani Singh
Jamia Millia Islamia, India

P. K. Singh
Magadh University, India

U. C. Kulshreshtha
Jawaharlal Neharu University, India

ABSTRACT

The ubiquitous presence of microplastics (MPs) in the environment and living organisms including humans has emerged as a significant global concern in the recent decades. Ambient MPs primarily originate from industrial activities and degradation of larger plastic debris, contributing to both coarse (PM10) and fine (PM2.5) particulate matter. Inhalation and ingestion of these particles pose significant health risks. MPs can affect plant growth, alter photosynthesis and modify genetic expression, thereby impacting agricultural productivity and ecosystem stability. MPs can also influence climate by disrupting the Earth's energy dynamics and carbon cycles. This chapter focuses on the source apportionment of MPs in coarse and fine particulate matter (PM) and examines their impacts on human health, plants life, and climate change while deliberating on the mitigation strategies.

DOI: 10.4018/979-8-3693-9163-1.ch002

1. INTRODUCTION

The microplastics (MPs) are emerging as a new air pollutant which is a matter of growing concern among researchers, medical professionals, and environmentalists. While the full extent of its impacts is still being studied, the micro-plastics have shown several adverse effects on the environment and human health. Inhalation or ingestion of microplastics may pose multiple health risks as these particles carry potentially harmful chemicals.

Plastic particles of a size range of 5 millimeters to 100 nanometers are usually defined as microplastics (GESAMP, 2016; Masura et al., 2015; Thompson et al., 2004). Particles smaller than 100 nanometres are called nanoplastics (NPs). MPs are synthetic, organic, insoluble, and non-biodegradable particles. The microplastics may be primary or secondary based on their sources. While the former is intentionally produced particles, the latter ones are produced by degradation of larger plastic particles (macroplastic) in the environment by natural processes. As per estimates, secondary MPs constitute 69-81% and the primary MPs constitute 15-31% of the oceanic MPs. Laundering of synthetic fabrics (35%), abrasion of tires (28%) and cosmetics (2%) are major sources of the secondary MPs.

The plastics are nonreactive, organic polymers of high molecular mass. Based on the response of temperature on the plastics, they can be classified as thermoplastics or thermosets. Thermoplastics are reversible as alteration in temperature can alter their forms. Some of the examples of thermoplastics are polypropylene (PP), polycarbonate (PC), polyarylsulfone (PSU), polystyrene (PS), polyethylene terephthalate (PET), polymethyl methacrylate (PMMA), polyvinyl chloride (PVC), polypropylene (PP), and polyamides (PA) etc. However, thermosets are a family of plastics that cannot be reversed by altering temperature as they undergo a chemical change, e.g., epoxy resins, vinyl ester, polyurethane (PUR), acrylic resin, silicone, and melamine resin etc.

In the past 20 years, MPs have emerged as a big environmental problem and drew researchers' attention. The microplastics present in the air pollute the environment far and wide when they settle. They pollute air, water, soil and harm plants, animals, and microorganisms. Researchers are trying to understand the long-term consequences of ambient microplastics, the risk associated with them and the sustainable practices of waste management to mitigate the impact.

1) Sources of microplastics: There are various sources of origin of MPs, like the degradation of larger plastic debris, synthetic fabrics, wear and tear of synthetic materials, tires, cosmetics, cigarette butts and industrial processes (Prata, 2018; Enyoh CE, 2019). As per an estimate, the global annual production of plastics has shot up from 2 million tons in 1950 to more than 450 million tons (Jones, 2023). Two-thirds of the plastics are short-lived which soon become waste and find their way into the ocean to marine life and then to the human food chain. However, despite all the environmental concerns, the production of plastics is increasing. The cumulative production of plastic has gone up to more than 9 billion tons, half of which has been discarded or gone to landfills. Some projections indicate that by 2025, about 11 billion tons of plastic would have accumulated in the environment (Jones, 2023). The microplastics can adsorb various toxic contaminants like pesticides, heavy metals and release them in the environment as they undergo degradation.

2) Atmospheric Transportation and Deposition: Microplastics can be carried over long distances by wind. They eventually settle from the air onto surfaces, including soil and water, leading to potential contamination of terrestrial and marine ecosystems. The microplastics travel around the globe and reach urban, rural, and remote areas. Contrary to popular understanding, oceans are not the 'final sink' for microplastics (Allen et al.6, 2022**).** MPs travel through atmospheric transport pathways at much faster speed than oceanic currents and can cover thousands of kilometers in a matter of days (Allen et al., 2022). Generally, plastics constitute less than 1 percent of anthropogenic aerosol depositing on the ground but in certain regions of the ocean it may constitute as much as 50 percent of the aerosol.

In the life cycle of production to dumping, the plastic items get degraded to MPs and carried away to different ecosystems in different locations and get bioaccumulated in living organisms, including human beings. Microplastics are almost found everywhere and on all the seven continents. A simplified microplastics life cycle is given in **Figure 1.**

3) Characterization of Microplastics: The characterization of MPs may include examination of size, morphology, composition, and their abundance. Different analytical methods available for MPs identification and characterizations are summarized in **Table 1.**

Table 1. Analytical methods for MPs identification and characterization

Methods	Instrumentation	References
Gas chromatography techniques	Desorption Gas Chromatography Mass Spectroscopy (TED GC-MS)	Duemichen et al. 2014; Dumichen et al. 2015; Dumichen et al. 2017; Eisentraut et al. 2018
	Pyrolysis Gas Chromatography (Py-GC-MS)	Fischer and Scholz- Bottcher 2017; Mintenig et al. 2018)
Spectroscopy methods	Focal Plane Array Fourier Infrared spectroscopy (FPA-FTIR)	Huppertsberg and Knepper 2018; Ivleva, Wiesheu, and Niessner 2017; Joachim et al. 2015; Kappler et al. 2016; Primpke et al. 2017; Simon, van Alst, and Vollertsen 2018.
	Raman Spectroscopy	Araujo et al. 2018; Lenz et al. 2015; Zhao et al. 2017.
	Enhanced Darkfield (EDF) Nanoscale Hyperspectral Imaging Spectroscopy (HIS)	Fournier et al. 2020

The gas chromatography technique has some limitations as it involves particle decomposition which prevents comprehensive characterization of MPs. The combined use of both FPA-FTIR and Raman spectroscopy methods can determine MP composition, size and morphology, with detection limits of 10 mm (Joachim et al. 2015; Vianello et al. 2019) and 2 mm (Ballent et al. 2016; Enders et al. 2015; Zada et al. 2018), respectively. Chen et al. 2024 observed average concentration of the fine plastic particles (FPPs) in Shanghai city as 5.6 μg/m^3 where the ratio of FPPs to PM2.5 was 13.2%.

Rahaman et al., 2021 collected and characterized the ambient MPs from fine particulate matters. They did outdoor, indoor as well as personal air sampling and collected fine PMs on Teflon and silver filters using OMNI FT ambient air samplers (BGI by MESA Labs, Butler, NJ) fitted with 5 L/min PM2.5 Impactor. The filters were extracted in methanol by 10 minutes sonication at 40kHz. Raman spectra were acquired of the extracted material for identification of PMs. A large variety of MP polymers were observed of size range below 2.5 μm to below 1 μm. Sources of these plastic polymers were mostly associated with packaging materials. Nylon was the leading polymer of the group **Figure 2**.

2. METHODS OF SAMPLING AND ANALYSIS

There is a wide range of methods of sampling, monitoring and analysis of MPs from a basic direct collection using brooms to advanced instrumental techniques. **Table 2** summarizes the key methods of sample collection of ambient MPs and related details.

Table 2. Methods of sample collection

Method	Equipment	Advantages	Limitations	References
Passive collection (Total Deposition)	Use of Brushes, Sweeping, Vacuum, NILU Collectors	Simple, standardized collection method, cheaper, No electricity required, very successful for remote locations	It only collects the deposited MPs from the ground or material surface. It does not give an idea about suspended MPs.	Deghani et al. (2017); Cai et al (2017); Dris et al. (2017); Abbar et al. (2017); Klien and Fischer (2019); Allen et al. (2019a);
Active methods of MPs collection	Active pumped samplers	Established method for atmospheric MPs.	It requires power supply. For comprehensive sampling of ambient MPs it is used in conjunction with passive sampling methods.	Hayward et al., (2010); Dommergue et al., (2019); Dris et al. (2017)

3. IMPACT OF MICROPLASTICS ON ENVIRONMENT:

Microplastics are one of the most damaging and long-lived legacies of plastic pollution which is an increasing threat to planetary health (UNEP 2021). Some of the key environmental impacts are described below:

1) **Ecosystem disruption**: Microplastics are ubiquitous and settle in all environmental compartments including air, soil and water. It enters into ecosystems and disrupts ecosystems adversely affecting various organisms (Roy et al. 2022).
2) **Effects on wildlife**: Animals may take microplastics from air, water and food leading to potential harm to their health. The existence of microplastics in the food chain may cause bioaccumulation as well (Llorca et al. 2020).

3) **Climate change:** Researchers are working on understanding the potential effects of MPs on climate. Major impacts of airborne microplastics on meteorological phenomena and earth's climate are as follows:
   a) **Radiative Forcing:** Atmospheric MPs may alter the radiative forcing resulting in the change in the Earth's energy balance. These tiny particles could disturb the climate dynamics by influencing the scattering and absorption of solar radiation.
   b) **Cloud Formation and Properties**: MPs might influence cloud formation as they act as cloud condensation nuclei (CCN) which can influence cloud cover, precipitation patterns, and regional climate.
   c) **Carrier of Pollutants**: Microplastics can adsorb and transport various pollutants in the atmosphere. After deposition, MPs may affect soil and water quality, contributing to changes in ecosystems and potentially impacting carbon cycling and greenhouse gas dynamics. According to Roy et al. (2020), MPs absorb chemicals and heavy metals which make them toxicologically more harmful for living organisms. They revealed that the older MPs show higher pollutant absorption capacity than newer ones. Apart from the age, the sorption behavior of MPs also depends on the particle size, hydrophobicity, hydrogen bonding and specific surface ratio (Yu et al., 2019).
   d) **Interaction with Microorganisms:** Microplastics in the air may also interact with microorganisms, forming complex microbial communities on their surfaces. These interactions can disturb biogeochemical cycles, especially carbon and nitrogen cycles.
   e) **Secondary Organic Aerosols:** Interaction between microplastics and organic compounds in the atmosphere could lead to the formation of secondary organic aerosols. These aerosols can impact air quality and atmospheric processes, with potential implications for climate.

4. IMPACTS ON HUMAN HEALTH

Ubiquitous microplastics carrying harmful chemicals cause serious health impacts. These can include respiratory issues, brain development and changes in human genetics. In addition to entering the food chain through seafood, living organisms can inhale microplastics from the air, ingest from water and absorb through the skin. Microplastics have been found in various human organs including the placenta

(UNEP) and are likely to impact women more. Different human health impacts have been shown in **Figure 3.**

According to Revel et al. (2018), humans are exposed to MPs/NPs primarily through ingestion and inhalation; the former being the most common way (Lehner et al., 2019). Inhalation is another major intake pathway of micro- and nano-plastics. It is estimated that humans can inhale up to 22,000,000 micro- and nano-plastics annually (Cox et al., 2019).

After MPs/NPs ingestion, translocation and absorption occur through the digestive system (Whelan et al., 2011). Plastic particles of polymers such as PS, PP, and PET were detected and quantified in human blood for the first time by Leslie et al., 2022. Other studies have shown that PS beads of 50, 80, and 240 nm, as well as microsized PP, can permeate the human placenta (Ragusa et al., 2021; Wick et al., 2010).

In an experimental study done by Nadanaciva et al. (2011), PS-NPs (44 nm) interiorized into human cells hindered cell viability, changed gene expression which resulted in morphological abnormalities and inflammation. Thus, MPs/NPs in the circulatory system may hinder normal circulation, leading to heart disease. MPs exposure also leads to cardiovascular disease such as hypertension and atherosclerosis. Persons with respiratory ailments experience more severity in their ailments after MPs exposure.

Bioaccumulation of MPs/NPs in the gut and liver leads to fatty liver and increased enzyme activities in liver (Lu et al., 2016). Large surface area and persistent nature of MPs/NPs in human tissues, causes oxidative stress and results in cytotoxicity with chronic inflammation, and increases the risk of cancer (Prata et al., 2020). Absorbed MPs/NPs may change the gut microbes' diversity and their function. It may cause disturbance in metabolism in the human body and may increase the risk of obesity, diabetes, and other chronic liver diseases (Weiss & Hennet 2017). Humans having exposure to airborne MPs develop many acute to chronic respiratory diseases like inflammation, irritation in airways to asthma. The adverse effects of MPs and NPs on the respiratory system range from irritation to pulmonary emphysema to autoimmune diseases to the onset of cancer in cases of chronic exposure (Faciola et al., 2019).

Jenner et al (2022) recorded region wise distribution of MP particles in human lungs as 0.8 ± 0.96 MP/g within the upper region, $0.41+ 0.37$ MP/g within the middle/lingular region, and 3.12 ± 1.30 MP/g within the lower region **(Figure 4).** While characterizing MPs found in human lungs, they observed 12 polymer types in the tissue samples with PP (23%) and PET (18%) showing the highest abundance **(Figure 5)**. The observed length and width of the MP particles in the tissues ranged between 12 - 24.75 μm and 4 - 88 μm respectively having mean length and width as 223.10 ± 436.16 μm and 22.21 ± 20.32 μm respectively. After entering into lungs on inhalation, MPs interfere with the biophysical function of pulmonary surfactants and damage alveoli and epithelial cells which induces epithelial cell apoptosis.

5. IMPACTS ON PLANTS:

Microplastics can adversely impact seed germination, root formation, growth of shoot and leaf, photosynthesis, and gene expression. It also affects the nutrient cycle as it affects soil microorganism and pollutes the soil by leachates released from the MPs. According to Roy et al. (2023), MPs might block the micropyle of crop seeds and inhibit the uptake of water and nutrients leading to decline in seed germination.

Root cracks facilitate the MPs to enter into newly developed roots and into the vascular system (Li et al., 2020). The same may be seen in **Figure 6A.** Entry of MPs into shoot and accumulation in leaves occurs through transpiration pull (Xu et al., 2022). MPs use the apoplastic pathway to enter the vasculature of the stems and leaves from the roots (Li et al., 2019). Roy et al (2022) explained the adverse effects of MPs on the plant growth as physical damage to xylem vessels. According to Wang et al. (2023c), PS MP caused reduction in leaves' fresh weight, number of leaves per plant, leaf surface area, and overall plant height. Liu et al. (2021) observed specific responses of plant tissue against MP stress in terms of shoot height and weight. **Table 3** summarizes the adverse effects of different MPs on plant growth.

There are some physiological and molecular changes behind the physical symptoms of plants due to MPs effects. These physiological symptoms include Photosynthesis, shoot physiology and root physiologies. MPs adversely affect the photosynthetic pigments, chlorophyll fluorescence, leaf gas exchange, ionic homeostasis, and redox regulation (Yang and Gao, 2022; **Figure 6B**). Reduction in photosynthesis in lettuce, pumpkins, and mangrove were noticed by Lian et al. (2021), Liu et al. (2021), and Chai et al. (2023) respectively. MPs reduce the chlorophyll content and block the transfer of electrons to PSII reaction centers, which subsequently affects ETC reaction in the PSII photosystem (Xu et al., 2022).

Disruption in the molecular structure of thylakoid membranes due to MPs inhibits electron transfer. Though the exact mechanism of the effect of MPs is not very clear, it might be due to functional groups associated with different MPs. Kalhor et al., 2018 explained that MPs with $-NH_2$ and $-SO_3H$ functional groups shut down the reaction center with a concomitant decrease in the number of electron receptors in PSII, leading to reduced transfer of electrons to QA-receptors in PSII (Wang et al., 2017), thus MPs reduce the overall efficiency of PSII.

Decline in photosynthesis does not occur only due to reduction in chlorophyll contents in the reaction center (Peng et al., 2021). It could also be associated with redox reactions and transfer of electrons in the reaction center of the PSII (Lu et al., 2020). Moreover, anthocyanins play a pivotal role in protecting leaves from photoinhibition as they absorb excess protons and through their antioxidant activity. Nevertheless, decreased biosynthesis of anthocyanin also reduces photosynthesis under MP stress (Wang et al., 2023b).

Table 3. Effects of different microplastics (MPs) on plant growth

MP type and concentration	Plant	Impact on plant	Research Reference
PE (0.0125-100 mg/L)	Maize	Impaired nitrogen uptake, Impaired water intake, Reduced overall growth	Urbina et al., 2020
PE (0.1% (w/w))	Maize	Reduced plant growth, Altered soil bacterial community	Fajardo et al.,2022
PS (0.01-10 mg/L)	Wheat	Reduced shoot to root biomass ratio, Impaired micronutrient contents	Lian et al., 2020
PS (25-400 mg/L)	Onion	Reduced root length, Cytotoxicity toxicity, Enhanced ROS Production	Maity et al., 2020
PVC (100- 200 mg/L)	Sweet potato	Reduced plant height, Reduced chlorophyll content and Reduced fresh biomass per plant.	Khan et al., 2023
PVC(100-1000 mg/L)	Spirodela polyrhiza	Significant reduction in leaf, Declined adventitious root elongation and Declined overall all growth	Wang et al.,2023b

Giorgetti et al., 2020 explained that MPs disturb the Reactive Oxygen Species *(ROS) production* and scavenging balance through increasing the ROS production beyond the capacity of plants to activate the antioxidant defense system, which consequently hampers the integrity of membranes, decreases mitotic cell division, and even causes cytogenetic anomalies in roots **(Figure 6C).**

MPs can also induce *alterations in gene expression* (Xu et al., 2020) in roots. The observed cytotoxic and genotoxic effects on the root growth may be possibly due to the alteration in gene expression under the influence of MPs. Li et al. (2021b) found that PAN MP treatment affects the expression of *CDC2* (the gene regulating the cell division cycle) and *CDK* (the coding gene of cyclin dependent kinase) resulting in increased root aging or even death. As these genes have a crucial role in cell cycle regulation, any change in these genes causes alteration in the root growth and development. A recent study reported that PS MP stress-induced negative effects on root and shoot growth were associated with the down regulation of genes involved in ROS scavenging and ionic homeostasis (Wang et al., 2023c). Thus, MP stress affects root physiology by altering gene expression. However, a variation in response may exist among plant species, which may further vary with size and type of MPs and other growth conditions, which needs to be further explored.

6. TREATIES AND REGULATIONS

A single comprehensive global treaty specifically dedicated to microplastics is yet to emerge. However, the issues of microplastics are increasingly being addressed within existing international agreements, conventions, and regulations. Some of the treaties/regulations are described below.

1) Basel Convention on the Control of Transboundary Movements of Hazardous Wastes and Their Disposal: The Convention aims at restricting the movements of hazardous waste between nations. It has been amended to include certain plastic wastes, including microplastics under its control regime.
2) Stockholm Convention on Persistent Organic Pollutants (POPs): The Convention aims at restricting the production and use of POPs, which can be associated with microplastics as they adsorb toxic chemicals. Efforts to control POPs indirectly contribute to reducing the impact of microplastics.
3) United Nations Environment Assembly (UNEA) Resolutions: The UN Environment Assembly (UNEA-5.2) passed an important resolution in 2022 to establish an Intergovernmental Negotiating Committee (INC) to draft a legally binding agreement to reduce plastic pollution.
4) Regional Agreements: Some regional agreements such as the Oslo-Paris Convention for the Protection of the Marine Environment of the North-East Atlantic (OSPAR Convention) have included measures to monitor and reduce marine litter, including microplastics. The European Union has set a target to reduce MPs by 30% by 2030.
5) National Regulations: Many countries have framed their own regulations and policies in recent years to control microplastics in various products and environments. The Netherlands was the first country to ban microplastics in 2014 followed by Austria, Luxemburg, Belgium, and Sweden. The USA followed in 2017. India notified the Plastic Waste Management Amendment (PUMA) Rules, 2021 to ban manufacture, import, sale and use of single use plastic items across the country from July 1, 2022.

7. RECOMMENDATIONS

Addressing ambient microplastics requires a collective approach involving governments, industries, researchers, and the public. Following key steps may be considered for addressing ambient microplastics:

1) Research and Monitoring: Conducting a long-term research will help in better understanding of the sources, transportation, and impacts of ambient microplastics on different environmental and human health aspects. Inclusion of microplastics as new pollutant in air quality index (AQI) may help monitor ambient microplastics levels and identify emerging hotspots.
2) Regulatory Frameworks: Strong regulatory frameworks at national and international levels can help address microplastics pollution comprehensively. This may include setting safe standards for ambient microplastics levels, regulating their emissions from various sources, and implementing product regulations to limit their use.
3) Pollution Prevention and Waste Management: Adopting sustainable waste management practices to reduce the release of plastics into the environment could be an important step. Promotion of recycling, circular economy approaches, and innovative technologies can help in prevention and cleaning of pollution.
4) Public awareness and education: Creating awareness among the public, policymakers, and industries about the environmental and health impacts of microplastics will be crucial in promoting behavioral changes and compliances.
5) Addressing the full life cycle of plastics: Airborne micro- and nano-plastics are small part of bigger phenomena. A systematic management covering production to use must be required to address the problem. The policy makers should support action at the global level for strict enforcement measures through legally binding agreement.

REFERENCES

Allen, D. (2022). Microplastics and Nanoplastics in the Marine-Atmosphere Environment. *Nature Reviews Earth & Environment,* 3(6), 393. https://doi.org/doi:10.1038/s43017-022-00292-x

Allen, S., Allen, D., Karbalaei, S., Maselli, V., & Waker, T. R. (2022). Micro(Nano) Plastics Sources, Fate, and Effects: What We Know after Ten Years of Research. *Journal of Hazardous Materials Advances*, 6, 1. DOI: 10.1016/j.hazadv.2022.100057

Amato-Lourenço, L. F., Galvao, L. D. S., Weger, L. A. D., Heimstra, P. S., Vijver, M. G., & Mauad, T. (2020). An Emerging Class of Air Pollutants: Potential Effects of Microplastics to Respiratory Human Health? *The Science of the Total Environment*, 749, 141676. Advance online publication. DOI: 10.1016/j.scitotenv.2020.141676

Chai, M., Li, R., Li, B., Wu, H., & Yu, L. (2023). Responses of mangrove (*Kandelia obovate* L.) growth, photosynthesis, and rhizosphere soil properties to microplastic pollution. *Marine Pollution Bulletin*, 189, 114827. DOI: 10.1016/j.marpolbul.2023.114827

Chen, Q., Gao, J., Yu, H., Su, H., Yang, Y., Cao, Y., Zhang, Q., Ren, Y., Holleret, H., Shi, H., Chen, C., & Liu, H. (2022). An emerging role of microplastics in the etiology of lung ground glass nodules. *Environmental Sciences Europe*, 34(1), 25. DOI: 10.1186/s12302-022-00605-3

Chen, Y., Jing, S., Wang, Y., Song, Z., Xie, L., Shang, X., Fu, H., Yang, X., Wang, H., Wu, M., Chen, Y., Li, Q., Zhang, Y., Wang, W., Zhang, L., Wang, R., Fang, M., Zhang, Y., Li, W., & Chen, J. (2024). Quantification and Characterization of Fine Plastic Particles as Considerable Components in Atmospheric Fine Particles. *Environmental Science & Technology*, 58(10), 4691–4703. DOI: 10.1021/acs.est.3c06832

Cox, K. D., Covernton, G. A., Davies, H. L., Dower, J. F., Juanes, F., & Dudas, S. E. (2019). Human Consumption of Microplastics. *Environmental Science & Technology*, 53(12), 7071. DOI: 10.1021/acs.est.9b01517

Delhi, P. I. B. (2022) Ban on identified Single Use Plastic Items from 1st July 2022. Ministry of Environment, forest, and climate change. https://www.unep.org/news-and-stories/story/microplastics-long-legacy-left-behind-plastic-pollution

Donkers, J. M., Hoppener, E. M., Grigorieve, I., Will, L., Melgret, B. N., Zan, B. V. D., Steeg, E. V. D., & Kooter, I. M. (2022). Advanced Epithelial Lung and Gut Barrier Models Demonstrate Passage of Microplastic Particles. *Microplastics and Nanoplastics*, 2(1), 6. DOI: 10.1186/s43591-021-00024-w

Enyoh, C. E. (2019). Microplastic pollution in Nigeria: a call for concern. 26th meeting of the Imo State chapter of the Chemical Society of Nigeria (CSN) held at Imo State University, Nigeria. https://doi.org/DOI: 10.13140/RG.2.2.21027.9190

Facciolà, A., Giuseppa, V., Ciarello, M. P., & Pietro A. D. (2021). Newly Emerging Airborne Pollutants. Breathing Plastic: The Health Impacts of Invisible Plastics in the Air - Center for International Environmental Law (ciel.org). doi:DOI: 10.3390/ijerph18062997

Gasperi, J., Wright, S. L., Dris, R., Collard, F., Mandin, C., Guerouache, M., Langlois, V., Kelly, F. J., & Tassin, B. (2018). Microplastics in Air: Are We Breathing It In? *Current Opinion in Environmental Science & Health*, 1, 1–5. DOI: 10.1016/j.coesh.2017.10.002

Jenner, L. C., Rotchell, J. M., Bennett, R. T., Cowen, M., Tentzeris, V., & Sadofsky, L. R. (2022). Detection of microplastics in human lung tissue using μFTIR spectroscopy. *Science of The Total Environment.* 831, 154907.

Jones, N. (2023). Microplastics Are Filling the Skies. Will They Affect the Climate? YaleEnvironment360 published at Yale school of environment. 2023. https://e360.yale.edu/features/plastic-waste-atmosphere-climate-weather

Kalhor, M. S., Aliniaeifard, S., Seif, M., Asayesh, E. J., Bernard, F., Hassani, B., & Li, T. (2018). Title: Enhanced salt tolerance and photosynthetic performance: Implication of γ-amino butyric acid application in salt-exposed lettuce (*Lactuca sativa* L.) plants. *Plant Physiology and Biochemistry*, 130, 157–172. DOI: 10.1016/j.plaphy.2018.07.003

Li, L., Zhou, Q., Yin, N., Tu, C., & Luo, Y. (2019). Uptake and accumulation of microplastics in an edible plant. *Kexue Tongbao*, 64(9), 928–934. DOI: 10.1360/N972018-00845

Li, W., Wufuer, R., Duo, J., Wang, S., Luo, Y., Zhang, D., & Pan, X. (2020). Microplastics in agricultural soils: Extraction and characterization after different periods of polythene film mulching in an arid region. *The Science of the Total Environment*, 749, 141420. DOI: 10.1016/j.scitotenv.2020.141420

Liu, K., Wang, X., Fang, T., Xu, P., Zhu, L., & Li, D. (2019). Source and potential risk assessment of suspended atmospheric microplastics in Shanghai. *The Science of the Total Environment*, 675, 462–471. DOI: 10.1016/j.scitotenv.2019.04.110

Liu, S., Wang, J., Zhu, J., Wang, J., Wang, H., & Zhan, X. (2021). The joint toxicity of polyethylene microplastic and phenanthrene to wheat seedlings. *Chemosphere*, 282, 130967. DOI: 10.1016/j.chemosphere.2021.130967

Llorca, M., Álvarez-Muñoz, D., Ábalos, M., Rodríguez-Mozaz, S., Santos, L. H., León, V. M., Campillo, J. A., Martínez-Gómez, C., Abad, E., & Farré, M. (2020). Microplastics in Mediterranean coastal area: toxicity and impact for the environment and human health.

Pascall, M. A., Zabik, M. E., Zabik, M. J., & Hernandez, R. J. (2005). Uptake of Polychlorinated Biphenyls (PCBs) from an Aqueous Medium by Polyethylene, Polyvinyl Chloride, and Polystyrene Films. [PubMed]. *Journal of Agricultural and Food Chemistry*, 53(1), 164–169. DOI: 10.1021/jf048978t

Peng, J., Feng, Y., Wang, X., Li, J., Xu, G., Phonenasay, S., Luo, Q., Han, Z., & Lu, W. (2021). Effects of nitrogen application rate on the photosynthetic pigment, leaf fluorescence characteristics, and yield of indica hybrid rice and their interrelations. *Scientific Reports*, 11(1), 7485. DOI: 10.1038/s41598-021-86858-z

Prata, J. C., da Costa, J. P., Lopes, I., Duarte, A. C., & Rocha-Santos, T. (2020). Environmental exposure to microplastics: An overview on possible human health effects. *The Science of the Total Environment*, 702, 134455. DOI: 10.1016/j.scitotenv.2019.134455

Rahman, L., Mallach, G., Kulka, R., & Halappanavar, S. (2021). Microplastics and nanoplastics science: Collecting and characterizing airborne microplastics in fine particulate matter. *Nanotoxicology*, 15(9), 1253–1278. DOI: 10.1080/17435390.2021.2018065

Roy, P., Mohanty, A. K., & Misra, M. (2022). Microplastics in ecosystems: Their implications and mitigation pathways. (Perspective). *Environmental Science. Advances*, 1(1), 9–29. DOI: 10.1039/D1VA00012H

Shi, M., Sun, Y., Wang, Z., He, G., Quan, H., & He, H. (2019). Plastic film mulching increased the accumulation and human health risks of phthalate esters in wheat grains. *Environmental Pollution*, 250, 1–7. DOI: 10.1016/j.envpol.2019.03.064

United Nation Environment Program. (2021). *From Pollution to Solution: A global assessment of marine litter and plastic pollution (unep.org)*. United Nation Environment Program.

United Nation Environment Program. (2023). Microplastics: The long legacy left behind by plastic pollution. Ocean and costs. United nations environment program. https://www.unep.org/news-and-stories/story/microplastics-long-legacy-left-behind-plastic-pollution

Wang, L., & Tanveer, M. (2023). Editorial to the special issue "Eco-physiological and molecular basis of stress tolerance in plants." *Biology (Basel)*, 12(3), 485. DOI: 10.3390/biology12030485

Wang, Y., Xiang, L., Wang, F., Redmile-Gordon, M., Bian, Y., Wang, Z., Gu, C., Jiang, X., Schäffer, A., & Xing, B. (2023c). Transcriptomic and metabolomic changes in lettuce triggered by microplastics-stress. *Environmental Pollution*, 320, 121081. DOI: 10.1016/j.envpol.2023.121081

Yang, C., & Gao, X. (2022). Impact of microplastics from polyethylene and biodegradable mulch films on rice (Oryza sativa L.). *The Science of the Total Environment*, 828, 154579. DOI: 10.1016/j.scitotenv.2022.154579

Yu, F., Yang, C., Zhu, Z., Bai, X., & Ma, J. (2019). Adsorption behavior of organic pollutants and metals on micro/nanoplastics in the aquatic environment. *Sci. Total Environ., 694*, 133643. https://doi.org/10.1016/j.scitotenv.2019.133643

Chapter 3
Status of Micro/Nano-Plastic (MNP) Pollution in the Aquatic System

Laxmi Kant Bhardwaj
https://orcid.org/0000-0001-7518-4199
Amity University, Noida, India

Prangya Rath
https://orcid.org/0000-0002-6764-1625
Amity University, Noida, India

Farhana Rehman
University of Technology, Sydney, Australia

Poonam Yadav
India Institute of Maize Research, Ludhiana, India

Tooraj Massahi
https://orcid.org/0000-0002-3602-2151
Research Center for Environmental Determinants of Health, Kermanshah University of Medical Sciences, Iran

Rashmi Mittal
Central Pollution Control Board, New Delhi, India

Harshita Jain
https://orcid.org/0000-0002-9544-2869
Amity University, Noida, India

ABSTRACT

Over the last few decades, due to the tremendous rise in plastic production, the contamination of micro/nano plastics (MNPs) has become widespread in aquatic systems around the globe. MNPs are tiny plastic particles with sizes < 5 mm/<100 nm. They are ubiquitous pollutants and can arrive in the aquatic environment from different sources. They persist in the environment for a long time. They can act as carriers for many organic pollutants, heavy metals, and pathogenic bacteria. They are an emerging environmental issue, and the scientific community is very

DOI: 10.4018/979-8-3693-9163-1.ch003

much concerned about this issue. Raman spectroscopy, Fourier transform infrared spectroscopy, pyrolysis gas-chromatography mass-spectrometry, scanning electron microscopy, etc. are advanced instrumental techniques used to analyze MNPs. This chapter focuses on the occurrence, sources, and impact of MNP on the health of aquatic organisms. This chapter also describes the identification and detection techniques of MNPs and suggests some recommendations for the elimination of MNPs from the environment.

1. INTRODUCTION

Plastics are synthetic organic polymers and can be made either from petrochemicals or renewable raw ingredients. They can be inexpensive, flexible, waterproof, and lightweight, and act as insulators. They are not biodegradable, but some plastics are biodegradable and can perish by hydrolysis or in the occurrence of ultraviolet (UV) light or by the action of microbes (Bhardwaj and Sharma 2021a; Bhardwaj 2022). Micro/nano plastics (MNPs) pollution pertains to the presence of small plastic particles, ranging from microplastics (MPs) (<5 mm) to nanoplastics (NPs) (<1 μm), in the environment. MNPs differ in colour, density, size, and type. They are naturally hydrophobic (Bhardwaj and Sharma 2021b).

MNPs are pervasive pollutants found in various environments worldwide, including oceans, rivers, soil, air, and even in organisms. They have been discovered in places you wouldn't expect, including beer, air, food, and tap water. They persist in the surroundings for a long time and can be ingested by biota (Peiponen et al. 2019). They can enter bottled water from the water distribution systems. They are present in marine environments from the surface to the seabed, where they are increased, transported, and accumulated (Bhardwaj 2023). These particles are also deposited on both urban beaches and natural sediments, posing a constant threat to marine organisms in all oceans of the planet.

MNPs can be classified as primary or secondary. Primary MNPs are purposefully manufactured; the main sources of their release are domestic and industrial activities. They are often made in small sizes, such as pellets, plastic powder, and microbeads, and are frequently found in cosmetics, face scrubbers, and personal care products (PCPs). These particles cause a substantial amount of MNP pollution since they are directly released into the environment (Dube and Okuthe 2023). Conversely, secondary MNPs are amorphous plastic fragments that come from the disintegration of bigger plastic objects including fishing nets and ropes, poly bags, and plastic bottles. These tiny particles are dangerous to the atmosphere and are produced after the breakdown of larger plastic particles (Choudhury et al. 2023).

Scientists stated that ~ $12{,}000\times10^9$ kgs of plastic would be present in the atmosphere by 2050 (Geyer et al. 2017). Koelmans et al. (2019) reported MNP particles in the aquatic environment in the range from 0.0 to 103×10^3 particles/m^3 and stated that it was due to a lack of proper waste management. The particles of polypropylene and polyethylene were reported in the Lijiang River (Zhang et al. 2021). MNP particles (30.3 $\pm$ 15.9 items/kg) were reported from Ciwalengke River (Yang et al. 2021). MNPs are an increasing hazard but due to a lack of data, our understanding of these particles is very limited. This chapter aims to focus on the occurrence, sources, harmful effects, and detection techniques of MNPs in the aquatic environment.

2. THE PRESENT CIRCUMSTANCES OF MICRO/NANO PLASTICS (MNPS) IN THE GLOBAL AQUATIC ENVIRONMENT

The increasing concentration of MNP particles in the aquatic environment is a very serious hazard to the aquatic ecosystem. The main source of this pollution is the growing manufacture and disposal of plastics globally. Several researchers have studied MNPs in the aquatic environment from 2021 to 2024 (Elgarahy et al. 2021; He et al. 2021; Lionetto et al. 2021; Junaid et al. 2023; Gao et al. 2023; Ouda et al. 2023 and Bhardwaj et al. 2024). Researchers have reported that MNP particles combine with different toxic pollutants such as heavy metals, and persistent organic pollutants (POPs), and produce toxic effects on the organisms. The ramifications are extensive and, therefore, require immediate ecological and physiological assessments. They have been reported to range from neurotoxicity to oxidative stress as well as genetic changes.

These environmental issues have been seen to be further compounded by MNPs' property to adhere with dangerous compounds like bisphenol-A (BPA) as well as bisphenol-S (BPS). Such characteristic features have emphasized the difficulties that are related to their pollutant-binding abilities. These have increased the requirement for creating as well as putting into practical usage of efficient techniques for the removal of such toxic compounds. Although potential benefits of nanotechnology-based approaches like adsorption, photocatalysis, and membrane filtration have been reported in such cases, still the field of analyzing such contaminants gets interrupted due to a lack of standardized procedures (Barros et al. 2023; Bhardwaj et al. 2023a). Wastewater treatment plants (WWTPs) have been observed to play an essential role that helps in mitigating the risk related to resistance gene transfer.

This is because MNPs have the ability to complicate matters when linked to antibiotics. The possibility that MNPs might be interacting with antibiotics to facilitate the propagation of resistance genes makes it even more challenging. As a result, WWTPs have now been seen as an essential control point (Kumar et al.

2022; Pourebrahimi and Pirooz 2023; Jain 2023). Such analytical recent studies have successfully highlighted the complexity and urgency towards this environmental concern. It has also highlighted the immediate requirement for a deeper understanding and a more strategic step for regulating MNPs in aquatic ecosystems. Different types of MNPs present in aquatic environments have been represented in Table 1.

Table 1. Occurrence of different types of Micro/Nano Plastics (MNPs) in aquatic environments

S. No.	Locations	Concentration of Micro/ Nano Plastics (MNPs)	Types of Micro/ Nano Plastics (MNPs)	References
1	Danjiangkou Reservoir, China	0.47 to 15.02×10^3 microplastic/m^3 in surface water and 15.0 to 40.0 microplastic/kg in wastewater	Fibers; polypropylene	Di et al. 2019
2	Wei River Basin, China	3.67 to 10.7 items /L	Fibers; polyethylene, polyvinylchloride, and polystyrene	Ding et al. 2019
3	Suzhou River, Huangpu River, and Yangtze Estuary, China	0.08 to 7.4 items/L	Fibers; polyester, rayon and polypropylene	Luo et al. 2019
4	Flemish River, Belgium		Foam, film, fiber, and fragment; ethylene-vinyl acetate, polypropylene, polyethylene, polyvinyl chloride, cellophane, polyvinyl acetate, and polyamides	Slootmaekers et al. 2019
5	Feilaixia Reservoir in the Beijiang River, China	0.56 ± 0.45 items/m^3	Foams, films, fragments, and fibers; polypropylene, polystyrene, expanded polystyrene, polyvinylchloride, polyethylene, and polyethylene	Tan et al. 2019
6	Pearl River along Guangzhou city and Pearl River Estuary, China	19,860.0 items/m^3 in urban and 8902.0 items/m^3 in the estuary	Films, fragments, and fibers; polyamide and cellophane	Yan et al. 2019
7	Poyang Lake, China	5.0 to 34.0 items/L for surface water and 54.0 to 506.0 items/kg for sediments	Fibers, films, pellets, and fragments; polypropylene and polyethylene	Yuan et al. 2019
8	Yangtze River, China	492000.0×10^{-6} microplastic/m^2	Polypropylene, polyethylene, polystyrene, nylon, polyoxymethylene, cellulose, and ethylene vinyl acetate	Xiong et al. 2019

continued on following page

Table 1. Continued

S. No.	Locations	Concentration of Micro/ Nano Plastics (MNPs)	Types of Micro/ Nano Plastics (MNPs)	References
9	Eight WWTPs in England and Wales (UK)	4.9 particles/L in raw water and 0.00011particles/L in potable water	Polyethylene, polyethylene terephthalate, polypropylene and polystyrene	Johnson et al. 2020
10	Yongjiang River, China	500.0 to 7700.0 items/m^3 in surface water and from 90.0 to 550.0 items/kg in sediments	Polypropylene and polyethylene	Zhang et al. 2020
11	Vistula River, Poland		Polyethylene, polypropylene, polyamide and polycarbonate	Rytelewska and Dąbrowska 2022

3. MICRO/NANO PLASTICS (MNPS): ORIGINS

The primary sources of the MNPs are particularly feedstocks containing polymeric particles, cosmetic materials, pellets, and plastic powders (Jain et al. 2022). Breaking down bigger plastic polymer particles is the main process of generating MNPs. In addition to stormwater runoff, combined sewer overflows, emissions from vehicles and wildfires, agricultural practices, atmospheric deposition, road markings and tyre wear, transportation, and maritime activities, and macroplastic fragmentation aspects that are directly into the water environment (Kumar et al. 2021; Gao et al. 2022). MNPs can cover large distances while airborne before deposition on land or water bodies (Kiran et al. 2022). Figure 1 presents the sources, formation, and transfer route. An extensive synopsis of MNPs' sources is provided below.

3.1. Industrial Activities, and Emissions from Vehicles & Wildfire

Microbeads are the most significant input stream for the generation of MNPs, with microbeads being used in toothpaste, cosmetics, as well as body scrubs. Polyethylene, polypropylene, and other polymers are frequently used to make these microbeads (Ojinnaka and Aw 2020). Washing clothing causes synthetic fibers to shed, which greatly increases MNP pollution. Fabrics made of polyester, nylon, and other synthetic materials release microfibers into wastewater (Cohen and Radian 2022). Abrasive blasting and grinding processes in industries release MNPs into the environment. These particles are frequently released from abrasives consisting of plastic or material containing plastic content (Wang et al. 2023). MNPs are used as fillers or additives in paints, coatings, and sealants. Weathering and degradation of these products release MNPs into the environment (Kundu et al. 2021). Medical equipment, pharmaceuticals, and healthcare products contain plastic components that can degrade into MNPs over time (de Souza et al. 2022). Exhaust emissions from vehicles and industries release MNPs into the atmosphere, which can subsequently be deposited onto land and water surfaces through dry and wet deposition processes (Dube and Okuthe 2023). Wildfires release large amounts of particulate matter, including MNPs, into the atmosphere.

3.2. Shipping & Maritime Activities, Sea Spray, and Recreational Activities

Shipping activities such as vessel maintenance, cleaning, and accidents (e.g., container spills) release MNPs into marine environments and impact coastal ecosystems and wildlife (Kiran et al. 2022). Ships' hulls and marine structures coated with antifouling paints release MNPs into aquatic ecosystems through leaching and abrasion (Wang et al. 2023). Breaking waves generate sea spray containing dissolved and suspended organic matter and include MNPs that are formed from degraded marine debris. Sea spray can be transported inland by winds and deposited onto terrestrial surfaces (Junaid and Wang 2022). Recreational activities such as boating, fishing, and beachgoing can contribute to MNP pollution through the release of plastic debris and recreational equipment containing plastic components (Gao et al. 2022).

3.3. Pellets, Resins, and Fragmentation of Macroplastics

Plastic pellets, also known as nurdles, are used to manufacture plastic products. Accidental spills and improper handling during production, transportation, and storage lead to the release of MNPs into the environment (Junaid and Wang 2022). Large plastic items such as bottles, bags, packaging materials, and fishing gear break down into smaller trashes due to UV radiation, mechanical abrasion, and weathering (Kiran et al. 2022).

3.4. Agricultural Activities, Landfills, and Atmospheric Deposition

Plastic films used in agriculture for mulching and greenhouse covering degrade over time, and release MNP particles into soil and water (Astner et al. 2022). These particles can accumulate in agricultural fields and nearby water bodies. Particles generated from road dust, agricultural activities, and industrial processes can contain

MNPs and contribute to atmospheric deposition (Caracci et al. 2023). Plastic waste disposed of in landfills undergoes degradation processes and releases MNPs into the surrounding environment through leachate and landfill gas emissions (Dube and Okuthe 2023). Improper disposal of plastic waste through littering and illegal dumping contributes to MNP pollution in terrestrial and aquatic ecosystems.

3.5. Tyre Wear, and Road Markings

Vehicle tyres contain synthetic rubber and other plastic additives. Wear and tear during driving releases tyre dust, which contains MNPs, into the environment (Kiran et al. 2022). Road markings contain plastic polymers that can degrade over time due to traffic and environmental factors and can release MNPs into the surrounding soil and water (Kumar et al. 2021).

3.6. Wastewater Treatment Plants (WWTPs), Combined Sewer Overflows, and Stormwater Runoff

MNPs from domestic and industrial sources are discharged into WWTPs. However, conventional treatment methods may not effectively remove all MNPs, and lead to release into receiving water bodies (Mohana et al. 2021). During heavy rainfall, combined sewer systems may overflow and can release untreated sewage containing MNPs directly into water bodies (Khan et al. 2022). Urban runoff and sewage effluents are significant pathways for the transport of MNPs from land-based sources to aquatic environments (Dube and Okuthe 2023). Rainwater sweeps plastic products off streets, avenues, and other urban locations and flows through the stormwater drains, into the river, lake, or sea (Meegoda and Hettiarachchi 2023).

Figure 1. Sources, formation, and transport route of MNPs in aquatic system

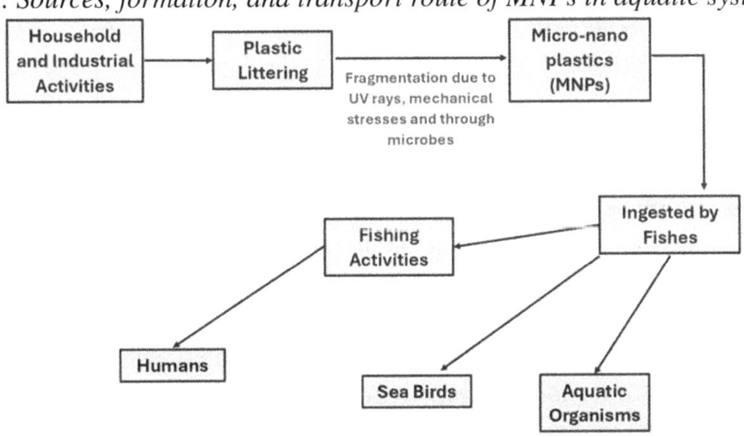

4. EFFECTS OF MICRO/NANO PLASTICS (MNPS) ON AQUATIC ORGANISM HEALTH

Most worrisome for the aquatic environment are marine nanoplastics, abbreviated as MNPs. The oddity in the research endeavors on the stressed aquatic ecosystems has shifted much emphasis toward MNP pollution in the last couple of years (Geremia et al. 2023; Massahi et al. 2024). The presence of MNP in aquatic habitats has raised some issues about its impacts on the health of aquatic animals. Some systematic research has revealed that MNPs are toxic to different types of water organisms including zooplankton, fish eggs and larvae, and water crustaceans. If ingested by water organisms, MNP particles may lodge themselves within the internal organs or become lodged in the digestive tract (GI) tract.

MNPs have a high surface-to-volume ratio and can adsorb the toxic metals and POPs that can enter the body of organisms after ingestion and may cause toxicity (Bhardwaj et al. 2019; Bhardwaj and Jindal 2020; Alam et al. 2023; Bhardwaj et al. 2023b). Consequently, pollutants can accumulate in the food chain with an even greater threat to the lives of aquatic animals (Peng et al. 2020). The MNP's physiochemical properties, including size, shape, content, and concentration, may influence their impacts on aquatic animals. Previous studies have demonstrated that MNPs elicit numerous physiological and biochemical effects in water-dwelling animals. Certain scientific papers suggest that MPs and NPs deposited in lower trophic level species can transport these compounds up to higher trophic level species where they may find their way to humans via the consumption of water fauna (Elizalde and Gómez 2021). Possible negative effects of MNPs on the physiological and morphological processes of metabolism food digestion and behavior. Hence the potential of MNPs to transform ecosystems or individual cells (Al-Thawadi 2020).

When exposed to MNPs, aquatic species such as fish, crabs, and mollusks experience changes in their growth, reproduction, and behavior (Kögel et al. 2020). Zebrafish are vulnerable to behavioural abnormalities, vasotoxicity, cytotoxicity, and developmental toxicity from MNPs. MPNs have an impact on phytoplankton, and negative effects include growth suppression, oxidative damage, and a reduction in photosynthetic activity. Major effects of MNP poisoning on zooplankton include slower growth, increased mortality, slower reproduction, and decreased fecundity. Ingested by zooplankton, MNPs can influence the migration of MNPs up the food chain and the connections between predators and prey. When MNPs are applied and exposed to sunlight, nano-sized titanium dioxide (TiO_2) becomes more detrimental to zooplankton and can enhance phytoplankton mortality (Sun et al. 2023).

The idea that MNPs biomagnify from lower trophic levels to higher trophic levels is currently not well supported by data. Some research has looked into the transfer of MNPs over a wide range of trophic levels in marine food chains, but no conclusive evidence of biomagnification has been found (Wang 2023). Data on the effects of MNPs on fish, which are higher trophic-level species, are currently scarce. Barría et al. (2020) have demonstrated that fish exposed to plastic particles smaller than 100 nm experience detrimental effects such as tissue accumulation, decreased locomotor, and foraging behaviours, implications on growth, the immune system, lipid metabolism, and neurotoxicity. Furthermore, histological analyses of fish exposed to MNPs have shown that these particles produce significant inflammatory reactions in the fish. Figure 2 shows the movement of MNP's toxicity from lower trophic levels to higher trophic levels.

Figure 2. The transfer of MNPs from lower trophic level to higher trophic level organisms

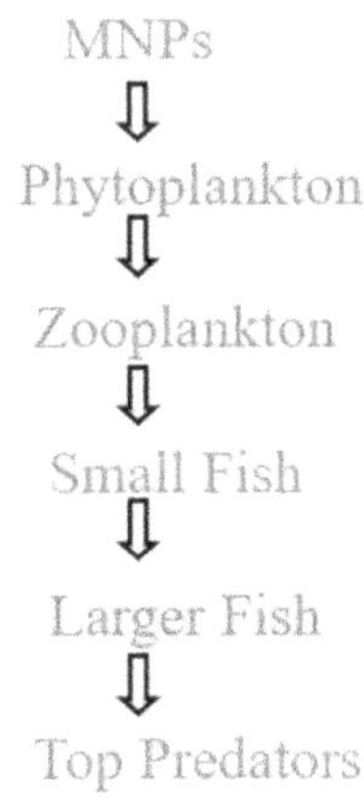

Other aquatic creatures MNPs can impact a variety of aquatic taxa, including mammals, ciliates, freshwater, and marine species. MNPs, for example, have the capacity to cause metabolic disturbances, increased absorption or translocation, inflammatory lesions, cytotoxicity, neurotoxicity, oxidative stress, and more (Rapa et al. 2023). Living organisms can easily access MNPs through the food chain because of their small size. Several harmful effects such as oxidative stress, disruption of the endocrine system, altered gene expression, neurotoxicity, abnormalities in reproduction, etc. have been reported in different studies and were due to the toxicity of MNP particles.

5. METHODS FOR THE DETECTION OF MICRO/NANO PLASTICS (MNPS)

The methods for the detection of MNPs are given below in detail.

5.1. Visual Identification

It is the basic method for the detection of MNP particles. We can detect only the shape, size, and color of the MNP by this method. However, only large MNP particles can be detected by this method (Kavaja et al. 2020).

5.2. Spectroscopic Method

This method is advanced and well-established from the visual identification method for the detection of MNP particles. In this method, the emission spectra of MNP particles are compared with the reference spectra. Tirkey and Upadhyay (2021) used Raman Spectroscopy for the identification of MNP particles in the aquatic environment. Fourier Transform Infrared Spectroscopy (FTIR) is used for the quantification of very small particles of MNPs. Particles can be detected by this method after the creation of vibrations. Micro-FTIR is used for the detection of very small particles of MNPs. Near Infrared Spectroscopy (NIR) is used to differentiate between plastic particles. NIR is a more advanced technique than FTIR. It can detect MNPs of size > 1 mm. Micro-NIR can identify MNP particles quickly (Pakhomova et al. 2020).

5.3. Pyrolysis Gas-Chromatography Mass-Spectrometry (Pyr-GC-MS) and Direct Pyrolysis Mass Spectrometry (DPMS)

Pyr-GC-MS is the most popular technique for the quantification of MNP on the basis of mass (Pico and Barcelo 2020). The plastic particles are pyrolyzed in an inert condition before entering the GC-MS. For this technique, only a small quantity of samples (0.35-7.0 mg) is needed. The temperature can reach up to 700°C. The samples are broken down into smaller particles after pyrolysis, and these particles are analyzed by GC-MS. To look into the degradation products of most polymers, use the DPMS approach. The polymer undergoes pyrolysis near the ion source, and the resulting primary chain cleavage products are promptly transported to the detector in order to get mass spectra. Thermally labile products are recognized in the absence of any subsequent reactions.

5.4. Nuclear Magnetic Resonance (NMR)

MNPs can be estimated by NMR based on their size or shape. In this technique, the signal is filtered to determine the number of MNPs in samples. Proton counts related to a particular resonance must be proportionate to signal intensities or integral values for quantitative NMR analysis to work. Using over 98% accuracy, qNMR offers accurate quantification. MNPs must be dissolved in a suitable deuterated solvent in order to perform NMR spectroscopy (Tirkey and Upadhyay 2021).

5.5. Scanning Electron Microscopy (SEM)

SEM produces an electron beam and makes pictures of MNPs with high resolution. The size and surface characteristics of MNPs can be identified by this technique. The appearance and topography of samples are determined by analyzing secondary electrons released by the signals created by the sample's interaction with the electron beam. Very small particles that are invisible to human eyes can be possible to identify quickly by using Energy-Dispersive X-ray Spectroscopy with Scanning Electron Microscopy (SEM-EDS) (Schwaferts et al. 2019).

5.6. Combined Method

This combined method is an advanced technique over a single technique because MNPs can be estimated more accurately by the combined method. The spectroscopic technique is applied with the microscopic technique for the estimation of MNPs. First MNP particles are estimated by spectroscopy and then the particles are counted by microscopy. This technique can increase the rate of analysis.

6. CONCLUSION AND RECOMMENDATIONS

MNPs are ubiquitous particles and are found in air, water, and soil. They affect more the aquatic environment. They persist in the environment for a long time and can harm aquatic organisms on cell and organ levels. They accumulate in the tissue of the lower trophic level organisms and move to the higher tropic level organisms. These particles can carry toxic chemicals and after ingestion, they cause harmful effects on humans also. Now, the researchers are worried about plastic pollution and are working on the mitigation of plastic pollution. The researchers suggest several recommendations to the regulatory body and government for the mitigation of plastic pollution. The efficient physical, chemical, and biological techniques should be developed for the removal of these tiny particles from the aquatic environment.

New guidelines should be made by the regulatory bodies to dump plastic garbage. People should be aware of waste management and the recycling of plastic waste. To mitigate MNP pollution, different sources of MNPs should be understood. More research is needed in the area of plastic pollution with sophisticated instruments. Advanced techniques should be developed for the estimation and identification of these particles.

ACKNOWLEDGEMENT

The authors are deeply grateful to Amity University, Noida for providing the platform to do this study.

REFERENCES

Al-Thawadi, S. (2020). Microplastics and nanoplastics in aquatic environments: Challenges and threats to aquatic organisms. *Arabian Journal for Science and Engineering*, 45(6), 4419–4440. DOI: 10.1007/s13369-020-04402-z

Alam, S., Bhardwaj, L. K., Mallick, R., & Rai, S. (2023). Estimation of heavy metals and fluoride ion in vegetables grown nearby the stretch of River Yamuna, Delhi (NCR), India. *Indian Journal of Environmental Protection*, 43(1), 64–73.

Astner, A. F., Hayes, D. G., O'Neill, H. M., Evans, B. R., Pingali, S. V., Urban, V. S., & Young, T. M. (2022). Forming micro-and nano-plastics from agricultural plastic films for employment in fundamental research studies. [Journal of Visualized Experiments] [PubMed]. *Journal of Visualized Experiments*, (185), e64112.

Barría, C., Brandts, I., Tort, L., Oliveira, M., & Teles, M. (2020). Effect of nanoplastics on fish health and performance: A review. [PubMed]. *Marine Pollution Bulletin*, 151, 110791. DOI: 10.1016/j.marpolbul.2019.110791

Barros, J., Kumar, S., & Seena, S. (2023). Does functionalised nanoplastics modulate the cellular and physiological responses of aquatic fungi to metals? [PubMed]. *Environmental Pollution*, 337, 122549. DOI: 10.1016/j.envpol.2023.122549

Bhardwaj, L., Sharma, S., Ranjan, A., & Jindal, T. (2019). Persistent organic pollutants in lakes of Broknes peninsula at Larsemann Hills area, East Antarctica. [PubMed]. *Ecotoxicology (London, England)*, 28(5), 589–596. DOI: 10.1007/s10646-019-02045-x

Bhardwaj, L. K. (2022). Evaluation of bis (2-ethylhexyl) phthalate (DEHP) in the PET bottled mineral water of different brands and impact of heat by GC–MS/MS. *Chemistry Africa*, 5(4), 929–942. DOI: 10.1007/s42250-022-00385-6

Bhardwaj, L. K. (2023). Occurrence of microplastics (MPs) in Antarctica and its impact on the health of organisms. *Maritime Technology and Research*, 6(2), 265418. DOI: 10.33175/mtr.2024.265418

Bhardwaj, L. K., & Jindal, T. (2020). Persistent organic pollutants in lakes of Grovnes Peninsula at Larsemann Hill area, East Antarctica. *Earth Systems and Environment*, 4(2), 349–358. DOI: 10.1007/s41748-020-00154-w

Bhardwaj, L. K., Kumar, D., & Kumar, A. (2023b). Phytoremediation Potential of Ocimum Sanctum: A Sustainable Approach for Remediation of Heavy Metals. In *Phytoremediation Potential of Medicinal and Aromatic Plants* (pp. 46-57). CRC Press. doi:DOI: 10.20944/preprints202308.0593.v1

Bhardwaj, L. K., Rath, P., & Choudhury, M. (2023a). A comprehensive review on the classification, uses, sources of nanoparticles (NPs) and their toxicity on health. *Aerosol Science and Engineering*, 7(1), 69–86. DOI: 10.1007/s41810-022-00163-4

Bhardwaj, L. K., Rath, P., Yadav, P., & Gupta, U. (2024). Microplastic contamination, an emerging threat to the freshwater environment: A systematic review. *Environmental Systems Research*, 13(1), 1–17. DOI: 10.1186/s40068-024-00338-7

Bhardwaj, L. K., & Sharma, A. (2021a). Estimation of physico-chemical, trace metals, microbiological and phthalate in PET bottled water. *Chemistry Africa*, 4(4), 981–991. DOI: 10.1007/s42250-021-00267-3

Bhardwaj, L. K., & Sharma, A. (2021b). Microplastics (MPs) in Drinking Water: Uses, Sources & Transport. https://doi.org/doi:10.20944/preprints202104.0498.v1

Caracci, E., Vega-Herrera, A., Dachs, J., Berrojalbiz, N., Buonanno, G., Abad, E., Llorca, M., Moreno, T., & Farré, M. (2023). Micro (nano) plastics in the atmosphere of the Atlantic Ocean. [PubMed]. *Journal of Hazardous Materials*, 450, 131036. DOI: 10.1016/j.jhazmat.2023.131036

Chaudhary, V., Thakur, N., Chaudhary, S., & Bangar, S. P. (2023). Remediation plan of nano/microplastic toxicity in food. [). Academic Press.] [PubMed]. *Advances in Food and Nutrition Research*, 103, 397–442. DOI: 10.1016/bs.afnr.2022.07.004

Cohen, N., & Radian, A. (2022). Microplastic textile fibers accumulate in sand and are potential sources of micro (nano) plastic pollution. [PubMed]. *Environmental Science & Technology*, 56(24), 17635–17642. DOI: 10.1021/acs.est.2c05026

de Souza, A. M., Santos, A. L., Araújo, D. S., de Barros Magalhães, R. R., & Rocha, T. L. (2022). Micro (nano) plastics as a vector of pharmaceuticals in aquatic ecosystem: Historical review and future trends. *Journal of Hazardous Materials Advances*, 6, 100068. DOI: 10.1016/j.hazadv.2022.100068

Di, M., Liu, X., Wang, W., & Wang, J. (2019). Corrigendum to" Pollution in drinking water source areas: Microplastics in the Danjiangkou Reservoir, China [Environ. Toxicol. Pharmacol. 65 (January) 2019 82-89] [PubMed]. *Environmental Toxicology and Pharmacology*, 66, 133–133. DOI: 10.1016/j.etap.2019.01.002

Ding, L., Mao, R., Guo, X., Yang, X., Zhang, Q., & Yang, C. (2019). Microplastics in surface waters and sediments of the Wei River, in the northwest of China. [PubMed]. *The Science of the Total Environment*, 667, 427–434. DOI: 10.1016/j.scitotenv.2019.02.332

Dube, E., & Okuthe, G. E. (2023). Plastics and Micro/Nano-Plastics (Mnps) in the environment: Occurrence, impact, and toxicity. [PubMed]. *International Journal of Environmental Research and Public Health*, 20(17), 6667. DOI: 10.3390/ijerph20176667

Elgarahy, A. M., Akhdhar, A., & Elwakeel, K. Z. (2021). Microplastics prevalence, interactions, and remediation in the aquatic environment: A critical review. *Journal of Environmental Chemical Engineering*, 9(5), 106224. DOI: 10.1016/j.jece.2021.106224

Elizalde-Velázquez, G. A., & Gómez-Oliván, L. M. (2021). Microplastics in aquatic environments: A review on occurrence, distribution, toxic effects, and implications for human health. [PubMed]. *The Science of the Total Environment*, 780, 146551. DOI: 10.1016/j.scitotenv.2021.146551

Gao, D., Junaid, M., Chen, X., Liao, H., Chen, G., & Wang, J. (2023). Interaction of micro (nano) plastics and bisphenols in the environment: A recent perspective on adsorption mechanisms, influencing factors and ecotoxic impacts. *Trends in Analytical Chemistry*, 165, 117132. DOI: 10.1016/j.trac.2023.117132

Gao, D., Liu, X., Junaid, M., Liao, H., Chen, G., Wu, Y., & Wang, J. (2022). Toxicological impacts of micro (nano) plastics in the benthic environment. [PubMed]. *The Science of the Total Environment*, 836, 155620. DOI: 10.1016/j.scitotenv.2022.155620

Geremia, E., Muscari Tomajoli, M. T., Murano, C., Petito, A., & Fasciolo, G. (2023). The Impact of Micro-and Nanoplastics on Aquatic Organisms: Mechanisms of Oxidative Stress and Implications for Human Health—A Review. *Environments (Basel, Switzerland)*, 10(9), 161. DOI: 10.3390/environments10090161

Geyer, R., Jambeck, J. R., & Law, K. L. (2017). Production, use, and fate of all plastics ever made. [PubMed]. *Science Advances*, 3(7), e1700782. DOI: 10.1126/sciadv.1700782

He, D., Zhang, Y., & Gao, W. (2021). Micro (nano) plastic contaminations from soils to plants: Human food risks. *Current Opinion in Food Science*, 41, 116–121. DOI: 10.1016/j.cofs.2021.04.001

Jain, H. (2023). Groundwater vulnerability and risk mitigation: A comprehensive review of the techniques and applications. *Groundwater for Sustainable Development*, 22, 100968. DOI: 10.1016/j.gsd.2023.100968

Jain, H., Yadav, V., Rajput, V. D., Minkina, T., Agarwal, S., & Garg, M. C. (2022). An eco-sustainable green approach for biosorption of methylene blue dye from textile industry wastewater by sugarcane bagasse, peanut hull, and orange peel: A comparative study through response surface methodology, isotherms, kinetic, and thermodynamics. *Water, Air, and Soil Pollution*, 233(6), 187. DOI: 10.1007/s11270-022-05655-0

Johnson, A. C., Ball, H., Cross, R., Horton, A. A., Jürgens, M. D., Read, D. S., Vollertsen, J., & Svendsen, C. (2020). Identification and quantification of microplastics in potable water and their sources within water treatment works in England and Wales. [PubMed]. *Environmental Science & Technology*, 54(19), 12326–12334. DOI: 10.1021/acs.est.0c03211

Junaid, M., Abbas, Z., Siddiqui, J. A., Liu, S., Tabraiz, S., Yue, Q., & Wang, J. (2023). Ecotoxicological impacts associated with the interplay between micro (nano) plastics and pesticides in aquatic and terrestrial environments. *Trends in Analytical Chemistry*, 165, 117133. DOI: 10.1016/j.trac.2023.117133

Junaid, M., & Wang, J. (2022). Interaction of micro (nano) plastics with extracellular and intracellular biomolecules in the freshwater environment. *Critical Reviews in Environmental Science and Technology*, 52(23), 4241–4265. DOI: 10.1080/10643389.2021.2002078

Kavya, A. N. L., Sundarrajan, S., & Ramakrishna, S. (2020). Identification and characterization of micro-plastics in the marine environment: A mini review. [PubMed]. *Marine Pollution Bulletin*, 160, 111704. DOI: 10.1016/j.marpolbul.2020.111704

Khan, N. A., Khan, A. H., López-Maldonado, E. A., Alam, S. S., López, J. R. L., Herrera, P. F. M., & Singh, L. (2022). Microplastics: Occurrences, treatment methods, regulations and foreseen environmental impacts. [PubMed]. *Environmental Research*, 215, 114224. DOI: 10.1016/j.envres.2022.114224

Kiran, B. R., Kopperi, H., & Venkata Mohan, S. (2022). Micro/nano-plastics occurrence, identification, risk analysis and mitigation: Challenges and perspectives. [PubMed]. *Reviews in Environmental Science and Biotechnology*, 21(1), 169–203. DOI: 10.1007/s11157-021-09609-6

Koelmans, A. A., Nor, N. H. M., Hermsen, E., Kooi, M., Mintenig, S. M., & De France, J. (2019). Microplastics in freshwaters and drinking water: Critical review and assessment of data quality. [PubMed]. *Water Research*, 155, 410–422. DOI: 10.1016/j.watres.2019.02.054

Kögel, T., Bjorøy, Ø., Toto, B., Bienfait, A. M., & Sanden, M. (2020). Micro-and nanoplastic toxicity on aquatic life: Determining factors. [PubMed]. *The Science of the Total Environment*, 709, 136050. DOI: 10.1016/j.scitotenv.2019.136050

Kumar, M., Chen, H., Sarsaiya, S., Qin, S., Liu, H., Awasthi, M. K., Kumar, S., Singh, L., Zhang, Z., Bolan, N. S., Pandey, A., Varjani, S., & Taherzadeh, M. J. (2021). Current research trends on micro-and nano-plastics as an emerging threat to global environment: A review. [PubMed]. *Journal of Hazardous Materials*, 409, 124967. DOI: 10.1016/j.jhazmat.2020.124967

Kumar, P., Inamura, Y., Bao, P. N., Abeynayaka, A., Dasgupta, R., & Abeynayaka, H. D. (2022). Microplastics in freshwater environment in Asia: A systematic scientific review. *Water (Basel)*, 14(11), 1737. DOI: 10.3390/w14111737

Kundu, A., Shetti, N. P., Basu, S., Reddy, K. R., Nadagouda, M. N., & Aminabhavi, T. M. (2021). Identification and removal of micro-and nano-plastics: Efficient and cost-effective methods. [PubMed]. *Chemical Engineering Journal*, 421, 129816. DOI: 10.1016/j.cej.2021.129816

Lionetto, F., Esposito Corcione, C., Rizzo, A., & Maffezzoli, A. (2021). Production and characterization of polyethylene terephthalate nanoparticles. [PubMed]. *Polymers*, 13(21), 3745. DOI: 10.3390/polym13213745

Luo, W., Su, L., Craig, N. J., Du, F., Wu, C., & Shi, H. (2019). Comparison of microplastic pollution in different water bodies from urban creeks to coastal waters. [PubMed]. *Environmental Pollution*, 246, 174–182. DOI: 10.1016/j.envpol.2018.11.081

Massahi, T., Amin, A. A., Meshabaz, R. A., Pirsaheb, M., Terry, L., Makhdoumi, P., Kianpour, S., Zamani, F., & Hossini, H. (2024). Microplastic occurrence and its potential role as a carrier for SARS-CoV-2 in health center wastewater treatment plant and surface water. *Journal of Sea Research*, 198, 102477. DOI: 10.1016/j.seares.2024.102477

Meegoda, J. N., & Hettiarachchi, M. C. (2023). A Path to a Reduction in Micro and Nanoplastics Pollution. [PubMed]. *International Journal of Environmental Research and Public Health*, 20(8), 5555. DOI: 10.3390/ijerph20085555

Mohana, A. A., Farhad, S. M., Haque, N., & Pramanik, B. K. (2021). Understanding the fate of nano-plastics in wastewater treatment plants and their removal using membrane processes. [PubMed]. *Chemosphere*, 284, 131430. DOI: 10.1016/j.chemosphere.2021.131430

Ojinnaka, D., & Aw, M. (2020). Micro and nano plastics: A consumer perception study on the environment, food safety threat and control systems. *Biomedical Journal of Scientific & Technical Research*, 32(2). Advance online publication. DOI: 10.26717/BJSTR.2020.31.005064

Ouda, M., Banat, F., Hasan, S. W., & Karanikolos, G. N. (2023). Recent advances on nanotechnology-driven strategies for remediation of microplastics and nanoplastics from aqueous environments. *Journal of Water Process Engineering*, 52, 103543. DOI: 10.1016/j.jwpe.2023.103543

Pakhomova, S., Zhdanov, I., & van Bavel, B. (2020). Polymer type identification of marine plastic litter using a miniature near-infrared spectrometer (MicroNIR). *Applied Sciences (Basel, Switzerland)*, 10(23), 8707. DOI: 10.3390/app10238707

Peiponen, K. E., Räty, J., Ishaq, U., Pélisset, S., & Ali, R. (2019). Outlook on optical identification of micro-and nanoplastics in aquatic environments. [PubMed]. *Chemosphere*, 214, 424–429. DOI: 10.1016/j.chemosphere.2018.09.111

Peng, L., Fu, D., Qi, H., Lan, C. Q., Yu, H., & Ge, C. (2020). Micro-and nano-plastics in marine environment: Source, distribution and threats—A review. [PubMed]. *The Science of the Total Environment*, 698, 134254. DOI: 10.1016/j.scitotenv.2019.134254

Pico, Y., & Barcelo, D. (2020). Pyrolysis gas chromatography-mass spectrometry in environmental analysis: Focus on organic matter and microplastics. *Trends in Analytical Chemistry*, 130, 115964. DOI: 10.1016/j.trac.2020.115964

Pourebrahimi, S., & Pirooz, M. (2023). Microplastic pollution in the marine environment: A review. *Journal of Hazardous Materials Advances*, 10, 100327. DOI: 10.1016/j.hazadv.2023.100327

Râpă, M., Darie-Ni ă, R. N., Matei, E., Predescu, A. M., Berbecaru, A. C., & Predescu, C. (2023). Insights into Anthropogenic Micro-and Nanoplastic Accumulation in Drinking Water Sources and Their Potential Effects on Human Health. [PubMed]. *Polymers*, 15(11), 2425. DOI: 10.3390/polym15112425

Rytelewska, S., & Dąbrowska, A. (2022). The Raman spectroscopy approach to different freshwater microplastics and quantitative characterization of polyethylene aged in the environment. *Microplastics*, 1(2), 263–281. DOI: 10.3390/microplastics1020019

Schwaferts, C., Niessner, R., Elsner, M., & Ivleva, N. P. (2019). Methods for the analysis of submicrometer-and nanoplastic particles in the environment. *Trends in Analytical Chemistry*, 112, 52–65. DOI: 10.1016/j.trac.2018.12.014

Slootmaekers, B., Carteny, C. C., Belpaire, C., Saverwyns, S., Fremout, W., Blust, R., & Bervoets, L. (2019). Microplastic contamination in gudgeons (Gobio gobio) from Flemish rivers (Belgium). [PubMed]. *Environmental Pollution*, 244, 675–684. DOI: 10.1016/j.envpol.2018.09.136

Sun, Y., Qian, Y., Geng, S., Wang, P., Zhang, L., & Yang, Z. (2023). Joint effects of microplastics and ZnO nanoparticles on the life history parameters of rotifers and the ability of rotifers to eliminate harmful phaeocystis. [PubMed]. *Chemosphere*, 310, 136939. DOI: 10.1016/j.chemosphere.2022.136939

Tan, X., Yu, X., Cai, L., Wang, J., & Peng, J. (2019). Microplastics and associated PAHs in surface water from the Feilaixia Reservoir in the Beijiang River, China. [PubMed]. *Chemosphere*, 221, 834–840. DOI: 10.1016/j.chemosphere.2019.01.022

Tirkey, A., & Upadhyay, L. S. B. (2021). Microplastics: An overview on separation, identification and characterization of microplastics. [PubMed]. *Marine Pollution Bulletin*, 170, 112604. DOI: 10.1016/j.marpolbul.2021.112604

Wang, H., Wang, Q., Lv, M., Li, J., Zhao, X., Song, Z., Wang, X., You, J., Wang, Y., & Chen, L. (2023). Marine micro (nano) plastics determination and its environmental toxicity evaluation. *Trends in Analytical Chemistry*, 168, 117332. DOI: 10.1016/j.trac.2023.117332

Wang, W. X. (2023). Environmental toxicology of marine microplastic pollution. Cambridge Prisms. *Plastics*, 1, e10. DOI: 10.1017/plc.2023.9

Xiong, X., Wu, C., Elser, J. J., Mei, Z., & Hao, Y. (2019). Occurrence and fate of microplastic debris in middle and lower reaches of the Yangtze River–from inland to the sea. [PubMed]. *The Science of the Total Environment*, 659, 66–73. DOI: 10.1016/j.scitotenv.2018.12.313

Yan, M., Nie, H., Xu, K., He, Y., Hu, Y., Huang, Y., & Wang, J. (2019). Microplastic abundance, distribution and composition in the Pearl River along Guangzhou city and Pearl River estuary, China. [PubMed]. *Chemosphere*, 217, 879–886. DOI: 10.1016/j.chemosphere.2018.11.093

Yang, L., Zhang, Y., Kang, S., Wang, Z., & Wu, C. (2021). Microplastics in soil: A review on methods, occurrence, sources, and potential risk. [PubMed]. *The Science of the Total Environment*, 780, 146546. DOI: 10.1016/j.scitotenv.2021.146546

Yuan, W., Liu, X., Wang, W., Di, M., & Wang, J. (2019). Microplastic abundance, distribution and composition in water, sediments, and wild fish from Poyang Lake, China. [PubMed]. *Ecotoxicology and Environmental Safety*, 170, 180–187. DOI: 10.1016/j.ecoenv.2018.11.126

Zhang, S., Wang, J., Yan, P., Hao, X., Xu, B., Wang, W., & Aurangzeib, M. (2021). Non-biodegradable microplastics in soils: A brief review and challenge. [PubMed]. *Journal of Hazardous Materials*, 409, 124525. DOI: 10.1016/j.jhazmat.2020.124525

Zhang, X., Leng, Y., Liu, X., Huang, K., & Wang, J. (2020). Microplastics' pollution and risk assessment in an urban river: A case study in the Yongjiang River, Nanning City, South China. *Exposure and Health*, 12(2), 141–151. DOI: 10.1007/s12403-018-00296-3

Chapter 4

Plant Stress Management Mitigating Micro/Nano Plastics Pollution

Suresh Waghamare
https://orcid.org/0009-0001-8909-2209
Rashtriya College of Pharmacy, Hatnoor, India

Poonam Kodag
Rashtriya College of Pharmacy, Hatnoor, India

ABSTRACT

Abstract Microplastics (MPs) and nanoplastics (NPs) are pervasive environmental pollutants, originating from plastic fragmentation and manufacturing processes. They enter ecosystems via wastewater discharge and littering, posing risks to wildlife and human health. In agriculture, MNPs disrupt soil health and crop productivity. Understanding their behavior is crucial for mitigation. Plant responses include nutrient uptake disruption and oxidative stress. Mitigation strategies include bio-augmentation, nanomaterials, and phytoremediation. Biochar and mustard plants show promise in MNP capture. Case studies highlight successful applications in compost for agriculture, emphasizing the need for further research and standardized monitoring for sustainable practices.

1. INTRODUCTION

Microplastics (MPs), nanoplastics (NPs), and tiny plastic fragments and particles less than 5 millimeters in size have become a pervasive environmental concern. Their presence in everything from the deepest ocean trenches to the highest moun-

DOI: 10.4018/979-8-3693-9163-1.ch004

tain peaks paints a concerning picture of plastic pollution's ubiquitous reach. Let's delve into the origins and rise of this environmental challenge.

The widespread adoption of plastic materials in the latter half of the 20th century revolutionized various industries. Plastics offered many advantages – they were lightweight, durable, and inexpensive. This led to a surge in plastic production and consumption, with global plastic production exceeding 380 million tons in 2015 according to a 2017 report by Plastics Europe, an industry association. The convenience and affordability of single-use plastic products like bags, bottles, and packaging materials further fueled this rise.

From Macro to Micro Plastic debris from various sources undergoes environmental fragmentation. Sunlight, wave action, and weathering break down larger plastic items into progressively smaller pieces, eventually generating MPs and NPs. This fragmentation process can be accelerated by mechanical forces like abrasion during transport or by biological processes as microbes break down the plastic. The resulting microplastics can further degrade into even smaller nanoplastics, raising concerns about the potential for plastic pollution to reach a truly nanoscopic scale. (Geyer et al., 2017)

- **Primary Microplastics:** These are manufactured directly in microscopic sizes, such as microbeads in cosmetics or plastic fibers released from synthetic clothing during washing. These tiny plastics enter the environment directly through product use or during the manufacturing process. A 2017 study published in Environmental Science & Technology Letters estimated that synthetic textiles could release up to 1.9 million fibers per wash.(Campanale et al., 2020)
- **Secondary Microplastics:** These arise from the fragmentation of larger plastic items, including plastic bags, bottles, and packaging materials. This fragmentation process is a major contributor to microplastic pollution, with estimates suggesting that secondary microplastics account for a larger proportion of the overall environmental microplastic burden. (Law & Thompson, 2014)
- **The Nanoplastics Enigma:** The origins of NPs are less well understood. Some might be directly manufactured as nanomaterials for various industrial applications, such as in electronics or cosmetics. However, a significant portion likely originates from the breakdown of larger microplastics or the fragmentation of plastic composites containing nanofillers. The small size and unique properties of nanoplastics raise additional concerns, as their behavior in the environment and potential interactions with biological systems are still being actively researched.(Sultana et al., 2022)

1.2. The Ubiquitous Reach of Microplastic Pollution

MPs and NPs easily enter the environment through various pathways, including wastewater discharge, littering, and atmospheric deposition. Wastewater treatment plants are not always efficient at capturing these tiny particles, allowing them to enter waterways. Littering and improper waste disposal contribute significantly, with plastic debris breaking down or being washed away into rivers and ultimately oceans. Wind and rain can transport microplastics over long distances, contaminating even remote regions like the Arctic or High Mountain ranges. A study published in Nature Geoscience reported finding microplastics in virtually all air samples collected worldwide. (Duis & Coors, 2016)

Tiny particles can absorb harmful pollutants onto their surfaces, potentially introducing them into the food chain as they are ingested by organisms. This bio-accumulation of toxins can have cascading effects on ecosystem health.(Soares et al., 2020)

1.3. Impact of Micro/Nano-Plastics on Plant Health

The widespread existence of tiny plastic pieces and particles known as micro-plastics (MPs) and nanoplastics (NPs) is becoming a serious worry for the health of plants. Through plastic degradation, wastewater irrigation, and agricultural activities, these toxins find their way into terrestrial ecosystems. MPs and NPs can negatively affect plant development once they are inside the soil matrix through a variety of methods. The main worry is that plant roots might be physically damaged directly. (Jia et al., 2023) MPs and NPs are encountered by roots when they take up water and nutrients. The plant's capacity to effectively absorb vital resources for growth and development may be hampered by these interactions, which may result in mechanical disruption of root structures. MPs and NPs can also change important soil characteristics like aeration and drainage, which can cause stressful root environments and perhaps jeopardize plant health. (Roy et al., 2023)

1.4 Importance of Controlling Agriculture's Micro/Nanoplastic Pollution

Because it may hurt crop yield, soil health, and food safety, controlling micro and nanoplastic contamination in agriculture is crucial. Over time, these microscopic plastic particles—which come from a variety of sources, including synthetic fibers, decomposed plastic debris, and plastic mulching—may build up in agricultural soils. Their presence can affect the availability of nutrients, water retention, and porosity in the soil, all of which have an impact on plant growth and development. Micro/

nanoplastics may also adsorb dangerous pollutants and agrochemicals, which might extend their stay in the environment and cause crops to absorb them. This could harm human health throughout the food chain.(Maddela et al., 2023)

Sustainable food production and environmental preservation depend on the implementation of efficient control solutions for micro- and nanoplastic contamination in agriculture. Agricultural systems may be made less susceptible to the introduction of micro- and nanoplastics by implementing strategies including utilizing biodegradable mulching materials, implementing precision farming techniques to limit plastic usage, and endorsing circular economy ideas for the management of plastic waste. Furthermore, boosting biodiversity in the soil, improving soil health through organic farming methods, and routinely checking soil quality can all help to lessen the negative effects of micro- and nanoplastics on agricultural sustainability. (Moeck et al., 2022) To fully address micro/nanoplastics pollution in agriculture, governments, researchers, farmers, and the general public must work together to create new solutions, enforce laws, and increase public awareness.

2. SOURCES OF ENVIRONMENTAL MICRO/NANOPLASTICS

The environment and maybe human health are seriously threatened by microplastics (MPs) and nanoplastics (NPs), which are tiny plastic fragments and particles smaller than five millimeters. Given their ubiquitous occurrence in the air, water, and soil, a complete grasp of their origins is necessary to successfully counteract this escalating issue. This research provides an in-depth examination of the primary environmental sources of microplastics and nanoplastics, as well as the ecological implications of these materials.

Fragmentation of Plastic Debris: A Cascade of Consequences

Environmental Degradation: A substantial source of MPs and NPs originates from the breakdown of larger plastic items. Sunlight, wave action, weathering, and biological activity can fragment plastic bags, bottles, packaging materials, and other debris into progressively smaller pieces. These fragments can further degrade into nanoplastics over time, amplifying the environmental burden. Biofilms forming on plastic debris in the ocean can accelerate this degradation by creating a microenvironment conducive to microbial breakdown (Bourdages et al., 2021). Textile Microfiber Shedding: Synthetic clothing releases microplastic fibers during laundering, polluting waterways through wastewater discharge. These fibers contribute to microplastic pollution in oceans and rivers. A 2017 study in Environmental Science & Technology Letters estimated that synthetic garments could release up to

1.9 million fibers per wash. Furthermore, these fibers can become airborne during tumble drying, potentially contaminating indoor environments and entering the human respiratory system through inhalation (Tiffin et al., 2022).

Microplastics find use in various industrial settings, including abrasive blasting materials, coatings for paints and varnishes, and additives in lubricants. Accidental spills or improper disposal of these materials can directly introduce microplastics into the environment. Industrial processes can also release microplastics into wastewater streams, potentially overloading treatment plants and exacerbating waterway contamination (Boucher & Friot, 2017).

2.1. Land-Based Sources: A Significant Contributor

Wastewater treatment plants are not foolproof in capturing microplastics. Tiny particles can evade treatment processes and be released into waterways, contributing to microplastic pollution in rivers, lakes, and ultimately the oceans. Upgrading wastewater treatment facilities with advanced filtration technologies specifically designed to capture microplastics is crucial for mitigating this source of environmental contamination. (Sun et al., 2019) Plastic mulch films used in agriculture can degrade over time, releasing microplastics into the soil. Additionally, fertilizers or biosolids applied to land might contain microplastic contaminants from sources like sewage sludge. These microplastics can be transported by rain or irrigation water, eventually reaching waterways. (Iqbal et al., 2023) The impact of microplastics on agricultural soil health and crop productivity is a growing research area, with potential concerns regarding reduced soil fertility and hindered plant growth.

2.2. Atmospheric Deposition: A Global Reach

Microplastics can become airborne through various processes, including windblown dust from landfills or wave action creating sea spray. These airborne microplastics can travel vast distances and contaminate even remote areas, highlighting the global scope of microplastic pollution. A 2019 study in Nature Geoscience reported finding microplastics in virtually all air samples collected worldwide. Microplastics deposited on land can also be washed away by rain into rivers and streams, ultimately reaching the oceans. (Enyoh et al., 2019)

2.3. Behavior and Fate of Micro/Nano-Plastics in Soil and Water Systems

Once released into the environment, micro/nanoplastics embark on a complex journey through soil and water systems. Their behavior and ultimate fate are influenced by various factors, including Smaller and less dense microplastics tend to remain suspended in water for longer durations, increasing their potential for transport over vast distances. Conversely, larger and denser particles may sink and accumulate in sediments. (K.L. et al., 2022) The surface characteristics of micro/nanoplastics, such as their charge and hydrophobicity (water repellency), determine how they interact with water molecules, soil particles, and organic matter. These interactions can influence their movement, attachment, and potential for aggregation (clumping together). (K.L. et al., 2022) Water flow velocity, turbulence, and temperature all play a role in the transport and distribution of micro/nanoplastics. For instance, prolonged periods of rain can move microplastics that have been collected in the soil and transfer them to bodies of surface water. (Ouyang et al., 2020) Their movement may be restricted by this connection, which may allow soil bacteria to degrade them (though this mechanism is not entirely understood). Microplastics can be consumed by earthworms and other soil invertebrates, which may hurt their health and help the particles spread throughout the soil food web.

2.4. Plant Uptake and Distribution of Micro/Nanoplastics

The invasion of micro/nanoplastics (MNPs) in the environment is also affecting the plants. They result in plastic particles that can be absorbed by plants through several pathways and can also move around in the plant leading to concerns with food safety and health of the plants.

- **Roots:** Plant roots serve as the primary defense against MNPs. The microplastics present in the soil remain absorbed and incorporated by root hairs or diffuse in through small fissures and openings or even sweat pores. Size, shape, and surface aspects of MNPs are determinants of plant absorption(Azeem et al., 2021)
- **Leaves:** Other MNPs in the atmosphere or the irrigation water may also attach themselves to leaf surfaces. Depending on the size and characteristics of the MNPs and the leaf structure, they may enter stomata (tiny pores) or become embedded in the waxy cuticle layer. (Azeem et al., 2021)

2.5. Translocation Within Plants

Once inside the plant, MNPs can be transported through the vascular system, primarily driven by the transpiration stream. This stream pulls water and dissolved nutrients from the roots to the leaves for photosynthesis. Microscopic MNPs, particularly nanoplastics, are small enough to be carried along with this flow, potentially reaching various plant tissues, including fruits and seeds. (Dehghanian et al., 2023)

- **Plant species:** Some plants are more or less able to take up and translocate MNPs.
- **MNP characteristics:** In addition to the aforementioned factors, the relative mobility of the MNPs within the plant is also affected by their size or shape and surface properties.
- **Environmental conditions:** Environmental factors such as soil moisture and temperature might also affect the rate of MNP uptake and MNP movement.
- **Disruption of plant growth and development:** MNPs can interfere with the uptake of nutrients, the transport of water, photosynthetic processes, and other activities that normally enable plant growth.
- **Effects on human health through the food chain:** The edible parts of plants also contain MNP and MNPs consumed by animals and humans can enter the food chain causing health concerns.
- **Impact on soil health:** The MNPs can also mingle with soil microorganisms and in effect interfere with important soil functions that are responsible for plant nutrition cycling and health.

3. EFFECTS OF MICRO/NANO-PLASTICS ON PLANT PHYSIOLOGY

3.1 Distribution and Uptake of Micro/Nano Plastics by Plants

The presence of micro/Nano-plastics (MNPs) in the environment as a new contaminant also brings forth new types of stress on the plants stimulating a series of cellular physiological responses. Expanding more in this part, the details of this process of plant response to MNPs are considered. Disruption of nutrient and photosynthetic processes. Root hairs and stomata placed over the MNPs serve as physical barriers blocking the uptake of soil nutrients and water respectively, which greatly limits nutrient absorption and water. Expand the areas in which MNPs can be overcome are the water and required micronutrient absorption necessary for plant growth and development. In addition, MNPs could reduce the light availability to the chloroplasts which are the known sites of photosynthesis. Besides, they may also inhibit the synthesis of chlorophyll further limiting the capacity of a plant to utilize sunlight energy in its growth processes. (Li et al., 2023)

Oxidative Stress and Cellular Damage

Exposure to MNPs can trigger the production of reactive oxygen species (ROS), highly reactive molecules that damage cellular components expand more this oxidative stress can lead to stunted growth, programmed cell death (apoptosis), and impaired plant function. Plants respond by activating antioxidant defense mechanisms to scavenge ROS. However, this is an energetically expensive process, diverting resources from growth and reproduction. (Yu et al., 2020)

Altered Gene Expression and Hormonal Imbalance

MNPs have the potential to disrupt gene expression patterns, impacting various physiological processes such as growth, stress response, and defense mechanisms. This disruption can lead to maladaptive responses and reduced resilience. Additionally, MNPs might interfere with plant hormone production or signaling pathways, resulting in developmental abnormalities and hindering a plant's ability to adapt to environmental fluctuations. (Abideen et al., 2022)

Reduced Seed Germination and Plant Growth:

MNPs can physically impede water uptake by seeds, hindering their germination and subsequent seedling establishment. The combined effects of disrupted nutrient uptake, compromised photosynthesis, and stress responses manifest as stunted plant growth, reduced biomass accumulation, and ultimately, lower crop yields.

Severity of the Response

The severity of these physiological responses hinges on several factors, including, Intrinsic differences in plant biology that lead to varying sensitivities to MNPs. Exclamation Some species exhibit a greater degree of tolerance than others. Exclamation. Size, shape, surface properties, and the chemical composition of MNPs can influence their interaction with plant tissues and the intensity of the response. Higher MNP concentrations and extended exposure periods generally lead to more pronounced physiological effects.(Zuverza-Mena et al., 2017)

3.2 Bimolecular and Cellular Mechanisms of Plant Stress Under Micro/Nano-plastics Pollution

The ubiquitous presence of micro/nano-plastics (MNPs) in the environment poses a novel and concerning stressor for plant life. This section delves into the biomolecular and cellular mechanisms underlying plant stress responses to MNP pollution, providing a deeper understanding of this nascent environmental threat.

MNPs, particularly nanoplastics due to their minute size, can physically disrupt plant cell membranes. This compromises their selective permeability, hindering essential functions like nutrient and water transport. The resulting cellular homeostasis dysregulation triggers stress response pathways. MNPs have the potential to interact with receptors on plant cell membranes, inadvertently triggering signaling cascades. These unintended pathways can alter gene expression and lead to the production of stress hormones. Such hormonal imbalances can disrupt plant growth, development, and defense mechanisms. (Jia et al., 2023b)

Plants possess a sophisticated defense system that includes the activation of stress-responsive genes. These genes encode proteins involved in antioxidant defense, detoxification, and repair mechanisms. However, the activation of these pathways requires significant energy and resources, potentially limiting plant growth and development. MNP exposure can disrupt the delicate balance of plant hormones, such as auxin, cytokinin, and ethylene. This hormonal imbalance can lead to stunted growth, developmental abnormalities, and a weakened defense system. (Sood et al., 2021)

By targeting specific pathways or enhancing plant defense mechanisms, we can potentially develop strategies to mitigate the detrimental effects of MNP pollution on plant health. Assessing food safety risks a deeper understanding of MNP uptake and translocation within plants is vital for ensuring food safety and consumer health. Informing Risk Assessment models comprehensive understanding of plant responses allows for the development of more accurate models to predict the ecological impact of MNP pollution. (Jia et al., 2023a)

4 COMPREHENSIVE STRATEGIES FOR MITIGATING MICROPLASTIC AND NANOPLASTIC (MNP) POLLUTION IN PLANT MANAGEMENT

The growing threat of microplastic and nanoplastic (MNP) pollution in the environment necessitates a multi-pronged approach to safeguard plant health and agricultural productivity. While directly managing MNPs within plants remains an area of active research, effective strategies exist at the agricultural and ecological levels. Here, we delve deeper into these strategies, exploring both preventative and responsive measures.

4.1. Upstream Prevention: Breaking the Plastic Cycle

However, the best strategy can be said to be based on stopping the creation of plastic in the first place. Promising policies that racemize using and sale of single-use plastics and the promotion of biodegradable plastics is also very essential. It is recommended to engage policymakers and other waste management structures to develop efficient waste collection systems as well as proper disposal, which would greatly minimize the general environmental impact of created MNPs. Further, promoting improved recycling systems that will effectively capture microplastics and recycle them instead of being dumped into the environment might also go a long way in the future. (Liu et al., 2024)

4.2. Fostering a Resilient Soil Ecosystem

Since plants can take up MNP directly, several barriers prevent this uptake such as healthy soil. Research indicates that the presence of organic matter additions such as compost or biochar may be capable of reducing the absorption of MNP. Its amendments improve nutrient release and recycling, promote the beneficial assembly of microbes that might immobilize MNPs, and formulate barriers that prevent MNP translocation to the plant roots. Implementing practices such as cover cropping and minimum tillage may also create more benefits to enhance the ability of soils to self-heal. (Verma et al., 2023)

4.3. Water Management for MNP Minimization

On the other occasion, contamination by MNP may be through irrigation water. Minimization of contamination from MNP through this route can therefore be achieved through irrigation with harvested rainwater or water from other sources with low MNP contents, particularly for sensitive crops. Likewise, evaluating more efficient irrigation techniques such as drip irrigation could reduce exposure of MNP to water and thus lessen the MNP load delivered to plant roots. Using pre-filtration systems under situations where the irrigation water sources contain high MNP concentrations may also be effective.

4.4. Strategic Plant Selection and Management Techniques

Studies indicate that some kinds of plants are more or less resistant to MNPs. Supposing that dead MNP settles at the base of these plants with the facility of tough cell wall, or the parts of the plant used for food are not<|reserved_special_token_258|> under the soil, then the said areas should select plants which should not be prone to MNP attack. It can be also useful to apply crop rotation measures with the help of MNP-tolerant species which will decrease the influence on agriculture productivity. Additionally, exploring the potential of hyperaccumulator plants, known for their ability to absorb and concentrate specific pollutants, might lead to phytoremediation techniques for MNP-contaminated soils in the future.(El-Saadony et al., 2022)

4.5. Public Awareness and Collaborative Action for Long-Term Sustainability

Fostering public awareness regarding MNP pollution is essential. Educating individuals and communities about the consequences of plastic usage and promoting responsible waste disposal practices can significantly reduce MNP influx into the environment. Collaborative efforts between policymakers, scientists, farmers, and consumers play a vital role in developing and implementing effective mitigation strategies and fostering a more sustainable future for our agricultural and ecological systems. Investment in research and development of bioplastics, biodegradable materials, and improved recycling technologies can pave the way for a more sustainable future.

5. ROLE OF PLANT STRESS MANAGEMENT IN MITIGATING MICRO/NANO-PLASTICS POLLUTION

Microplastics (MPs) and nanoplastics (NPs), insidious environmental contaminants, pose a growing threat to ecosystems worldwide. Expanding their minuscule size and pervasiveness necessitates innovative solutions for remediation. Enter phyto remediation, a bioremediation technique that harnesses the power of plants to cleanse the environment.expand_morethis eco-friendly approach offers a promising avenue for tackling MP/NP pollution, promoting a sustainable path forward.

Table 1. Mechanisms and benefits of phytoremediation for MP/NP remediation

Sr.No	Topic	Detail	Reference
1.	**Plants as Nature's Filters: Unveiling the Mechanisms of MNP Capture**	**Role of Plants:** Plants act as biological filters with root hairs trapping tiny plastic particles. Some plants absorb specific types of plastics.	(Jia et al., 2023a)
		Selective Absorption: Certain plant species, like ryegrass, show a preference for absorbing specific plastics like polyethylene. This can help in creating targeted phytoremediation strategies.	
		Microbial Muster: The rhizosphere, full of microbes, aids in MP/NP remediation. Some microbes can break down plastic polymers, enhancing biodegradation.	
2.	**Beyond Green: The Broader Benefits of Phytoremediation for MNP Remediation**	**Sustainable Ally:** Phytoremediation uses sunlight, reducing environmental impact and addressing plastic pollution.	(Yan et al., 2020)
		Holistic Ecosystem Improvement: Plants improve soil health by promoting microbial communities, nutrient cycling, and soil structure, fostering a thriving ecosystem.	
		Aesthetics with a Purpose: MP/NP-accumulating plants can turn contaminated areas into green spaces, enhancing public well-being.	
3.	**Challenges and the Road Ahead: Optimizing Plant-Based Remediation**	**Scaling Up for Impact:** Developing strategies to break down larger plastics for plant uptake is crucial. This might involve mechanical preprocessing or using specific microbes.	(Ali et al., 2021)
		Safe Plastic Disposal: The ways of effectively destroying or reducing the scope of captured plastics within the plant or via composting are also crucial. This consists of using specific enzymes or regulating plant metabolism in some way.	
		Long-Term Effects: This paper pointed out the deficit of the long-term impact studies of MP/NP accumulation on plant health and the possible transfer of toxicities in the food chain for efficient phytoremediation.	

5.4 Improving Plant Resistance to Micro/Nanoplastics Using Biotechnological Approaches

Microplastics (MPs) and nanoplastics (NPs) are relatively new pollutants that have a propensity for entering the environment and negatively impacting plants. Although phytoremediation presents a good approach to achieving this in capturing these offenders, it is also crucial to arm plants with enhanced tolerances for effectiveness. Strengthening plants' ability to cope with the effects of exposure to MP/NPs appears as one of the promising areas addressed with the help of biotechnological interventions. As it is seen, when exposed to MPs/NPs, plants are subjected to a four-pronged attack. Because of MPs/NPs presence in the soil, root growth is restricted, and therefore a plant cannot get water and nutrients. This narrows the plants' growth and results in low yields. To some extent, microplastic fibers can pick up essential nutrients such as phosphorus and nitrogen then become unavailable for plant uptake. This worsens nutrition depletion status further straining the plants and consequently malformation. MP/NP exposure in plants induces the formation of detrimental reactive oxygen species that occur in the plants. Some of these ROS cause damage to plant cells and tissues therefore compromising the plant's defense. (Jia et al., 2023a)

5.5. Biotechnology Offers a Multifaceted Toolkit to Enhance Plant Stress Tolerance, Paving the way for a more Resilient Future:

Genetic engineering can be used to introduce genes from other organisms that offer resistance to specific stressors. For instance, genes encoding for enzymes that degrade plastic polymers could be introduced into plants, potentially promoting biodegradation of captured MPs/NPs within the plant itself. However, meticulous consideration of potential ecological risks and rigorous regulatory oversight are crucial for the responsible application of transgenic technologies.

Plants possess an arsenal of genes responsible for stress tolerance. Researchers are actively identifying and manipulating the expression of these genes to create plants better equipped to handle the challenges posed by MPs/NPs. This might entail strategies such as promoting root growth under stressful conditions or over-expressing genes that strengthen antioxidant defense mechanisms. Rather than adding whole new genes, this technique makes use of the plant's existing stress response mechanisms, thereby providing a more tailored and environmentally sound approach.

A phytoremediation and tolerance enhancement phytoremediation strategy taken together with an enhancement in plant tolerance provides a stronger solution. The ability of the plants to capture MPs/NPs and at the same time being able to withstand the pressures for a long, this will enable the plants to alleviate the adverse impacts of plastic pollution in the long run. For example, breeding plant varieties with root architecture that allows it to avoid MPs/NPs in the soil intercalation at the same with stronger stress tolerance mechanisms would be a breakthrough. Sustainable Practices cutting down the utilization of plastic at its root level continues to be important. Reducing people's uses of plastic products, manufacturing them sustainably as well as the proper ways in which they can be disposed of when they are not reusable is also essential in lessening the amount of plastics that end up in the environment. A switch to biodegradable or compostable single-use plastics along with better waste management policies can go a long way in cutting the amount of strain on natural ecosystems. Integrated Ecosystem Concept MP/NP pollution is easily controlled if the ecosystem it belongs to has a natural capability to overcome different stress factors. Simply, encouraging programs such as those of biodiversity, of sustainable land usage, and of maintaining and improving the well-being of the earth's soil can go a long way in making the environment more prepared to handle shocks when they occur. Soil having a healthier microbial population is competent to decompose organic matter inclusive of MPs/NPs; at the same time soil is competent to provide plants with required nutrients, uptake, and root proliferation.

Exciting opportunities exist for increasing plant stress tolerance to MPs/NPs through biotechnological treatments. However prudent development and research are essential. To make sure that the introduction of transgenic plants or the modification of the rhizosphere microbiome does not upset the ecological balance, a careful assessment of the possible unintended consequences is necessary. For example, considerable consideration needs to be given to the possible impacts of transgenic plants on creatures that are not their intended target or the unexpected outcomes of releasing new microbial strains into the ecosystem. Establishing confidence and guaranteeing responsible development and implementation necessitates transparent communication and public involvement. (Jain et al., 2023)

5.4. Integrating Sustainable Agriculture Practices with Micro/Nano-plastics Management

Sustainable agricultural practices not only promote environmental well-being but also lay a robust foundation for MNP resilience. Reducing on-farm plastic use through practices such as mulching with organic materials instead of plastic films, opting for reusable containers, and implementing efficient irrigation systems that minimize plastic tubing dependence, significantly diminishes the potential for plastic breakdown into MNPs within the agricultural environment. (Moeck et al., 2022) Soil with a rich organic matter content and a diverse microbial community act as a natural filter, potentially degrading MNPs and hindering their migration. Practices like cover cropping, composting, and no-till farming actively promote soil health, fostering a more resilient ecosystem with enhanced MNP mitigation potential. Healthy soil with a thriving microbial community can also break down organic matter more efficiently, potentially including MNPs, while also supporting plant growth by providing essential nutrients and promoting root development.

This practice allows pastures to recover between grazing periods, promoting healthy plant growth and robust root development. A vigorous root system acts as a more effective barrier against MNP migration within the soil profile. Additionally, rotational grazing can help maintain healthy soil structure, which can further improve drainage and reduce the potential for MNPs to be transported deeper into the soil profile by water runoff. Biochar a charcoal-like material derived from organic waste, possesses a high surface area and can effectively adsorb MNPs within the soil. Strategic biochar application acts as a filter, capturing MNPs and preventing them from entering the food chain. Research is ongoing to optimize biochar production methods and application strategies for maximizing MNP capture efficiency in agricultural settings. Life cycle assessments need to be conducted to ensure the sustainability of biochar production methods and to evaluate potential downstream effects, such as the long-term fate of captured MNPs within the biochar. (Yuan et al., 2022)

Microorganisms living within the soil contribute to nutrient cycling and the breakdown of organic matter. The use of bioremediation in tackling MNPs has been implicated as having the potential to serve as a long-term solution by presenting an innovative method of introducing specific microbial strains that are competent in breaking down plastic polymers. The discovery and subsequent utilization of these 'plastic-eating' microbes can therefore transform how agricultural MNP management is done. Nevertheless, it is vital to identify other potential side-effects of the treatments explored on the soil microorganisms as a whole. The introduction of new microbial strains to the soils may have bad impacts on the soils and the plants because it may alter the equilibration of the existing microbial population. More studies must be conducted and carried out for the said approach to be safe and effective. (Pérez-Reverón et al., 2022)

6. CASE STUDIES OF EFFECTIVE MANAGEMENT OF MICRO/NANOPLASTIC POLLUTION IN AGRICULTURAL ENVIRONMENTS

Even though the field of MNP management in agriculture is still in its infancy, several innovative projects provide insightful information for the future. These examples demonstrate a range of possible ways, emphasizing the significance of customizing tactics to particular situations and combining MNP management with sustainable farming methods.

6.1 Case Study 1: Biochar for MNP Capture in Strawberry Fields (Europe)

- **The Challenge:** Strawberry cultivation often relies on plastic mulching films, a potential source of MNP contamination. Traditional removal and disposal of these films can be cumbersome and expensive, raising concerns about both environmental impact and economic feasibility. (Yang et al., 2019)

- **The Strategy:** Researchers in Europe experimented with a novel approach – applying biochar derived from agricultural waste to strawberry fields before laying down a biodegradable mulch film. Biochar a charcoal-like material, has a high surface area and the potential to adsorb MNPs.
- **The Results:** The experiment yielded encouraging results. The biochar effectively captured MNPs leaching from the biodegradable mulch film, preventing them from entering the soil. This approach offers a potential two-fold solution: mitigating MNP contamination while promoting sustainable waste management practices by utilizing agricultural byproducts for biochar production. However, further research is needed to assess the long-term fate of captured MNPs within the biochar and perform a life cycle assessment to evaluate the overall environmental and economic sustainability of this approach for large-scale implementation. Here, the biodegradability of the mulch film itself becomes another crucial factor. Ideally, the entire system, from biochar application to mulch film degradation, should contribute to a minimized environmental footprint.

6.2. Case Study 2: Mustard Plants for MNP Phytoremediation in Greenhouses (China)

- **The Challenge:** Greenhouse agriculture often relies on intensive plastic use, creating a potential risk for MNP contamination. Traditional methods for removing MNPs from contaminated greenhouse soils can be disruptive and expensive, involving complete soil replacement or harsh chemical treatments. In addition to interfering with current agricultural productivity, these practices give rise to worries about possible environmental repercussions from chemical usage or the disposal of tainted soil. (Du et al., 2020)
- **The Approach:** Indian mustard plants are a widely available option that researchers in China looked at for MNP phytoremediation in greenhouse settings. The capacity of mustard plants to absorb MNPs was assessed while they were cultivated in soil polluted with MNPs.
- **The Findings:** The investigation demonstrated a possible use for phytoremediation. The ability of the mustard plants to function as natural filters in greenhouses polluted with MNPs was demonstrated by the efficient accumulation of MNPs within their root tissues. This method offers a viable and economical way to clean up polluted soils while also producing useful biomass that might be used for biofuel generation or composting. To investigate safe techniques for removing or breaking down the collected MNPs within the plant material, more study is necessary. Further research is needed to determine how MNP accumulation affects plants over the long term and whether it

affects the plants' viability for composting or producing biofuel. It's important to comprehend this process's potential for MNPs to infiltrate the food chain. Restrictions on the use of this biomass or alternate remediation techniques may be required if mustard plants exhibit a considerable MNP absorption.

6.3. Case Study 3: Compost Amendments for Enhanced MNP Degradation (Global)

- **The Challenge:** Composting is a popular method for handling organic waste in farming and has several advantages for the environment. When compost is spread to agricultural areas, there are worries about how MNPs may affect plant health and the environment due to their possible presence. MNPs may not be sufficiently broken down by conventional composting techniques, which increases the possibility that they may enter the food chain. (Swart et al., 2022)
- **The Approach:** To tackle this problem, researchers are looking at creative ways. Examining the possibility of employing particular compost additions to improve MNP breakdown in compost heaps is one strategy that shows promise. These changes might incorporate plastic-degrading enzymes or microorganisms recognized for their capacity to break down plastic chains. (Cai et al., 2023)
- **The Promise:** This strategy has the potential to reduce the hazards of MNP contamination connected to the use of compost in agricultural contexts. By encouraging the breakdown of MNPs in the compost itself, scientists intend to produce a compost product that is safer and better for the environment to use in farming. Studies in this field are continuing, and more research is required to assess the safety and efficacy of different compost amendment techniques. Furthermore, life cycle analyses are required to guarantee the environmental sustainability of these modifications during both manufacturing and use. It is essential to have a comprehensive strategy that takes into account the environmental effects of every step of the process, from producing amendments to applying compost.

7. RECOMMENDATIONS FOR POLICY AND PRACTICE IN MICRO/NANO-PLASTICS POLLUTION MANAGEMENT

The subtle encroachment of microplastics (MPs) and nanoplastics (NPs) into agricultural environments presents a serious risk to environmental health and global food security. To tackle this issue, an all-encompassing strategy that combines strong legislative frameworks with sustainable farming methods is required. This text provides important guidelines for decision-makers in agriculture and paves the road for a more sustainable future.(Lakhiar et al., 2024)

- **Consistent Monitoring Protocols:** It's critical to establish consistent procedures for monitoring MNP in agricultural soils. This makes it possible to do thorough evaluations of contamination levels, which influence focused mitigation plans.
- **Promoting Microplastic Reduction at the Source:** Lawmakers ought to encourage the creation, manufacturing, and application of compostable or biodegradable substitutes for agricultural plastics such as irrigation tubing and mulching films. Further encouraging their wider adoption might be incentives for embracing these sustainable alternatives.
- **Microplastic Waste Management Infrastructure:** Legislation mandating the proper collection and responsible disposal of agricultural plastic waste is vital to prevent plastic debris from entering the environment and fragmenting into MNPs. Supporting the development of efficient recycling infrastructure for agricultural plastics fosters a circular economy approach.
- **Microbial Inoculant Regulation (if applicable):** If research on "plastic-eating" microbes for MNP degradation proves successful, robust regulatory frameworks are needed. These frameworks should ensure the safety and efficacy of introducing these microbes into agricultural ecosystems, with thorough evaluations of potential unintended consequences on the soil microbiome and antibiotic resistance risks.
- **Transparency and Public Engagement**: Open communication regarding MNP contamination risks and ongoing mitigation efforts is essential. Educating the public about responsible plastic use practices and fostering collaboration between policymakers, researchers, and agricultural practitioners is crucial for building trust and achieving long-term success.

7.1. On the Farm Implementing Sustainable Practices for MNP Reduction

- **Embrace Integrated Pest Management (IPM):** Promoting IPM practices that minimize reliance on synthetic pesticides can indirectly reduce MNP contamination. Many pesticides come in plastic containers, and promoting alternative pest control methods can help reduce overall plastic use in agriculture.(Lakhiar et al., 2024)
- **Promote Organic Mulching Alternatives:** Encouraging the use of organic mulching materials like straw, wood chips, or cover crops instead of tradi-

tional plastic mulching films significantly reduces the risk of MNP contamination. Research and development of effective, commercially viable alternatives are crucial for widespread adoption. (Pérez-Reverón et al., 2022)

- **Optimize Irrigation Systems:** Upgrading irrigation infrastructure to minimize leaks and promote water efficiency can significantly reduce plastic use associated with irrigation tubing and related components. Investing in drip irrigation or other water-saving technologies can decrease reliance on plastic within irrigation systems.
- **Composting with Caution:** Composting agricultural waste offers numerous environmental benefits, but the potential presence of MNPs necessitates caution. Research is needed to develop guidelines for safe composting practices that minimize the risk of MNPs entering the final compost product used for soil amendment. This might involve exploring techniques for separating MNPs from compostable materials before the composting process or developing methods to ensure their degradation within the compost pile itself.
- **Embrace Rotational Grazing:** Implementing rotational grazing practices allows pastures to recover between grazing periods, promoting healthy plant growth and robust root development. A healthy root system acts as a more effective barrier against MNP migration within the soil profile. (Yang et al., 2019)

8. CONCLUSION AND FUTURE PROSPECTS

Microplastics (MPs) and nanoplastics (NPs), tiny plastic particles less than 5 millimeters in size, have become a significant environmental concern due to their widespread presence across various ecosystems, from the deep ocean trenches to the highest mountain peaks. The challenge posed by MPs and NPs is complex, affecting marine and terrestrial environments, agricultural systems, and potentially human health. This conclusion synthesizes the key findings and outlines prospects for addressing this pressing issue.

Key Findings

The rapid increase in plastic production and consumption, driven by the benefits of plastics such as being lightweight, durable, and inexpensive, has led to extensive plastic pollution. The breakdown of larger plastic debris through environmental and biological processes results in the formation of MPs and NPs. These particles are found everywhere, entering ecosystems through multiple pathways like wastewater discharge, littering, and atmospheric deposition. MPs and NPs pose severe threats to both marine and terrestrial ecosystems. They can be ingested by wildlife, causing physical harm, and introducing harmful pollutants into the food chain. Although the impact on human health is still under investigation, there are concerns about the potential ingestion and inhalation of these particles, with possible links to endocrine disruption, inflammation, and certain cancers. In agricultural soils, MPs and NPs can damage plant roots, alter soil properties, and disrupt essential soil processes. These contaminants may be taken up by plants and transported within their vascular systems, raising concerns about food safety and broader ecological impacts.

Future Prospects

Addressing the pervasive issue of MPs and NP pollution requires a comprehensive approach involving research, policy, and community engagement. Here are some prospective actions and research directions:

1. **Enhanced Waste Management and Recycling**:- Improve plastic waste management and recycling infrastructure to reduce the amount of plastic entering the environment. Promote the use of biodegradable and compostable materials to minimize persistent plastic waste.
2. **Advanced Filtration Technologies**:- Upgrade wastewater treatment plants with advanced filtration technologies designed to capture MPs and NPs, preventing their release into waterways.
3. **Agricultural Practices:-** Implement sustainable agricultural practices, such as biodegradable mulching materials and precision farming techniques to minimize plastic use. Enhance soil health through organic farming practices and promote soil biodiversity to mitigate the impact of MPs and NPs.
4. **Research and Monitoring:** Conduct comprehensive research to understand the behavior, fate, and ecological impact of MPs and NPs in various environments. Develop standardized methods for detecting and quantifying MPs and NPs in environmental samples to facilitate consistent monitoring.
5. **Policy and Regulation: -** Enforce stricter regulations on plastic production, use, and disposal to limit the introduction of MPs and NPs into the environment. Encourage international cooperation and the development of global policies to address the transboundary nature of plastic pollution.
6. **Public Awareness and Education: -** Raise public awareness about the sources and impacts of MPs and NP pollution through education and outreach programs. Promote community participation in plastic pollution prevention initiatives and encourage sustainable consumption behaviors.

7. **Innovative Solutions**: - Explore innovative solutions, such as developing new materials and technologies to replace conventional plastics and reduce reliance on single-use plastics. Investigate the potential of bioremediation and other natural processes to degrade MPs and NPs in the environment.
8. **Enhanced Phytoremediation Techniques**: - Further research into plant species effective at capturing and degrading MNPs is crucial. Genetic engineering of plants for enhanced MNP tolerance and degradation could unlock new phytoremediation potentials.
9. **Microbial Interventions:** - Harness the power of microbial communities in the rhizosphere for MNP degradation. Further studies should focus on identifying and safely deploying these microbes in agricultural settings.
10. **Sustainable Agricultural Practices:**-Develop and refine sustainable farming practices that naturally mitigate MNP contamination—such as biochar application, organic mulching, and rotational grazing. Integrating these practices into standard farming operations can provide a long-term solution to MNP pollution.
11. **Policy and Regulatory Frameworks:**- Establish global and regional policies that incentivize the reduction of plastic use in agriculture, alongside robust waste management infrastructure. Regulatory oversight for new technologies, including genetically modified organisms (GMOs) and microbial inoculants, will ensure that interventions are safe, effective, and environmentally sound.
12. **Public Education and Collaboration:-** Build public awareness and foster collaboration among scientists, policymakers, and agricultural practitioners to drive the adoption of these strategies. Continuous education on the impact of MNPs and the importance of sustainable practices can empower communities to contribute to a cleaner, healthier environment.

REFERENCES

Abideen, Z., Hanif, M., Munir, N., & Nielsen, B. L. (2022). Impact of Nanomaterials on the Regulation of Gene Expression and Metabolomics of Plants under Salt Stress. *Plants*, 11(5), 691. DOI: 10.3390/plants11050691

Ali, S. S., Elsamahy, T., Koutra, E., Kornaros, M., El-Sheekh, M., Abdelkarim, E. A., Zhu, D., & Sun, J. (2021). Degradation of conventional plastic wastes in the environment: A review on the current status of knowledge and future perspectives of disposal. *The Science of the Total Environment*, 771, 144719. DOI: 10.1016/j.scitotenv.2020.144719

Azeem, I., Adeel, M., Ahmad, M. A., Shakoor, N., Jiangcuo, G. D., Azeem, K., Ishfaq, M., Shakoor, A., Ayaz, M., Xu, M., & Rui, Y. (2021). Uptake and Accumulation of Nano/Microplastics in Plants: A Critical Review. *Nanomaterials (Basel, Switzerland)*, 11(11), 2935. DOI: 10.3390/nano11112935

Boucher, J., & Friot, D. (2017). *Primary microplastics in the oceans: A global evaluation of sources*. IUCN International Union for Conservation of Nature., DOI: 10.2305/IUCN.CH.2017.01

Bourdages, M. P. T., Provencher, J. F., Baak, J. E., Mallory, M. L., & Vermaire, J. C. (2021). Breeding seabirds as vectors of microplastics from sea to land: Evidence from colonies in Arctic Canada. *The Science of the Total Environment*, 764, 142808. DOI: 10.1016/j.scitotenv.2020.142808

Cai, Z., Li, M., Zhu, Z., Wang, X., Huang, Y., Li, T., Gong, H., & Yan, M. (2023). Biological Degradation of Plastics and Microplastics: A Recent Perspective on Associated Mechanisms and Influencing Factors. *Microorganisms*, 11(7), 1661. DOI: 10.3390/microorganisms11071661

Campanale, C., Massarelli, C., Savino, I., Locaputo, V., & Uricchio, V. F. (2020). A Detailed Review Study on Potential Effects of Microplastics and Additives of Concern on Human Health. *International Journal of Environmental Research and Public Health*, 17(4), 1212. DOI: 10.3390/ijerph17041212

Dehghanian, Z., Asgari Lajayer, B., Biglari Quchan Atigh, Z., Nayeri, S., Ahmadabadi, M., Taghipour, L., Senapathi, V., Astatkie, T., & Price, G. W. (2023). Micro (nano) plastics uptake, toxicity and detoxification in plants: Challenges and prospects. *Ecotoxicology and Environmental Safety*, 268, 115676. DOI: 10.1016/j.ecoenv.2023.115676

Du, J., Guo, Z., Li, R., Ali, A., Guo, D., Lahori, A. H., Wang, P., Liu, X., Wang, X., & Zhang, Z. (2020). Screening of Chinese mustard (Brassica juncea L.) cultivars for the phytoremediation of Cd and Zn based on the plant physiological mechanisms. *Environmental Pollution*, 261, 114213. DOI: 10.1016/j.envpol.2020.114213

Duis, K., & Coors, A. (2016). Microplastics in the aquatic and terrestrial environment: Sources (with a specific focus on personal care products), fate and effects. *Environmental Sciences Europe*, 28(1), 2. DOI: 10.1186/s12302-015-0069-y

El-Saadony, M. T., Saad, A. M., Soliman, S. M., Salem, H. M., Desoky, E.-S. M., Babalghith, A. O., El-Tahan, A. M., Ibrahim, O. M., Ebrahim, A. A. M., Abd El-Mageed, T. A., Elrys, A. S., Elbadawi, A. A., El-Tarabily, K. A., & AbuQamar, S. F. (2022). Role of Nanoparticles in Enhancing Crop Tolerance to Abiotic Stress: A Comprehensive Review. *Frontiers in Plant Science*, 13, 946717. DOI: 10.3389/fpls.2022.946717

Enyoh, C. E., Verla, A. W., Verla, E. N., Ibe, F. C., & Amaobi, C. E. (2019). Airborne microplastics: A review study on method for analysis, occurrence, movement and risks. *Environmental Monitoring and Assessment*, 191(11), 668. DOI: 10.1007/s10661-019-7842-0

Geyer, R., Jambeck, J. R., & Law, K. L. (2017). Production, use, and fate of all plastics ever made. *Science Advances*, 3(7), e1700782. DOI: 10.1126/sciadv.1700782

Iqbal, B., Zhao, T., Yin, W., Zhao, X., Xie, Q., Khan, K. Y., Zhao, X., Nazar, M., Li, G., & Du, D. (2023). Impacts of soil microplastics on crops: A review. *Applied Soil Ecology*, 181, 104680. DOI: 10.1016/j.apsoil.2022.104680

Jain, R., Gaur, A., Suravajhala, R., Chauhan, U., Pant, M., Tripathi, V., & Pant, G. (2023). Microplastic pollution: Understanding microbial degradation and strategies for pollutant reduction. *The Science of the Total Environment*, 905, 167098. DOI: 10.1016/j.scitotenv.2023.167098

Jia, L., Liu, L., Zhang, Y., Fu, W., Liu, X., Wang, Q., Tanveer, M., & Huang, L. (2023). Microplastic stress in plants: Effects on plant growth and their remediations. *Frontiers in Plant Science*, 14, 1226484. DOI: 10.3389/fpls.2023.1226484

K.L., P., K.R., R., Joseph, C. J., M.S., I., Srinivas, R., & S., H. (2022). Fate, transport and degradation pathway of microplastics in aquatic environment—A critical review. *Regional Studies in Marine Science*, 56, 102647. DOI: 10.1016/j.rsma.2022.102647

Lakhiar, I. A., Yan, H., Zhang, J., Wang, G., Deng, S., Bao, R., Zhang, C., Syed, T. N., Wang, B., Zhou, R., & Wang, X. (2024). Plastic Pollution in Agriculture as a Threat to Food Security, the Ecosystem, and the Environment: An Overview. *Agronomy (Basel)*, 14(3), 548. DOI: 10.3390/agronomy14030548

Law, K. L., & Thompson, R. C. (2014). Microplastics in the seas. *Science*, 345(6193), 144–145. DOI: 10.1126/science.1254065

Li, X., Wang, R., Dai, W., Luan, Y., & Li, J. (2023). Impacts of Micro(nano)plastics on Terrestrial Plants: Germination, Growth, and Litter. *Plants*, 12(20), 3554. DOI: 10.3390/plants12203554

Liu, X., Li, M.-C., Lu, Y., Li, Z., Liu, C., Liu, Z., Mei, C., & Wu, Q. (2024). Cellulose nanofiber-coated delignified wood as an efficient filter for microplastic removal. *Progress in Natural Science*, 34(1), 162–171. DOI: 10.1016/j.pnsc.2024.02.010

Maddela, N. R., Ramakrishnan, B., Kadiyala, T., Venkateswarlu, K., & Megharaj, M. (2023). Do Microplastics and Nanoplastics Pose Risks to Biota in Agricultural Ecosystems? *Soil Systems*, 7(1), 19. DOI: 10.3390/soilsystems7010019

Moeck, C., Davies, G., Krause, S., & Schneidewind, U. (2022). Microplastics and nanoplastics in agriculture—A potential source of soil and groundwater contamination? *Grundwasser*. Advance online publication. DOI: 10.1007/s00767-022-00533-2

Ouyang, W., Zhang, Y., Wang, L., Barceló, D., Wang, Y., & Lin, C. (2020). Seasonal relevance of agricultural diffuse pollutant with microplastic in the bay. *Journal of Hazardous Materials*, 396, 122602. DOI: 10.1016/j.jhazmat.2020.122602

Pérez-Reverón, R., Álvarez-Méndez, S. J., Kropp, R. M., Perdomo-González, A., Hernández-Borges, J., & Díaz-Peña, F. J. (2022). Microplastics in Agricultural Systems: Analytical Methodologies and Effects on Soil Quality and Crop Yield. *Agriculture*, 12(8), 1162. DOI: 10.3390/agriculture12081162

Roy, T., Dey, T. K., & Jamal, M. (2023). Microplastic/nanoplastic toxicity in plants: An imminent concern. *Environmental Monitoring and Assessment*, 195(1), 27. DOI: 10.1007/s10661-022-10654-z

Soares, M. D. O., Matos, E., Lucas, C., Rizzo, L., Allcock, L., & Rossi, S. (2020). Microplastics in corals: An emergent threat. *Marine Pollution Bulletin*, 161, 111810. DOI: 10.1016/j.marpolbul.2020.111810

Sood, M., Kapoor, D., Kumar, V., Kalia, N., Bhardwaj, R., Sidhu, G. P. S., & Sharma, A. (2021). Mechanisms of Plant Defense Under Pathogen Stress: A Review. *Current Protein & Peptide Science*, 22(5), 376–395. DOI: 10.2174/1389203722666210125122827

Sultana, A., Zare, M., Thomas, V., Kumar, T. S. S., & Ramakrishna, S. (2022). Nano-based drug delivery systems: Conventional drug delivery routes, recent developments and future prospects. *Medicine in Drug Discovery*, 15, 100134. DOI: 10.1016/j.medidd.2022.100134

Sun, J., Dai, X., Wang, Q., Van Loosdrecht, M. C. M., & Ni, B.-J. (2019). Microplastics in wastewater treatment plants: Detection, occurrence and removal. *Water Research*, 152, 21–37. DOI: 10.1016/j.watres.2018.12.050

Swart, B., Pihlajamäki, A., John Chew, Y. M., & Wenk, J. (2022). Microbubble-microplastic interactions in batch air flotation. *Chemical Engineering Journal*, 449, 137866. DOI: 10.1016/j.cej.2022.137866

Tiffin, L., Hazlehurst, A., Sumner, M., & Taylor, M. (2022). Reliable quantification of microplastic release from the domestic laundry of textile fabrics. *Journal of the Textile Institute*, 113(4), 558–566. DOI: 10.1080/00405000.2021.1892305

Verma, K. K., Song, X.-P., Xu, L., Huang, H.-R., Liang, Q., Seth, C. S., & Li, Y.-R. (2023). Nano-microplastic and agro-ecosystems: A mini-review. *Frontiers in Plant Science*, 14, 1283852. DOI: 10.3389/fpls.2023.1283852

Yan, A., Wang, Y., Tan, S. N., Mohd Yusof, M. L., Ghosh, S., & Chen, Z. (2020). Phytoremediation: A Promising Approach for Revegetation of Heavy Metal-Polluted Land. *Frontiers in Plant Science*, 11, 359. DOI: 10.3389/fpls.2020.00359

Yang, X., Zhang, S., Ju, M., & Liu, L. (2019). Preparation and Modification of Biochar Materials and their Application in Soil Remediation. *Applied Sciences (Basel, Switzerland)*, 9(7), 1365. DOI: 10.3390/app9071365

Yu, Z., Li, Q., Wang, J., Yu, Y., Wang, Y., Zhou, Q., & Li, P. (2020). Reactive Oxygen Species-Related Nanoparticle Toxicity in the Biomedical Field. *Nanoscale Research Letters*, 15(1), 115. DOI: 10.1186/s11671-020-03344-7

Yuan, S., Liang, J., Zhang, Y., Han, H., Jiang, T., Liu, Y., Zhang, Y., Wang, W., & Dong, X. (2022). Evidence from Thermal Aging Indicating That the Synergistic Effect of Glyoxal and Sodium Sulfite Improved the Thermal Stability of Conformational Modified Xanthan Gum. *Polymers*, 14(2), 243. DOI. 10.3390/polym14020243

Zuverza-Mena, N., Martínez-Fernández, D., Du, W., Hernandez-Viezcas, J. A., Bonilla-Bird, N., López-Moreno, M. L., Komárek, M., Peralta-Videa, J. R., & Gardea-Torresdey, J. L. (2017). Exposure of engineered nanomaterials to plants: Insights into the physiological and biochemical responses-A review. *Plant Physiology and Biochemistry*, 110, 236–264. DOI: 10.1016/j.plaphy.2016.05.037

Chapter 5
Soil Physio-Biochemical Alterations Due to Micro/Nano-Plastic Pollution

Suresh Waghamare
https://orcid.org/0009-0001-8909-2209
Rashtriya College of Pharmacy, Hatnoor, India

ABSTRACT

The ubiquitous presence of micro/nano-plastics in soil raises growing concerns about their impact on soil health and ecosystem function. This review comprehensively examines the physio-biochemical alterations caused by MNPs in soil, focusing on physical, chemical, and biological changes. We analyze how MNPs affect soil texture, structure, and water dynamics, potentially leading to erosion, altered hydrological regimes, and reduced agricultural productivity. Furthermore, we explore the interactions between MNPs and soil contaminants, highlighting their potential to mobilize harmful chemicals and pose risks to human health through food chain contamination. The review then delves into the crucial role of the soil microbiome and its disruption by MNPs. We detail how MNPs physically and chemically inhibit microbial activity, potentially impacting nutrient cycling, decomposition, and plant-microbe interactions. Additionally, we investigate the consequences of MNP exposure on plant physiology and growth, emphasizing potential reductions in photosynthesis, germination, and yield.

DOI: 10.4018/979-8-3693-9163-1.ch005

1 INTRODUCTION

1.1 Importance of Soil

Soil health is a critical factor in sustaining and supporting ecosystems, playing a vital role in maintaining the overall health of our planet. Healthy soil is essential for a variety of reasons that impact both natural environments and human societies. Soil serves as a medium for nutrient cycling, providing essential elements such as nitrogen, phosphorus, and potassium that are crucial for plant growth. Nutrient-rich soil supports the growth of diverse plant species, forming the foundation of terrestrial ecosystems. Healthy soil is a habitat for a vast array of microorganisms, insects, and other small organisms. These organisms contribute to the soil's fertility, help decompose organic matter, and play a crucial role in supporting biodiversity by forming complex and interconnected food webs. Soil acts as a natural filter for water, removing impurities and contaminants as water percolates through it. Additionally, soil plays a crucial role in regulating water flow, preventing floods by absorbing and storing water during periods of heavy rainfall and releasing it gradually during drier periods. Soil is a significant reservoir for carbon. Healthy soils store carbon in organic matter, contributing to climate change mitigation by helping to reduce the concentration of greenhouse gases in the atmosphere. This process, known as carbon sequestration, is vital in the fight against global warming. For agriculture, soil health is synonymous with crop productivity. Fertile and well-structured soils provide the necessary support for plant roots to access nutrients and water, leading to improved crop yields. Sustainable farming practices that prioritize soil health contribute to long-term food security. The structure and composition of healthy soil help prevent erosion. Soil erosion, caused by factors like wind and water, can degrade landscapes and lead to the loss of fertile topsoil. Healthy soil with good structure and organic matter is more resistant to erosion. Healthy soils enhance the resilience of ecosystems to climate change. They contribute to improved water retention during droughts, reduced vulnerability to extreme weather events, and overall stability in the face of changing environmental conditions. Soil health indirectly affects human health by influencing the quality of crops and water sources. Nutrient-rich soil produces healthier and more nutritious crops, contributing to the overall well-being of populations that rely on agriculture for sustenance.

1.2 Plastic Pollution of Micro/Nano-Plastics in the Environment

The public's awareness of plastic pollution in terrestrial and marine habitats has grown as a result of extensive and well-reported media coverage, particularly in light of the possible dangers of tiny resin pellets. For instance, it was claimed that the plastic industry's unintentional spilling during handling and transportation resulted in the accumulation of over 1000 tons of resin pellets on New Zealand beaches, seriously contaminating the marine and coastal eco-systems. The amount of plastic debris entering terrestrial and marine environments has stabilized and even decreased in terms of plastic pellets since 1990, thanks to strong laws and regulations that forbade the dumping of marine debris and strengthened the raw material recycling management in the plastic industries. Still, Large-scale environmental microplastics events have become more common in recent years as a result of the decomposition of the environment's pre-existing plastic trash. Plastic debris having a diameter or length of less than 0.5 cm is commonly referred to as MPs. (Lebreton, 2011; Lebreton, 2011)

Plastic debris in the environment can be categorized into four groups based on their sizes: MPs (less than 0.5 centimeter), mesoplastics (0.5–5 cm), macroplastics (between 5 and 50 cm), and megaplastics (more than 50 cm). (Randhawa, 2023) Depending on where they come from, MPs can be divided into primary and secondary categories microplastics (MPs) are mostly released into the environment through direct contact with goods that include MPs. Examples of these products include plastic microbeads added to toothpaste, cosmetics, and hand and face cleansers, as well as pre-production MP pellets.

Secondary microplastics are plastic pieces that are left over after bigger plastic goods, like clothes, rope, and packaging, break down. (Wua, 2019)

Due to their enduring presence in the environment and possible harm to aquatic life, fauna, ecosystems, and public health, microplastic contamination has garnered significant scientific interest in recent times. Although there have been a number of critical reviews on microplastics (MPs) and their environment, the majority of them concentrate on how MPs affect the marine environment. MPs can originate from different sources, interact with a variety of environmental media (such as freshwater, marine water, groundwater, sediments, and soils), and go through a number of different transport and transformation pathways. Apart from the maritime domain, MPs have been consistently identified in freshwater soils and the atmosphere.

Plastic pollution, a global environmental crisis, is characterized by the pervasive accumulation of plastic waste across various ecosystems, posing profound threats to both the environment and human health. The ubiquity of Micro plastics tiny plastic particles measuring less than 5mm, has further intensified this crisis. Originating from the breakdown of larger plastic items, microbeads, and synthetic textiles, Micro and nano plastics have infiltrated virtually every corner of the planet, from the depths of oceans to the air we breathe. The alarming rise of plastic pollution is evident in the persistence of plastic waste, which takes centuries to decompose, leading to widespread contamination of land, water bodies, and natural habitats.

This environmental challenge has dire consequences for ecosystems, causing harm to marine life through ingestion and entanglement, disrupting food chains, and even potentially impacting human health as microplastics enter the food chain. Microplastics, being small and resilient, have become ubiquitous in various environments. They are transported across vast distances by wind and water currents, entering ecosystems far removed from their original sources. The widespread distribution of Micro and nano plastics raises concerns about their bioaccumulation in organisms and potential health risks. The ubiquity of Microplastics not only poses challenges in terms of environmental impact but also complicates the task of removal due to their small size and ability to evade conventional waste management systems. As research continues to uncover the intricate web of consequences associated with plastic pollution and MNPs, urgent and coordinated efforts on a global scale are imperative to address this crisis, protect ecosystems, and secure the well-being of both the planet and its inhabitants.

1.3 Define Micro/Nano-Plastics, Categorize Their Sources, and Describe Their Pathways to Soil Contamination

1.3.1 Definition of Microplastics

Microplastics are microscopic particles of plastic that are usually smaller than 5 millimeters. They can be divided into two categories: primary microplastics, which are produced at this size, and secondary microplastics, which are produced when bigger plastic products break down. They can be in the shape of pellets, fibers, pieces, or microbeads, among other forms.

1.3.2 Categorization of Microplastics Sources

Microplastics come from many different places, and classifying them makes it easier to comprehend the different ways they might get into the environment. Microplastics primarily come from the following sources:

A. Primary Microplastics:
   - Microbeads: Tiny plastic particles that are purposefully produced to be used in toothpaste and exfoliating scrubs, among other personal care products
   - Microfibers: Tiny plastic threads released during the washing of synthetic textiles, such as polyester and nylon.

B. Secondary Microplastics:

- Fragmentation: Breakdown of larger plastic items due to weathering, UV radiation, and physical forces. This includes pieces from plastic bottles, bags, and other debris.
- Disintegration: natural processes causing plastic waste in the environment to break down into smaller bits

1.3.3 Pathways to Soil Contamination

Microplastics can find their way into soils through various pathways, creating concerns about their impact on terrestrial ecosystems. The primary pathways for soil contamination by microplastics include

1.3.3.1. Agricultural Practices

Agricultural plastics, commonly used in various farming practices, have the potential to contribute to soil contamination through several mechanisms. Here are the primary ways in which agricultural plastics can contaminate soil:

Table 1. Source of microplastics

Source of Microplastics	Description	Reference
Plastic Mulching	Covering the soil surrounding plants with tiny plastic sheets. Over time, these plastic films degrade due to exposure to sunlight, temperature changes, and mechanical stressors.	(Mulvaney, 2011)
Introduction of Microplastics	Plastic mulching involves thin plastic films that, over time, break down into smaller particles, introducing microplastics into the soil.	(Mulvaney, 2011)
Plastic Film Residue	Agricultural plastic films used for weed control and soil moisture retention may not completely decompose, instead fragmenting into persistent plastic residue in the soil.	(Mulvaney, 2011)
Plastic Residue from Crop Covers	Some farmers abandon plastic covers used to protect crops after harvest, leading to the degradation and release of plastic particles into the soil.	(Barnes, 2020)
Deterioration of Irrigation Components	Plastic drip lines and hoses degrade over time due to UV radiation and weathering, releasing microplastics into the surrounding soil.	(Barnes, 2020)
Plastic Seed Coatings	Seeds coated with polymer materials to enhance properties can wear off during planting and germination, contributing microplastics to the soil.	(Huerta-Lwanga, 2016)
Trash Disposal Procedures	Inadequate agricultural waste management, such as improper disposal of plastic films, containers, or equipment, directly causes macro- and microplastic contamination of the soil.	(Huerta-Lwanga, 2016)

continued on following page

Table 1. Continued

Source of Microplastics	Description	Reference
Trash Disposal Procedures	Inadequate agricultural waste management, such as improper disposal of plastic films, containers, or equipment, directly causes macro- and microplastic contamination of the soil.	(Huerta-Lwanga, 2016)
Plastic Pesticide Containers	Inappropriate disposal of plastic containers for fertilizers and pesticides can contribute to soil pollution if not cleaned and disposed of correctly, especially with leftover chemicals and plastic particles.	(Hess, 2012)
Plastic-Based Fertilizer Coatings	Controlled-release fertilizers use plastic coatings that break down over time, releasing plastic particles and residual nutrients into the soil.	(Zhan, 2018)

1.3.3.2 Wastewater Irrigation

Wastewater irrigation, the practice of using treated or untreated wastewater to irrigate agricultural fields, can introduce microplastics into the soil. Microplastics in wastewater can originate from various sources, including domestic, industrial, and urban runoff. Here's how microplastics may play a role in soil contamination through wastewater irrigation:

- **Wastewater Composition:-** Wastewater from households may contain microplastics derived from personal care products (e.g., microbeads in cosmetics) and other household items. Industrial discharges can contribute microplastics from manufacturing processes and plastic-containing products. (Gouin, 2019)
- **Ineffective Filtration in Treatment Plants:** While wastewater treatment plants aim to remove contaminants, including microplastics, not all facilities are equipped to effectively filter out these small particles. As a result, treated wastewater may still contain microplastics. (Carr, 2016)
- **Use of Treated Wastewater for Irrigation:** Treated or partially treated wastewater, which may contain residual microplastics, is often used for irrigation to supplement water resources for crops.
- **Microplastics in Irrigation Water:** When the soil is being watered, microplastics in the irrigation water may be carried to the soil. When the soil surface gets wet with water that contains microplastics, this transfer takes place.
- **Accumulation in Soil:** Microplastics, once in the soil, may undergo sorption and retention processes. This means they can adhere to soil particles and accumulate in certain soil layers.

- **Possible Uptake by Plants:** Research into the possibility of microplastics being absorbed by plant roots is still ongoing. According to certain research, microplastics may build up in plant tissues and eventually find their way into the food chain.
- **Effect on Soil Health:** Microbial activity and soil structure may be impacted by the buildup of microplastics in the soil. Such modifications' long-term effects are currently being investigated.
- **Environmental Risks:** Plastic particles in soil may affect species that live there and so disturb soil ecosystems. Research on the potential ecological dangers posed by microplastics in soil is ongoing.

1.3.3.4 Atmospheric Deposition

The process of microplastics being carried by the air and deposited on land surfaces, including soil, is known as atmospheric deposition. This route adds to the microplastic pollution of the soil. The following is how soil contamination by microplastics is influenced by air deposition:

- **Atmospheric microplastics' sources:** - The atmosphere's microplastics can come from a number of places, such as weathering causes bigger plastic objects to break down into smaller pieces. microplastics released from synthetic fabrics when they are worn and washed. Microplastics are transported by air from open landfills, industrial sites, and urban areas. (Allen, 2019) (Dris, 2017)
- **Wind and Air Currents:** -Microplastics are lightweight and can be carried over long distances by wind and air currents. They become suspended in the atmosphere, creating a potential for widespread distribution. (Zhang, 2023)
- **Settling onto Soil Surfaces:**-Eventually, microplastics suspended in the air settle onto the Earth's surface. This settling can occur through gravitational forces, precipitation, or atmospheric turbulence. (Geyer, 2015)
- **Infiltration into Soil:** - Microplastics that have settled onto the soil surface can be further transported into the soil through processes such as rainfall, snowmelt, or direct contact with the ground. (Napper, Assessment of microplastic pollution on land and ocean., 2016)
- **Adhesion to Soil Particles:** -Microplastics may adhere to soil particles due to various physicochemical interactions. This adhesion process contributes to the retention of microplastics in the soil. (Li J. S., Sorption behaviors of microplastics in soil and water: A critical review., , 2020)
- **Impact on Soil Properties:** - The buildup of microplastics in the soil can change properties of the soil, like water absorption capacity and porosity.

These alterations may have an effect on the composition of the soil and the availability of nutrients for plants. (Zhang, 2020)

- **Transportation through Runoff:** -Microplastics in the soil may be transported to other locations through surface runoff during rainfall events. This runoff can carry microplastics to nearby water bodies, contributing to aquatic contamination. (Horton, 2017)
- **Potential Curiosity by Plants: -** Studies suggest that microplastics in the soil may be taken up by plant roots. This raises concerns about the potential transfer of microplastics from soil to plants and, subsequently, into the food chain. (Liu, 2022)
- **Long-Range Transport:** -Microplastics deposited on soil surfaces can persist for extended periods. They may be re-suspended into the atmosphere through wind erosion, contributing to long-range transport and further deposition. (Brahney, 2020) Concerns concerning potential ecological effects on plants, soil-dwelling creatures, and general soil health are raised by the discovery of microplastics in the soil. Research on the long-term effects of microplastic contamination in soil is still underway. (Rillig, 2020)

2. PHYSICAL AND CHEMICAL ALTERATIONS IN SOIL

2.1 Adsorption and Aggregation

The increasing presence of microplastics in soil environments poses a new and significant challenge with important consequences for soil health. These tiny plastic particles, similar to their metallic counterparts engage with soil particles in complex ways, potentially disturbing the fundamental aspects of healthy soil, such as its pore structure, aeration, and water dynamics. Because of their high surface area-to-volume ratio, Microplastics readily stick to soil particles through various physical and chemical interactions. This sticking can serve as a binding agent, encouraging the clumping together of individual soil particles into larger aggregates. While controlled aggregation can improve soil structure in certain cases, uncontrolled aggregation induced by Microplastics can significantly change how pore spaces are distributed within the soil. (Li J. Y., 2022) This altered pore distribution brings about various consequences. Large pores, essential for air and water drainage, can become clogged or reduced in size due to aggregation, impeding gas exchange and potentially causing waterlogging and reduced oxygen levels for plant roots and soil microbes. However, excessive aggregation can also impede water movement within the soil, restricting water accessibility for plants even in situations where the soil has an adequate total water content. Microplastics, on the other hand, may increase the number of micropores, which are tiny pores essential for water retention. Having a thorough understanding of MP-soil interactions requires an all-encompassing approach.

The nature and degree of these interactions are largely determined by the original texture and organic matter content of the soil, as well as by factors including MP size, surface charge, and composition. (Wang, 2022) Furthermore, the total impact of microplastics is highly dependent on their concentration within the soil ecosystem. Developing sustainable soil management techniques and assessing the possible advantages and disadvantages of plastic use require a full understanding of these relationships. It is possible to reduce the negative impacts of microplastic contamination and protect the long-term health of our vital soil ecosystems by carefully taking these aspects into account.

2.3 Soil Texture and Structure

The growing presence of microplastics in soil environments presents a new and worrisome challenge. These tiny plastic fragments disrupt the intricate balance between soil texture and structure, potentially increasing the risk of erosion. This section examines how Microplastics influence soil aggregation, bulk density, and stability, ultimately affecting a soil's ability to resist erosive forces.

2.3.1 Dual effects: mp-induced aggregation and its ramifications

with their adhesive properties, can serve as binding agents, encouraging the formation of larger soil aggregates. While controlled aggregation can improve soil structure by enhancing aeration and drainage, excessive MP-induced aggregation can be harmful. These tightly bound aggregates can impede water infiltration and gas exchange, thereby hindering plant growth. Furthermore, because of their fragility, they are more likely to be disturbed by wind and rain, which increases soil erosion and separation. The degree to which this impact manifests itself is mostly determined by the texture of the soil; sandy soils are especially susceptible because of their bigger pores and lack of binding substances such as clay.

2.3.2 The Bulk Density Puzzle: Compaction versus Loss of Cohesion?

Bulk density, the mass of soil solids per unit volume, is another crucial property influenced by Microplastics. Their presence can impact bulk density in contrasting ways. In some cases, Microplastics may fill pore spaces between soil particles, leading to increased bulk density, which can hinder water infiltration and root penetration, thus affecting plant growth and ecosystem health. Conversely, certain Microplastics, like plastic fragments, may create new pathways within the soil, potentially reducing bulk density. However, this looser structure can also increase erosion risks by reducing the soil's resistance to detachment forces.

2.3.3 Stability Dynamics: A Fragile Balance at Risk

Soil stability, or its resistance to breakdown by external forces like water and wind, is complex and influenced by various factors, including microplastics. While controlled MP-induced aggregation might enhance stability, excessive aggregation or loose, MP-influenced structures can decrease stability, making the soil more prone to erosion.

2.3.4 The Erosion Equation: Disruption of a Fragile Balance

Increased aggregation of a fragile nature, alongside altered bulk density and reduced stability, makes soil particles more susceptible to detachment by wind and water forces, leading to accelerated erosion. This affects the general health of the ecosystem and agricultural productivity by removing vital nutrients and depleting critical topsoil. Developing mitigation solutions requires an understanding of the complex effects of microplastics on soil characteristics. We can endeavor to reduce the negative effects of microplastics pollution and guarantee the ongoing health and resilience of our soil resources by taking into account soil texture, MP properties, and the application of suitable land management measures. (Wang, et al., 2022)

2.4 Chemical Sorption and Release

Microplastics pervasive pollutants in the environment, introduce a new and concerning aspect to soil chemistry. These minuscule particles possess a remarkable ability to both attract and release various harmful environmental contaminants, including pesticides and heavy metals. This section delves into this intriguing yet troubling phenomenon, focusing on its implications for the movement and availability of contaminants.

2.4.1 Increased Sorption: A Potential Trap for Contaminants

Microplastics have an excellent surface area to volume ratio, which makes them effective sorbents. Through various physical and chemical interactions, they are able to readily attract and bind a wide range of environmental pollutants. Weak van der Waals forces, electrostatic attractions, or even the formation of particular chemical bonds between the contaminant and functional groups on the MP surface are examples of these interactions. The particular contaminant, its chemical make-up, and the intrinsic qualities of the MP itself all influence the degree of sorption. (Rai, et al., 2023)

2.4.2 The Desorption Predicament: Trojan Horses or Temporary Reservoirs?

Although adsorption initially seems advantageous by isolating contaminants from the available pool, there is more to the story. Microplastics can act as Trojan horses, transporting adsorbed contaminants deeper into the soil. More worrisome, though, is that, under some circumstances, they may leak toxins that were previously bound into the environment. Pollutants that have been sequestered may release when desorption is triggered by variations in pH, salinity, or the presence of competing organic materials. Furthermore, the slow release of pollutants that have previously been adsorbed might also be attributed to the breakdown of microplastics over time.

2.4.3 Bioavailability Concerns: A Threat to Ecosystem Well-being

The health of ecosystems is seriously threatened by microplastics' capacity to both attract and discharge pollutants. Pesticides and heavy metals that have been adsorbed can become more accessible, which means that plants and soil creatures can absorb them more readily and eventually include them in the food chain. This condition may have a negative impact on plant health, disturb vital soil microbial communities, and may endanger the health of higher trophic levels. (Muhammad Sajjad, 2022)

2.5 Altered Ph and Salinity

A novel and worrisome threat to soil ecosystems' general health is the pervasive presence of microplastics in them. Recent study suggests that microplastics have the capacity to alter the pH and salinity of soil, which could have an impact on microbial activity and nutrient availability, two essential components of soil health, even if the precise mechanisms are still being investigated.

2.5.1 Fragile Balance: Potential Perturbations in Soil pH

The pH of the soil is essential for controlling microbial populations and nutrient solubility. Microplastics can affect soil pH in a variety of ways, depending on their chemical composition and surface characteristics. Certain microplastics may produce basic or acidic components during their degradation, which could lead to localized pH changes. Furthermore, the breakdown of organic acids resulting from the adsorption of organic materials onto Microplastics might cause a decrease in pH. On the other hand, some microplastics may act as buffers, reducing the effects of pH variations that occur naturally. However, more research needs to be done to determine how microplastics generally affect the pH of soil.

2.5.2 The Complexity of Salinity: A Two-Sided Impact

The concentration of soluble salts in the soil determines its salinity, which has a major impact on soil health. As microplastics break down, they can directly contribute to rising salinity, particularly if they contain leftover salts or industrial additives. Furthermore, microplastics have the ability to obstruct natural drainage channels, which could result in the buildup of salt from irrigation or precipitation. However, there are some positive effects of microplastics on salinity as well. Microplastics have the ability to function as barriers in some situations, preventing the leaching of necessary salts and possibly promoting plant development in arid areas.

2.5.3 Ripple Effects: Disrupted Microbial Activity and Nutrient Availability

Microplastic-induced changes in soil pH and salinity can have a broad impact on soil health. Changes in pH have a significant impact on microbial activity, which is essential for processes like decomposition and the cycling of nutrients. Significant pH changes can change the functionality and makeup of the microbial community, which affects the availability of nutrients for plants. Elevated salinity levels have the potential to exacerbate stress on soil microorganisms, impeding their vital processes and hence restricting plant development and output.

3 DISRUPTION OF SOIL MICROBIOME

3.1 Soil Microbiome

The complex and varied population of microorganisms known as the soil microbiome is frequently disregarded yet is vital to the health of terrestrial ecosystems. This intricate network, which is made up of bacteria, fungi, archaea, protozoa, and viruses, is essential to the cycling of nutrients, breakdown, and the stimulation of plant development. Major contributors to the breakdown of organic materials and the release of nutrients necessary for plant growth are bacteria. Plant roots and fungi, especially mycorrhizae, develop symbiotic interactions that improve plant resistance to environmental challenges and help the plant absorb nutrients. The nitrogen fixation and nutrient cycling that archaea support are essential to preserving soil fertility. Protozoa control the number of bacteria in an environment, whereas viruses affect the variety and metabolism of microbes. The complex network of relationships found in the soil microbiome affects global biogeochemical cycles in addition to maintaining the health of ecosystems. In the face of environmental concerns like climate change and land degradation, sustainable agriculture and ecosystem management depend on comprehending and utilizing the potential of these unsung heroes.

3.2 Physical and Chemical Disruption

The widespread presence of Microplastics in soil ecosystems is raising significant concerns regarding their long-term impact on environmental health. These minuscule plastic particles pose a substantial threat to the crucial element of healthy soil – its diverse and dynamic microbial community. This section explores the various ways in which Microplastics disrupt microbial activity and diversity, ultimately affecting ecosystem function.

Physical barriers created by Microplastics hinder microbial mobility within the soil matrix. These obstacles obstruct the intricate network of pores essential for microbial exploration, limiting access to vital nutrients and potential food sources. Moreover, certain Microplastics with sharp edges and rigid structures can cause physical harm to microbial cells, further compromising their survival and function.

3.3 Loss of Biodiversity and Symbiotic Interactions

The increasing prevalence of microplastics in soil ecosystems poses a new and worrying challenge to their long-term health. These tiny plastic particles pose a significant threat to the fundamental aspect of healthy soil its diverse and active microbial community. This intricate network of microorganisms plays a crucial role in vital processes such as nutrient cycling, decomposition, and supporting plant growth. This section explores the various ways in which microplastics can disrupt these critical microbial communities, ultimately affecting interactions between plants and microbes and the essential process of nitrogen fixation.

Soil microbial communities display remarkable diversity, with each member serving a specific ecological function. Microplastics, due to their extensive surface area, act as tiny absorbers, capturing essential nutrients and organic matter that usually support microbial growth and function. This disturbance in the delicate balance of the microbial network may result in the decline of certain populations and a decrease in overall biodiversity. Additionally, some microplastics may release harmful chemicals or possess inherent antimicrobial properties, further stressing or directly inhibiting the growth of beneficial microbes.

Nitrogen is a fundamental nutrient for plant development, and the process of nitrogen fixation by soil microbes is vital for maintaining soil fertility. Any disturbance to this process can significantly impact plant health and agricultural productivity. By affecting the microbial communities responsible for nitrogen fixation, microplastics can create a bottle neck in nitrogen availability. With reduced accessible nitrogen in the soil, plant growth and agricultural yields may face substantial limitations.

3.4 Emerging Pathogens and Antibiotic Resistance

The pervasiveness of microplastics in soil ecosystems presents a novel and worrisome challenge that may have consequences for public health. Although the precise processes are still under investigation, new research suggests that microplastics may act as agents in the transmission of diseases and bacteria resistant to antibiotics. This emphasizes how urgently this new threat needs to be thoroughly investigated.

A major worry stems from the ability of microplastics, owing to their extensive surface area, to encourage the formation of biofilms. These dense communities of microorganisms provide an ideal environment for horizontal gene transfer, a process where bacteria exchange genetic material, including genes responsible for antibiotic resistance. Within these biofilms, antibiotic-resistant bacteria can thrive, potentially rendering existing antibiotics ineffective against a broader array of pathogens.

Microplastics can serve as carriers for harmful pathogens by attracting them to their surfaces. This shielding not only shields the pathogens from environmental stressors such as UV radiation and desiccation but also increases their mobility within the soil structure. This heightened persistence and mobility significantly heighten the risk of pathogen transmission to plants and potentially to humans via contaminated agricultural products.

Antibiotics are frequently used in common agricultural techniques to treat animal illnesses. Antibiotic-resistant bacterial strains linked to microplastics may experience selective pressure to develop and survive as a result of interactions between residual drugs and microplastics in the soil. This may eventually cause the diversity and prevalence of antibiotic-resistant bacteria in the soil ecosystem to rise alarmingly. Previous studies suggest that microplastics may upset the delicate microbial community balance in soil. Healthy soil has natural defenses against the spread of pathogens, which may be weakened by this disruption. Microplastics may have detrimental effects on beneficial microbes that prey on or compete with pathogenic bacteria, so reducing competition and fostering an environment that favors disease multiplication.

4 IMPACT ON PLANT PHYSIOLOGY AND GROWTH:

4.1 Uptake and Translocation

Concerns are raised by the growing amount of microplastics in soil ecosystems about their potential to enter the food chain through plant absorption and movement to aboveground portions of the plant. This calls for a thorough investigation of the processes controlling MP absorption by roots and their subsequent transport throughout the plant. Plant roots operate as the first point of contact between microplastics in the soil environment due to their vast network of root hairs. Microplastics' ability to be absorbed is greatly influenced by their size and surface properties. Because of their greater surface area, smaller microplastics are more readily absorbed by passive diffusion into root cells. Furthermore, any lesions or fissures in plant roots could make it easier for larger microplastics to enter. According to recent research, some microplastic types may cause particular cellular absorption processes in plant roots, which could facilitate the ingestion of the microplastics. (Koycheva, 2021) (Roy, 2023)

Microplastics may begin their upward journey through the plant's vascular system, specifically the xylem, once they penetrate root cells. Transporting water and dissolved nutrients from the roots to the plant's aboveground components is the function of this specialized tissue. Some microplastics can be passively carried through the xylem of the water flow due to their small size. Additionally, studies indicate that microplastics may attach themselves to xylem walls or travel through vesicles of xylem sap, which are specialized structures that aid in the movement of fluid throughout plants. (Azeem, 2021)

Microplastics have the ability to accumulate in a variety of aboveground plant tissues, including leaves and fruits, if they are effectively transported through the xylem. Concerns about possible food crop contamination and its effects on human health are raised by this. However, a number of variables, such as the type of plant, the size and properties of the microplastics, and the current environmental circumstances, affect how much translocation and accumulation occurs. (Azeem, 2021)

4.2 Stress Responses and Reduced Photosynthesis

The emergence of Microplastics in soil ecosystems presents a fresh challenge to plant well-being. Beyond the risk of physical damage, these minute particles can trigger various detrimental effects within plant cells, ultimately affecting essential physiological functions. This section explores how exposure to Microplastics leads to oxidative stress in plants, causing disruptions in processes vital for their survival such as photosynthesis, respiration, and nutrient absorption. Plants perceive Microplastics as a potential hazard, prompting the production of reactive oxygen species (ROS), which are highly reactive molecules capable of damaging cellular components. Numerous factors contribute to this oxidative stress response. Microplastics, with their distinctive surface properties, can directly generate ROS upon interacting with plant cells. Furthermore, they may interfere with the natural antioxidant defense mechanisms within plants, leaving them vulnerable to damage inflicted by ROS. (Jasińska A, 2022)

Photosynthesis, the fundamental process by which plants harness sunlight to produce energy, is particularly susceptible to oxidative stress induced by it. Damage to chloroplasts, the cellular organelles responsible for photosynthesis, can significantly impede light absorption and energy transfer, resulting in reduced photosynthetic efficiency. (Pignattelli, 2021) Consequently, plants face challenges in producing the essential carbohydrates necessary for growth and development.

Cellular respiration, the process by which plants convert sugars into usable energy, is another essential function impacted by MP exposure. Oxidative stress can harm mitochondria, the cellular structures responsible for respiration, disrupting the electron transport chain, a critical component of energy production. Consequently, ATP (adenosine triphosphate) production, the primary energy currency within cells, decreases. With limited energy resources, plant growth and development are significantly affected.

Exposure to microplastics can cause disruptions to the complex network of channels and transporters that are essential for nutrient uptake within plant roots. Damage brought on by oxidative stress may impede the performance of these transporters, making it more difficult for the plant to efficiently take up vital nutrients from the soil. (Pignattelli, 2021) This nutrient deficiency further restricts plant growth and productivity.

4.3 Reduced Germination and Stunted Growth

The growing amount of microplastics in soil ecosystems poses a serious threat to agricultural productivity and plant health. This section looks at how microplastics negatively affect root growth, seed germination, and eventually plant biomass and yield as a whole.Plant growth in its early stages is especially susceptible to exposure to MP. There are several ways in which microplastics might physically prevent seed germination. They might stick to the seed coat and form a barrier that prevents the absorption of oxygen and water, which are essential for germination. Furthermore, it might interfere with the complex hormonal signaling pathways that initiate seed germination, which would impede the sprouting process. (Campanale, 2020)

Plants utilize their roots as an essential pathway for the uptake of water and nutrients. Root development is considerably impeded by the presence of microplastics in the soil. Smaller may penetrate root hairs, impairing their functionality and reducing the surface area that is accessible for the absorption of nutrients and water. They may also change the structure of the soil, which would compact the soil and decrease its pore space, which would hinder root growth and make it more difficult for supplies to reach the soil. (Joos, 2022)

Reduced overall plant biomass and yield are the combined effects of limited root growth and impaired seed germination. Throughout their life cycle, plants exposed to microplastics frequently show stunted growth, with less development of their roots and shoots. Agricultural production is impacted as a result of the reduction in the overall amount of plant material produced. For instance, crops grown in microplastic contaminated soil may produce less fruits, vegetables, or grains, costing farmers a substantial amount of money.

Microplastics have detrimental effects on plant growth that go beyond simple physical limitations. Due to their weaker root systems, stunted plants are more vulnerable to external stresses including disease, drought, and insect infestations, which can further reduce their production. Furthermore, microplastics may interfere with the complex relationships that plants have with helpful soil bacteria, which could have an impact on the cycling of nutrients and the general health of the soil.

4.4 Altered Quality and Safety of Food Crops

The widespread presence of microplastics in the environment has raised concerns about their possible introduction into the food chain. This section explores the likelihood of fruits and vegetables being contaminated with MPs and discusses the potential implications for human health. Fruits and vegetables can become contaminated with MPs through various pathways. One significant concern is root uptake, where small MPs may be directly absorbed by plant roots, potentially accumulating in parts that are consumed. Additionally, MPs might adhere to the surfaces of root vegetables or penetrate through micro-cracks in the skin. Another potential route for contamination is foliar uptake, where microplastics from airborne dust or irrigation water can deposit onto leaves, particularly affecting leafy greens. Once inside the plant, the fate of MPs is still being investigated. While some studies suggest limited movement to edible parts, others indicate the possibility of MPs accumulating in fruits and vegetables. The extent of this accumulation depends on various factors such as the type of MP, the plant species, and the specific part of the plant (e.g., leaves versus seeds). Furthermore, the potential for biomagnification, where microplastics concentrations increase at higher levels of the food chain, requires further examination.

The potential health implications of consuming MP-contaminated fruits and vegetables are not fully understood. Although the immediate health risks of ingesting small quantities of MPs are uncertain, potential concerns include:

- **Physical Harm:** MPs with sharp edges might inflict internal injuries within the digestive tract. (Chen, 2022)
- **Disruption of Gut Microbiota:** MPs may disrupt the delicate equilibrium of gut microbes, potentially affecting digestion and overall gut health.
- **Chemical Release:** Certain MPs might harbor harmful additives or breakdown products that could leach into food and be ingested. (Do Atn, 2022)

5. EMERGING RESEARCH AND FUTURE DIRECTIONS

Microplastics (MPs) are ubiquitous in soil settings and have become a major scientific issue. Although preliminary studies have illuminated possible effects on soil health and plant productivity, significant gaps in information prevent us from fully evaluating the long-term ramifications. This emphasizes the necessity of moving toward rigorous research projects that decipher the complex relationships between MPs, particular soil types, varied soil microbial populations, and different plant species.

5.1 Tailoring Approaches to Soil Specifics

The properties of the soil have a major influence on how MPs behave and interact with one another in ecosystems. For instance, compared to sandy soils, clay-rich soils might have a higher capacity for Microplastics sorption. Gaining a comprehensive understanding of the ways in which soil composition, texture, and organic matter content affect Microplastics transport, persistence, and potential effects on soil biota is essential. To effectively estimate Microplastics fate and transport, research efforts should focus on developing robust models that take soil heterogeneity into account. (Yu, 2021)

5.2 Navigating the Microbial Maze

Soil microbial communities boast immense diversity, with each member contributing uniquely to ecological functions. To comprehensively assess the overall impact of MPs on soil health and biogeochemical cycles, deeper insights into interactions between different MP types and specific microbial populations are essential. Rigorous research should delve into the susceptibility of various microbial taxa to Microplastics induced stress or disruption, alongside exploring the potential of certain microbial communities to degrade specific Microplastics. This species-specific understanding will inform targeted interventions aimed at preserving or enhancing soil microbial functionality amidst Microplastics pollution. (Du, Liang, Li, & Gong, 2020)

5.3 Plant Diversity and Varied Responses:

Different reactions are shown by plants to environmental stresses, such as exposure to MP. Certain plant species may have innate defense systems, while others may be more vulnerable to Microplastics uptake and transfer. Research should look at the unique susceptibilities and potential tolerance mechanisms of various plant species in response to Microplastics exposure in order to assure food safety and reduce risks. By utilizing this information, crop management strategies that are specifically designed to reduce the hazards of MP contamination in agricultural settings can be developed. (hen G, 2022)

5.4 A Holistic Approach to Bridging Knowledge Gaps

To close the existing information gaps, research on the interactions between MPs, certain soil types, varied microbial populations, and a variety of plant species must be prioritized. This all-encompassing strategy will enable: (Rai, et al., 2023)

- Development of accurate models predicting Microplastics fate and transport in diverse soil environments.
- Identification of microbial populations vulnerable to Microplastics exposure and those capable of biodegradation.
- Determination of plant species susceptible to Microplastics uptake, informing targeted crop management strategies.
- Exploration of the long-term implications of Microplastics exposure on soil health, ecosystem function, and food safety.

6. CONCLUSION AND FUTURE PERSPECTIVE

6.1. Conclusion

The widespread occurrence of microplastics in soil habitats is a complex problem that has significant effects on plant physiology, soil health, and ecosystem stability. MPs cause substantial physical and chemical changes when they interact with soil particles, which affect water dynamics, pore structure, and soil aggregation. Plant growth and agricultural production are eventually impacted by these changes, which also have an impact on critical soil characteristics as bulk density, stability, and susceptibility to erosion. Microplastics (MPs) operate as both adsorbents and potential sources of hazardous compounds; the chemical interactions between MPs and soil pollutants further complicate the situation. Microplastics are a concern to soil chemistry, microbial health, and the larger ecosystem because of their capacity to draw in and release pollutants, such as pesticides and heavy metals. . The disruption of soil pH and salinity by Microplastics exacerbates these issues, impacting microbial activity and nutrient availability.

The physical and chemical presence of microplastics has a significant impact on the microbial populations in the soil, which are essential for the cycling of nutrients and plant health. There are significant worries regarding the resilience of the soil ecosystem and public health due to the ensuing decrease in microbial variety and activity as well as the possible rise of antibiotic-resistant diseases. From the standpoint of plant physiology, microplastics obstruct vital functions including nutrient intake, photosynthesis, root growth, and germination. Concerns regarding food safety and human health are raised by the possibility that MPs will be absorbed by plants and build up in edible components. Agricultural yields may be jeopardized by the physical presence of MPs in the soil matrix and the stress reactions that accompany it, which can result in decreased biomass and stunted development in plants.

6.2. Future Perspective

Addressing the challenge of MPs in soil ecosystems necessitates a comprehensive and interdisciplinary research approach. Future studies should focus on several key areas to enhance our understanding and develop effective mitigation strategies:

1. Tailoring Approaches to Soil Specifics
    a. Research need investigate how soil composition, texture, and organic matter content influence Microplastics behavior and interactions.
    b. Goal Develop robust models that account for soil heterogeneity to predict MP fate and transport more accurately.

2. Navigating the Microbial Maze
    a. Research need explore the interactions between different MP types and specific soil microbial populations.
    b. Goal Identify microbial taxa susceptible to Microplastics-induced stress and those with potential to degrade MPs, informing targeted interventions to preserve soil microbial functionality.
3. Varieties of Plants and Their Reactions
    a. Research Need Examine the differential responses of various plant species to MP exposure, identifying innate defense mechanisms and vulnerabilities.
    b. Goal Develop crop management strategies that mitigate Microplastics contamination risks and ensure food safety in agricultural settings.
4. A Holistic Approach to Bridging Knowledge Gaps
    a. Research Need Prioritize interdisciplinary research on Microplastics interactions with specific soil types, diverse microbial populations, and a range of plant species.
    b. Goal Create comprehensive frameworks to understand the complex dynamics of Microplastics pollution, guiding sustainable soil management practices and policy decisions.

By integrating these research directions, we can work towards mitigating the negative impacts of microplastic contamination and protecting the long-term health and resilience of our vital soil ecosystems. This will not only preserve ecosystem services but also safeguard agricultural productivity and food security for future generations.

REFERENCES

Allen, S. A. (2019). Atmospheric microplastics in remote and high-altitude ecosystems. *Nature*, 569(7756), 641–645.

Azeem, A. M., Adeel, M., Ahmad, M. A., Shakoor, N., Jiangcuo, G. D., Azeem, K., Ishfaq, M., Shakoor, A., Ayaz, M., Xu, M., & Rui, Y. (2021). Uptake and Accumulation of Nano/Microplastics in Plants: A Critical Review. *Nanomaterials (Basel, Switzerland)*, 11(11), 1–16. DOI: 10.3390/nano11112935

Barnes, J. K. (2020). Accumulation and fragmentation of plastic debris in marine environments. *Environmental Pollution*, 269(110915), 1–17.

Brahney, J. H. (2020). Microplastics in the marine environment. *Environmental Science & Technology*, 54(19), 11490–11514.

Campanale, M. C., Massarelli, C., Savino, I., Locaputo, V., & Uricchio, V. F. (2020). Detailed Review Study on Potential Effects of Microplastics and Additives of Concern on Human Health. *International Journal of Environmental Research and Public Health*, 17(4), 1–26. DOI: 10.3390/ijerph17041212

Carr, S. A. (2016). Removal of microplastics by wastewater treatment plants: Does it really matter? *Water Research*, 92, 90–99.

Chen, L. Y., Li, Y., Liu, S., Junaid, M., & Wang, J. (2022). Effects of micro(nano) plastics on higher plants and the rhizosphere environment. *The Science of the Total Environment*, 807, 1–9. DOI: 10.1016/j.scitotenv.2021.150841

Do Atn, H. Y. (2022). Leaching of microplastic-associated additives in aquatic environments: A critical review. *Environ Pollut.*, 1–14.

Dris, R. G. (2017). Synthetic microfiber pollution in airborne dust: Potential for human inhalation exposure. *Marine Pollution Bulletin*, 120(1-2), 49–53.

Du, C., Liang, H., Li, Z., & Gong, J. (2020). Pollution Characteristics of Microplastics in Soils in Southeastern Suburbs of Baoding City, China. *nternational Journal of Environmental Research and Public Health*, 1-14.

Geyer, R. K. (2015). Leaked: The story of lost and mismanaged plastic from plastic production to plastic pollution. *Environmental Science & Technology*, 49(22), 12925–12932.

Gouin, T. M. (2019). Microplastics in commercial laundry detergents: Washing clothes as a source of microfibre pollution. *Microplastics and Nanoplastics*, 1(4), 259–267.

hen G, L. Y. (2022). Effects of micro(nano)plastics on higher plants and the rhizosphere environment. *Sci Total Environ.*, 1-13.

Hess, S. M. (2012). Pesticide container disposal practices in developing and newly industrialized countries: A review. *Waste Management (New York, N.Y.)*, 32(5), 899–912.

Hong Yu, Y. Z. (2022). Microplastics as an Emerging. *Frontiers in Environmental Science*, 1–18.

Horton, T. L. (2017). Microplastic in terrestrial environments: Evaluating the evidence for source-to-sink fluxes. *Marine Pollution Bulletin*, 114(1-2), 201–216.

Huerta-Lwanga, H. R.-O. (2016). The use of seed coating technologies to enhance water uptake and germination of soybean seeds. *Industrial Crops and Products*, 93, 155–167.

Jasińska, A. R. S., Różalska, S., Rusetskaya, V., Słaba, M., & Bernat, P. (2022). Microplastic-Induced Oxidative Stress in Metolachlor-Degrading Filamentous Fungus Trichoderma harzianum. *International Journal of Molecular Sciences*, 23(21), 1–12. DOI: 10.3390/ijms232112978

Joos, D. T., & De Tender, C. (2022). The importance of soil life and how it is influenced by (micro)plastic pollution. *Computational and Structural Biotechnology Journal*, 20, 1–21. DOI: 10.1016/j.csbj.2022.03.041

Koycheva, M. L.-S., Mihaylova, L. V., Todorova, M. N., Balcheva-Sivenova, Z. P., Alipieva, K., Ferrante, C., Orlando, G., & Georgiev, M. I. (2021). Leucosceptoside A from Devil's Claw Modulates Psoriasis-like Inflammation via Suppression of the PI3K/AKT Signaling Pathway in Keratinocytes. *Molecules (Basel, Switzerland)*, 26(22), 1–16. DOI: 10.3390/molecules26227014

Lebreton, L. C. (2011). Future trends in plastic pollution and debris flow into the marine environment. *Environmental Science & Technology*, 45(6), 9052–9059.

Li, J. S. (2020). Sorption behaviors of microplastics in soil and water: A critical review. *The Science of the Total Environment*, 715, 1–17.

Li, J. Y., Yu, S., Yu, Y., & Xu, M. (2022). Effects of Microplastics on Higher Plants: A Review. *Bulletin of Environmental Contamination and Toxicology*, 109(2), 1–18. DOI: 10.1007/s00128-022-03566-8

Liu, L. Y. (2022). Uptake of microplastics by rice (Oryza sativa L.) from soil with or without fertilizer amendment. *Chemosphere*, 310(135809), 1–8.

Muhammad Sajjad, Q. H. (2022). Microplastics in the soil environment: A critical review. *Environmental Technology & Innovation*, 1–24.

<bok>Mulvaney, R. L. (2011). Biodegradation of plastic mulches. *In Handbook of bioplastics and biodegradable polymers.*, (pp. 563-585).</bok>

Napper, I. E. (2016). Assessment of microplastic pollution on land and ocean. *Environmental Pollution*, 212, 150–159.

Pignattelli, B. A. (2021). Stress Effect Induced by Microplastics Coupled with Acid Rain. *Annals of Agricultural & Crop Sciences*, 1–12.

Rai, M., Pant, G., Pant, K., Aloo, B., Kumar, G., & Singh, H. (2023). icroplastic Pollution in Terrestrial Ecosystems and Its Interaction with Other Soil Pollutants: A Potential Threat to Soil Ecosystem Sustainability. *Resources*, 1–14.

Rai, M., Pant, G., Pant, K., Aloo, B., Kumar, G., Singh, H., & Tripathi, V. (2023). Microplastic Pollution in Terrestrial Ecosystems and Its Interaction with Other Soil Pollutants: A Potential Threat to Soil Ecosystem Sustainability. *Resources*, 12(6), 1–14. DOI: 10.3390/resources12060067

Randhawa, J. (2023). Advanced analytical techniques for microplastics in the environment: A review. *Bulletin of the National Research Center*, 47(1), 1–14. DOI: 10.1186/s42269-023-01148-0

Rillig, M. C. (2020). Microplastic pollution in soil: Are plants and soil biota at risk? *Environmental Science & Technology*, 54(16), 9409–9419.

Roy, T. D., Dey, T. K., & Jamal, M. (2023). Microplastic/nanoplastic toxicity in plants: An imminent. *Environmental Monitoring and Assessment*, 195(1), 1–35. DOI: 10.1007/s10661-022-10654-z

Wang, C., Tang, J., Yu, H., Wang, Y., Li, H., & Xu, S. (2022). Microplastic Pollution in the Soil Environment: Characteristics, Influencing Factors, and Risks. *Sustainability*, 1–14. DOI: 10.3390/su15010001

Wang, Q. W., Wang, Q., Adams, C. A., Sun, Y., & Zhang, S. (2022). Effects of microplastics on soil properties: Current knowledge and future perspectives. *Journal of Hazardous Materials*, 424, 1–36. DOI: 10.1016/j.jhazmat.2021.127531

Wua, J. (2019). Environmental occurrences, fate, and impacts of microplastics. *Ecotoxicology and Environmental Safety*, 184, 1–16. DOI: 10.1016/j.ecoenv.2019.109612

Yu, J. &.-L. (2021). Micro plastics in soil ecosystem – A review of sources, fate, and ecological impact. *Plant, Soil and Environment*.

Zhan, M. L. (2018). Biodegradable, slow-release fertilizer prepared with superabsorbent polymer composites and controlled-release coatings. *Journal of Applied Polymer Science, 136(4), 46068*, 1-12.

Zhang, G. X. (2023). Global dust cycle transports microplastics from arid and semiarid regions to marine environments. *Nature Geoscience*, 16(3), 346–352.

Zhang, Z. C. (2020). Microplastics alter soil aggregate stability and microbial community composition. *Environmental Science & Technology*, 54(10), 5905–5914.

Chapter 6
Strategies to Mitigate Microplastic (MP) and Nanoplastic (NP) Pollution in Soil

Hema Kandpal
G.B. Pant University of Agriculture and Technology, India

R. K. Srivastava
G.B. Pant University of Agriculture and Technology, India

ABSTRACT

The mismanagement of plastic waste has led to the environmental pollution and bioaccumulation of these plastic particles into living organisms through various food chains and food webs in the ecosystems. The breakdown of plastic into micro-plastic (MP) and nanoplastic (NP) intensifies its adverse impact on soil fertility, food security and living organisms. MP and NP can enter the soil through various pathways such as plastic mulching, compost and sewage sludge and atmospheric deposition. These particles migrate through various ecosystems. Small sizes limit the efficiency of detection of these particles and thus require urgent, sustainable solutions to mitigate MP and NP pollution. These particles can adsorb harmful chemicals like pesticides, PAHs, PCBs and heavy metals on their surface. Therefore, this chapter discusses the sources, impact and mitigation strategies for MP and NP pollution. Moreover, this chapter enhances understanding of the environmental threats by MP and NP pollution and provides sustainable approaches to mitigate their detrimental effects. .

DOI: 10.4018/979-8-3693-9163-1.ch006

1. INTRODUCTION

The widespread discovery of plastic pollution in diverse regions, such as the cryosphere, deep ocean floor and atmosphere where human activity is almost negligible has drawn attention to plastic pollution. Microplastics are particles that have a small size, ranging from 1 µm - 5 mm (Kiran et al., 2022). On the other hand, nanoplastics have a size smaller than microplastics, measuring < 1 µm. These are created through large-scale plastic waste breaking down into smaller fragments. Additional sources of microplastics (MP) and nanoplastics (NP) include particles from coating abrasions and pieces from tires etc. (Scaria et al., 2023). The study by Rai et al. (2021) explained that plastic waste released into the environment mostly consists of macroplastic ranging from 200µm-100µm. These plastic particles slowly deteriorate and broken down into tiny particles by action of various factors such as mechanical forces, ultraviolet radiation, etc. It is very challenging to remove such small-size MNPs (Micronanoplastics) from the environment. The MNPs present in the environment often consist of small low-density polystyrene and polyethylene particles which are hydrophobic (de Costa et al., 2016). These small particles impact living systems and grow increasingly harmful because of their tiny size and long-lasting properties. Plastic consumption has become an essential aspect of societies because of its unique structural characteristics and widespread availability at a low cost (Banerjee & Srivastava, 2012).

Globally, the annual production of plastic waste is nearly 400 metric tons. The low biodegradability of plastic polymers results in a prolonged disintegration time, leading to their breakdown into smaller pieces and increasing their persistence in nature (Chamas et al., 2020). Microplastics are formed due to the gradual degradation of macroscopic plastic fragments caused by mechanical, photochemical and biological processes. Based on the diameter of plastic particles, they are referred to as micro, macro, meso and nanoplastics i.e., 1-1000µm >20 mm, 1-10 mm and 0.001-1µm respectively (Bermúdez & Swarzenski, 2021). Just 15-20% of the microplastics (MP) that enter the environment originate from primary MPs, which are produced in large quantities by household applications such as abrasive cleaning agents, detergents, toothpaste, plastic powder for personal care products, exfoliating face scrubs and synthetic clothing. Hartmann et al. (2017) stated that MPs can effectively adsorb hydrophobic organic compounds like nonylphenol ethoxylates, polychlorinated biphenyls (PCBs), polybrominated diphenyl ethers and polycyclic aromatic hydrocarbons (PAHs). This ability is due to the hydrophobic nature, mobility, strong hydrophobicity and high stability of MPs (Liu et al., 2021).

NPs cause greater risks than MP because they can enter the cell membranes of living organisms (Huang et al., 2021). The quantity of MP and NP emitted per person varies significantly depending on factors such as sustainable waste management practices, wealth and population size and location, etc. Additionally, composts derived from biosolids, mixed municipal solid waste (MSW) and non-source-separated waste also contribute to plastic pollution in agroecosystems. Secondary microplastics are expected to be produced due to the physical breakdown of plastics caused by fragmentation and abrasion during mixing and transportation. These MNPs change the physiochemical properties of soil and transfer into the biotic system from one organism to another.

The study of terrestrial systems in terms of micro and nanoplastic pollution has been ignored compared to the water. The plastic once entered into the soil system mixes with the organic matter of the soil that can't be removed easily and it ends up in the soil for more than 100 years for further production of nanoplastic (Wahl et al., 2021). Recent studies explore the interaction of MNPs in the soil but lack analytical methods to study the MNP in a heterogeneous mixture of soil. New regulations for plastic in soils are needed because the major resources such as soil and water are contaminated with MNPs. This chapter focuses on MPs and NPs occurrence and transport in soil and MNPs interaction and impact on soil and strategies to mitigate MNP's impact in the soil ecosystem.

2. MICROPLASTIC (MP)

The concept of microplastic was first introduced by Richard Thompson. Microplastics (MP) are plastic particles with a size of less than 5mm. MP is composed of a polymeric chain made up of hydrocarbon and additives that contain heteroatoms like tetrabromobisphenol A, polybrominated diphenyl ethers and phthalates (Ramasamy & Palanisamy, 2021). The plastic undergoes fragmentation due to thermal, chemical, mechanical and physical stresses present in the environment and forms MP. Microplastics are classified into two categories: primary MP and secondary MP. MP are utilized as cosmetics and cleansers in their primary form. MP in secondary form is derived from the fragmentation of plastic materials. MP exists in various forms, including fibers, beads and irregular shapes and sizes. They have a high surface area and can absorb various pollutants like heavy metals and POPs (Persistent Organic Pollutants) on their surface. Microplastics primarily enter the environment through tires, synthetic textiles and plastic pellets and slowly degrade and transform from microplastic to nanoplastic (Ng et al., 2018).

3. NANOPLASTIC (NP)

Plastic waste production is continuously increasing and accumulating in the environment. The MNPs are used in textiles, electronic items, cosmetic products, and laboratories due to their unique properties such as small size, durability, hydrophobic nature, strength, high surface area and insulation and anti-corrosive nature. The plastic waste slowly decomposes into subfractions of size less than 100 nm which is known as nanoplastics (NP) (Ramasamy & Palanisamy, 2021). The NP is produced by various processes such as photodegradation, biodegradation, oxidation, mechanical force, and hydration, etc. (Yu and Flury, 2020). The introduction of functional groups in the plastic particles by these processes causes an increase in the hydrophobicity of NP. The NPs also have toxicological impacts which is a great cause of concern. With their small size, they can penetrate the biological membranes of organisms (Haldar et al., 2023). They can interact with soil particles and can alter their physiochemical properties such as pH, salinity, etc.

4. SOURCES OF MP AND NP POLLUTION IN SOIL

Plastic use in modern agriculture and irrigation systems has increased in the past few decades. The mulching film, nets, pipes for irrigation, fertilizers with polymers and greenhouses made of plastic are multiple sources of MNP for the soil ecosystem. MNP particles in soil are generated from larger-sized plastic materials like bottles, packaging materials and bags which break down into tiny fragments with time by degradation processes, mechanical abrasion and photooxidation (Surendran et al., 2023). MP undergoes a process of slow fragmentation into smaller particles on exposure to environmental conditions such as sunshine, temperature changes, etc. Plastic mulch is used to prevent weed production and moisture loss from agriculture fields is a common contributor to the MNP in soil. Plastic mulch is used for every crop cycle in the fields, and it leaves its polymer residue in the soil on degradation. Those residues due to their non-biodegradable nature persist in agricultural land for a long time. The waste disposal on open dump sites plays an important part in spreading the waste in the terrestrial ecosystem. MNPs can also enter the soil system from irrigation water (Khalid et al., 2023). Other sources of MNP in soil are biosolids generated from sludge. Ziajahromi et al. (2016) found that around 95% of MP that pass through WWTP end up in biosolids. Biosolids are organic matter recovered from wastewater treatment plants that are used as fertilizer in agriculture (Roy et al., 2022). They improve soil water holding capacity and permeability and provide nutrients such as phosphate but also contaminate fields with microplastic and nanoplastic. MNPs are non-soluble, and they are widely distributed in the cropland with time. Mohaierani and Karabatak (2020) found that the yearly increase of MP resulting from the use of biosolids in important areas such as Canada, the United States, and Australia, the European Union, China, is roughly 1,518, 21,249, 1,241, 26,042, 13,660 tonnes respectively. Studies have shown that MNPs can move from one cropland to an aquatic ecosystem through runoff (Rehm et al., 2021). In addition, atmospheric deposition adds MP and NP into soil ecosystems, when airborne particles from urban and industrial areas settle on the ground surface (Wright et al., 2023). Atmospheric transmission is a rapid method of pollution transportation that explains why MNP pollution is found even in remote areas. Atmospheric MNP can potentially impact surface climate and consequently, the health of ecosystems through its effects on surface albedo, cloud formation and radiative forcing (Aeschlimann et al., 2022). Research indicates that several natural systems, such as oceans, land and freshwater are interrelated and transfer significant quantities of plastic through the atmosphere. The use of water contaminated with microplastic is also a major source of pollution in the soil ecosystem. So, advancement in treatment technology is needed which can remove MNPs from water. Bottom ash, a residual material produced after the incineration of solid waste for its intended use in managing waste, is found to be an unforeseen contributor to MP in the environment. Bottom ash can be utilized in agriculture as a means to enhance soil fertility. Incineration is well recognized as a significant method for disposing of plastic waste (Yang et al., 2021). Although the high temperatures generated by incinerators, not all plastic items are completely burned, leading to an accumulation of plastic remains in the bottom ash. Reuse of bottom ash in the fields might release MNPs into the environment (Shen et al., 2021).

5. EFFECT OF MNP ON SOIL ECOSYSTEM

The MNPs can affect the growth of plants and reduce agriculture productivity. They can accumulate in edible plants passing through the soil. MNPs can also change microbial population and their diversity in the soil ecosystem (Lima et al., 2023). The MNP present in soil favors a few microbes and changes their dynamic and abundance. This change can be a threat to the ecological process which depends on microbes. The MNP distribution in croplands has increased the risk of food contamination with microplastic. Their adsorption and transportation into various ecosystems have further increased the risk of contamination (Liu et al., 2021).

5.1. Impact on Soil

MP adversely affects the physical structure of soil such as water holding capacity, aeration and permeability, reduced aeration and decreased water permeability and water holding capacity (Sajjad et al., 2022). MP also affects the biodiversity, microbial activity, nutrient transfer, nitrogen cycling and organic carbon content of soils leading to reduced plant growth and productivity. Nevertheless, the outcome also depends on the quantity of MP present in the soil. The presence of phthalic acid esters (PAEs) in agricultural plastics mulch has been recognized as a factor in fruit and vegetable growth (Bansal et al., 2022). Biodegradable plastic is often seen as a substitute for traditional plastic. However, research has demonstrated that biodegradable mulch plastic, specifically manufactured from polybutyrate adipate-co-terephthalate (PBAT) and polyhydroxybutyrate (PHB) can generate microplastics (MP) and nanoplastics (NP) during the decomposition process (Table 1). MNPs in the soil can form homogenous aggregates with each other. Zhang et al. (2019) investigated the effect of polyester microfiber (PMF) on soil. They found that the interaction between polyester microfibers and fine soil particles led to the increased production and stability of macro-aggregates. When MPs/NPs accumulate in the soil, they form hetero-aggregates with a variety of particles. These MNPs can move into deep soil and contaminate the groundwater. Hou et al. (2021) reported that polyethylene microplastic inhibits soil enzyme activity, changes the physiochemical characteristics of soil, and impacts microbial diversity. Polystyrene nanoparticles can harm the soil microbiome by inhibiting the enzyme activity that plays a vital role in the phosphorus, nitrogen, and carbon. These enzymes include β-glucosidase, alkaline phosphatase, cellobiohydrolase and leucine-aminopeptidase (Burke et al., 2014; Awet et al., 2018). The use of additives such as dyes or plasticizers enhances the ecological toxicity of MP. For plastic production, various additive components are usually integrated into the polymer backbone. These additives, such as colors, photostabilizers, plasticizers and antioxidants can leak and be released into soil habitats because they are not strongly bound to the polymer (Boston et al., 2023). Phthalate esters, which have been extensively detected in soils are likely to be absorbed and accumulated by plants. Consequently, MNPs can affect soil characteristics, including the physical, biological, and chemical properties of the soil. Additionally, this can affect the growth and development of plants, particularly in agroecosystems.

Table 1. List of different types of nano and microplastics with their impact on the soil

Plastic form	Size	Rate	Results	References
Functionalized Polystyrene Nanoplastics	71±6nm and 55nm ±7 nm	0.3 g/kg and 1.0 g/kg	An increased accumulation of reactive oxygen species and an effect on the growth of plants and the development of seedlings were observed.	Sun et al., 2020
Low-density polyethylene (LDPE)	<2 mm	0.1, 0.3 and 1.0%, m/m	An impact on microbial communities is observed and can potentially disrupt global carbon and nitrogen cycles.	Gao et al., 2021
LDPE fragments	0.3 to 400 μm	+5, +10 and +15 g m−2)	The introduction of MP had a significant impact on both the composition and number of microarthropod and nematode communities.	Lin et al., 2020
Polyester fibers, Polypropylene (PP), polyethylene high density (PEHD) and Polyamide (PA)	8 μm, 2-3mm, 2–3 mm, 15–20 μm,	2.0% of soil fresh weight	The plant characteristics and function were most significantly affected by polyester fibers and polyamide beads.	Machado et al., 2019
Polylactic acid microplastics	20-50 μm	2%	No significant effects on the overall composition and diversity of bacterial communities. However, it affects the interactions between species.	Chen et al., 2020

5.2. Impact on Plants

Plant roots are affected by MP and NP contaminants in the soil. These particles are taken in by plants through their roots and move into the shoot (Campanale et al., 2022). The limit for restricting particles based on size is observed to be between 40 and 50 nanometers (Taylor et al., 2020). Thus, plastic particles can be absorbed and moved within plant tissues similar to nanomaterials. MNPs stick to germinating seeds in soil and block their pores which inhibits nutrients and water uptake by seed. So, MP can slow down crop germination. They can probably infiltrate the food chain through consumable plants or crops. MNP can also induce blockages in the root hairs of plants and inhibit the absorption of vital nutrients (Iqbal et al., 2020). They can also affect the respiration ability of roots. Sometimes MNPs induce changes in cell wall structure and increase the pore size allowing more MNPs to enter inside the root cell. Chemicals in MNP cause plant toxicity through changes in metabolites, hormonal abnormalities, and other effects. The additives such as di(2-ethylhexyl) phthalate used for making polyvinyl chloride can cause toxicity in the soil. Additionally, they decrease the bioavailability of crucial minerals like phosphorus and nitrogen. Air-deposited MNP can obstruct light, reducing photosynthetic efficiency on leaf surfaces. It has been found that MNPs in soil lead to a decrease in water retention capacity, which subsequently results in a reduction in oxygen levels in soil aggregates. The MNPs also change the nutritional status of plants and influence the leaf attributes and root thickness in plants. Larger NP particles can accumulate in the cytoplasm, while smaller ones (about 30 nm) may even penetrate the nucleus, disrupting chromatin structure and function (Haldar et al., 2022). In response to these changes, plant cells release reactive oxygen species that cause changes in the defense system of the cell. NP might potentially cause a genotoxic impact. The ingestion of NP may potentially yield beneficial outcomes for plant growth. According to Lian et al. (2020), wheat roots elongated by 89% to 123% over the control group when exposed to polystyrene NP at concentrations ranging from 0.01 to 10 mg/L ($p < 0.01$). The exposure of NP increased the development of wheat seedlings. There was a significant impact on rhizosphere bacterial populations due to the presence of plastic mulch films, indicating that plastic residues affect both living and non-living organisms (He et al., 2021). Pignattelli et al. (2019) found polyvinyl chloride (PVC) to be the most toxic of all plastic polymers. In the study, PVC causes oxidative stress in the plant *Lepidium sativum.*

5.3 Impact on Health

Plastic often gets lost in the environment through various sources. The size and type of plastic entering the environment depends on its source. The targeted use of plastic in agriculture such as plastic mulch to prevent moisture loss and increase crop yields leaves its residue in the fields (Allouzi et al., 2021). Recollection of the plastic mulch residue is not possible. Similarly, tire wear once lost in the environment cannot be recollected. In-vitro studies have observed that the MNPs can cross the cell membrane and induce inflammation and oxidative stress in human cells (Wang et al., 2022). PVC is considered one of the most hazardous plastics in terms of toxicity due to its carcinogenic monomer and various dangerous additives (Blackburn & Green, 2022). Microscopic particles can pass across the cell membrane and be absorbed by endocytosis (Wang et al., 2022). The ingestion of food contaminated with MP and the exposure to phthalates in toys have led to the development of congenital illnesses, malignancies, and adverse effects on the neurological and reproductive systems. The consumption of microplastics by avian species, particularly young birds, hinders the absorption of nutrients, impairs growth, and affects reproductive systems causing a significant threat to their survival. A study by Cox et al. (2019) examined the quantity of MP ingested by individuals through their average food intake. The study revealed that the mean yearly intake of MP in the American populace varied between 39,000 and 52,000 particles. MP can transport potentially harmful substances such as pesticides, fungicides, and pathogens (Blackburn & Green, 2022). MP can also spread fungal disease in humans. Additionally, the mycobiomes found in plastic waste from municipalities have been identified as potential sources of fungal pathogens (Gkoutselis et al., 2021). MP not only attracts unique fungal communities but also gathers certain human infections, such as cryptococcal and Phoma-like organisms. The size of plastic particles significantly influences the bioaccumulation characteristics and toxicity. The act of breathing in aerosols containing nanoparticles and the subsequent entry of these nanoparticles into the capillary blood system allows for the distribution of this pollutant throughout the human body. The consumption of contaminated water and food is likely the primary route of exposure to NPs.

5.4 Impact on the Surrounding Environment

The MNPs present in soil can move to the surrounding environment such as air, and water by processes like erosion, runoff, and wind. The lightweight plastic microfibre released from synthetic textiles can migrate to the air by the action of wind. The MPs available in soil can penetrate the soil surface and can reach groundwater through infiltration (Bostan et al., 2023). The MNPs present in soil can be a source of pollution to its surrounding environment. MNP also inhibits Isocitrate dehydrogenase activity in the organisms that is essential to maintain redox reactions therefore oxidative damage occurs in the organism. As discussed in the above section MNPs affect microbial population, nutrient cycling, and soil fauna and can enter into the food chain. The impact and risk are also regulated by the size and concentration of MNPs.

6. INTERACTION OF MNP WITH SOIL

MNP present in the soil is a habitat for microbes. Microbes grow over MPs and form a Plastisphere and high accumulation of microbes on the MP surface can change the decomposition of organic matter present in the soil. The ecological process of nutrient recycling can be affected by this. MP also acts as a vector for microbes. The organisms present in soil interact with MNPs and create various types of responses. Some plastics release chemicals such as phthalates which remove the microbial population from soil. These tiny pieces of MP can change microbial diversity (De Souza Machado et al., 2019). The MNPs with large surface area and hydrophobic nature can absorb the toxic chemicals present in soil and transport them. The polystyrene particles exhibit a primarily spherical morphology, characterized by a smooth surface and a diameter that varies between 10 nm and a maximum of 25 μm. These particles can be either unmodified or modified functional groups of the surface, such as amino groups, sulfate groups and carboxyl. These alterations to the surface impact the electric charge and the water-repelling properties of the particles. Upon discharge into the environment, these particles will undergo additional surface alterations through interactions with inorganic ions, dissolved organic matter and extracellular polymeric compounds (Yu & Flury, 2021). Their destiny and movement in soil or sediments will thereafter be influenced by their altered surface characteristics (Figure 1). Secondary MP and NP mostly consist of polyvinyl chloride, polyethylene terephthalate, polyester, and polyethylene. Secondary micro- and nanoplastics differ from primary particles in terms of their shape irregularity and the diversity and heterogeneity of their surface properties. Parkinson et al. (2022) investigated the interaction of nanoparticles with *Arabidopsis thaliana* and observed the successful movement of nanoparticles into the root of the plant.

The surface characteristics of secondary MP and NP undergo alterations as a result of weathering, particularly when exposed to oxygen, water and light. Weathering causes the particles' surface to gain O-containing functional groups, such as C=O, COOH and OH. This makes the particles more water-attracting (hydrophilic) and frequently negatively charged. An eco-corona can be created when microbes and microbial exudates interact with plastic surfaces, resulting in a coating of biomolecules such as exoproteome and extracellular polymeric compounds (Yu & Flury, 2021). Soil is the direct source of penetration of microplastic into the food web. The movement of MNPs is influenced by soil minerals and soil pH, as they collectively impact the surface charges of PSNPs (Polystyrene nanoparticles) and soil. Furthermore, the transportation of PSNPs showed a high level of sensitivity to both the ionic strength and the type of cation, as observed by Wu et al. (2020). The earthworm and other microarthropods also enhance the movement of MNPs in the soil system (Heinze et al., 2021).

Figure 1. Interaction of micro and nano plastics in soil ecosystem

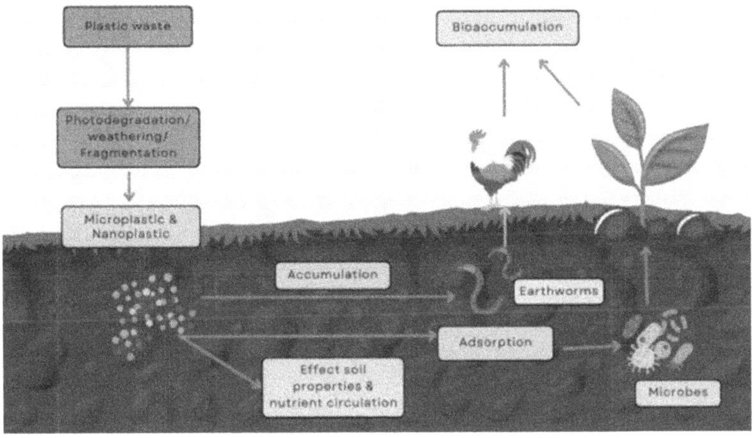

7. STRATEGIES TO REDUCE MP/NPS

Although the ban on the use of plastic can improve future contamination of MNPs to some extent but complete ban on plastic products is hard to achieve. Plastic has solved many problems and consumption has increased very rapidly in the past few decades. They are not sustainable due to their non-biodegradable nature but are lightweight, easy to carry and cheap. The first approach to reduce MNP pollution should be to prevent the dispersal of uncollected primary sources of pollution. An efficient collection system with sorting of plastic waste according to its characteristics should be installed (Mitrano & Wohlleben, 2020). Adopting the use of biodegradable plastic and the application of green chemistry in the field of polymer synthesis can be a mitigation measure for plastic pollution (Fig 2). Waste management and a circular economy approach can help to deal with the plastic problem. Restriction on the use of single-use plastic and recycling to a maximum extent can lead to plastic waste management. With the spread of microplastic into remote zones and glaciers it is not a local problem whereas it is a transboundary issue. It should be tackled locally as well as internationally.

Figure 2. Remediation strategies for MP/NP pollution

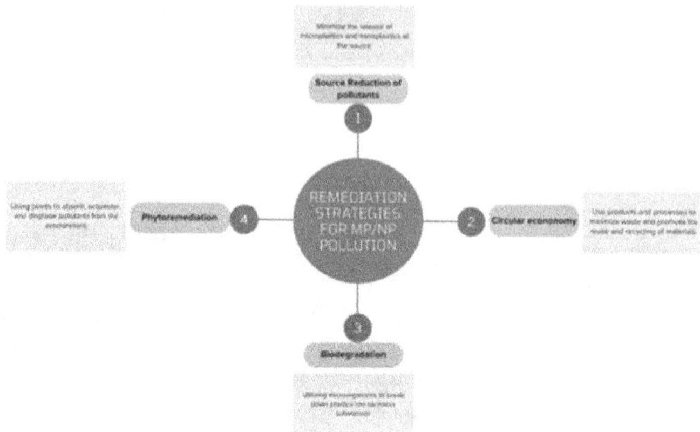

7.1. Source Reduction

The consumption of plastic is more than the capacity for the management of waste. Plastic is becoming a burden on the environment as well as the economy. A study by Borrelle et al. (2020) estimated that annually 19-23 million tons of waste is released into the environment. The restriction at the source level and control on the manufacturing of plastic products can prevent further damage by plastic pollution. Single-use plastic is also a problematic waste in almost all around the world. It is short-lived and more durable. It cannot be easily recycled so it is necessary to reduce its production from the source level. Government interference to stop the manufacturing of short-lived plastic products and promote alternative biodegradable products to replace them can lead to responsible consumption of plastic (Sarkar et al., 2022). Cutting down the production of plastic and replacing it with cardboard, paper and other sustainable materials can reduce MNP pollution. Raising awareness among farmers to prevent the use of plastic products such as plastic mulch, plant protection nets, etc can lead to behavioral change. Regulations to improve the collection efficiency of waste through extended producer responsibility can help in waste management. Implementation of rules and regulations can help in the reduction of pollutants.

7.2. Circular Economy

To achieve zero plastic waste in the waste management system circular economy can be a solution (Kiran et al., 2022). CE cantered on keeping the product in the value chain. The circular economy also addresses all the stages of plastic from its production to its end. Most MNPs come from the breakdown of plastic released into the environment so, reducing the plastic waste will reduce the MNP's pollution. Recycling plastic not only reduces pollution but also lowers further resource consumption for the production of plastic. Recycling of plastic is performed by open and closed-loop methods. The open-loop system utilizes plastic waste as raw material to produce a new plastic product however in closed-loop recycling the plastic is reused again and again and it remains unchanged (Shamsuyeva & Endres, 2021). For reduction in MNP pollution, it is necessary to reduce the loss of secondary microplastic from the value chain of any product. A major part of plastic is lost due to mishandling of plastic at its end phase. The value of plastic can be preserved by redirecting it to the circular economy. In plastic management, efficient sorting of waste can prevent its loss in the environment for which mobile apps and other technologies can be used for monitoring the loss of plastic from the value chain. Plastic has the potential to generate energy when it undergoes thermochemical conversions. Pyrolysis of plastic waste to form fuel can be an effective strategy for plastic waste management. This approach can also prevent the leaching of plastic into various ecosystems. Extended Producer Responsibility (EPR) is another measure to reduce plastic pollution by preventing the loss of products. This approach makes the producer legally responsible for determining the fate of its product and motivates them to produce environment-friendly products that are biodegradable. It addresses the production phase as well as the end phase of plastic. It is an important measure for the circular economy (Syberg et al., 2022).

7.3 Biodegradation

The MNPs persist in nature for a long duration but some microbes can break the plastic into non-toxic compounds by releasing enzymes that act on MNPs (Kiran et al., 2022). The MNP surface is suitable for the growth of microbes. The microbes grow and consume the MNP as a source of energy by converting it into CO_2, N_2, CH_4, H_2O, and H_2S. *Ideonella sakaiensis 201-F6* is a recently discovered bacterial strain that has the potential to convert polyethylene terephthalate (PET) into terephthalic acid and ethylene glycol (Yoshida et al., 2016). Bacterial strains of *Bacillus gottheilii* and *Bacillus cereus* have been identified from some mangrove sediments that exhibit the potential to degrade PET, polystyrene, polypropylene, and polyethylene (Auta et al., 2017). Microbes undergo four major steps in the process of degrading plastic: microbial film production, biodeterioration, bio-fragmentation and mineralization (Rai et al., 2021). Some studies have also concluded that fungus can also consume polyethylene as a source of energy. The evidence is collected by studying biomass fluctuation in fungi when plastic is provided as a source of energy (Paco et al., 2017). Fungus such as *Agaricus bisporus*, *Claviceps fusiformis* and

Schizophyllum commune secret some adhesive hydrophobins which attach to the hydrophobic surface of MNP. Fungi can also release some digestive enzymes from its hyphae through the exocytosis process. These enzymes help in the conversion of plastic into CO_2 and H_2O. The degradation also depends on chemical structure, water absorption, molecular weight, and other properties of MNP. Moreover, abiotic factors such as pH, salinity, temperature and surrounding environment also influence the process of degradation. Hyper thermophilic phase during composting can degrade MNPs through the activity of bacteria. Degradation also depends on the chemical and physical properties of MNPS. The type of plastic also determines the microbe's ability to degrade it. Not all types of plastic can be degraded by microbes. For example, ascomycetes, a group of fungi can break LDPE plastic (Ameen et al., 2015). Arbuscular mycorrhizal fungi which remain in symbiotic association with roots of plants also prevent MNPs pollution. It helps restrict the movement of MNPs by releasing hyphal exudates that cause NP to combine and form larger particles (Li et al., 2024). Earthworms are indicator organisms for monitoring the change occurring in the soil ecosystem. Worms help in the transportation of MNPs through their excretion. Studies have also shown that soil fauna is affected by MNPs present in soil which leads to decreased or increased mobility, change in metabolic functions, and oxidative stress in organisms. Sheng et al. (2021) study has shown that exposure of worms to MNP increases the production of enzymes such as peroxidase, catalase, and lipid peroxidase. The degradation of MNPs can be increased by introducing a genetic approach. The genetic material of bacterial strains that have the potential to degrade plastic can be transferred to other bacterial strains through genetic engineering (Anand et al., 2023). A similar application was observed by Moog et al. (2019) by introducing PET hydrolyzing enzymes into *Phaeodactylum tricornutum* which increases the efficiency of pollutant removal. Aliphatic polyesters, including polylactic acid, polybutylenesuccinate and polycaprolactone as well as natural biopolymers like polyhydroxyalkanoates and cellulose are among the biodegradable polymers that can be used for the synthesis of BP (Table 2). Biodegradable polymers are characterized by the presence of heteroatoms (namely oxygen, nitrogen, and sulfur) that are present in the polymer backbone. Heteroatoms function as reactive sites for enzyme activities, leading to significant decreases in the molecular weight of the polymer within a very brief timeframe, spanning from days to many years (Ng et al., 2018). These mechanisms induce the degradation of the polymer's structure into smaller molecular fragments, which can be assimilated by microbial cells and subsequently mineralized to generate CO_2, water, and biomass in aerobic conditions whereas CH_4 and CO_2 in anaerobic conditions (Sajjad et al., 2022).

Table 2. Structures of common biodegradable and non-biodegradable plastic

Non-Biodegradable Polymer	Biodegradable polymer
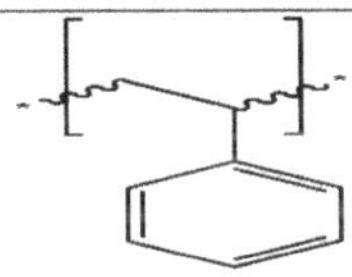 Polystyrene	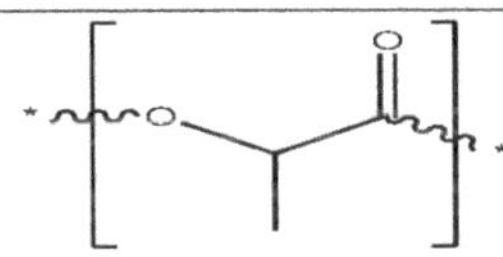 Polylactic Acid
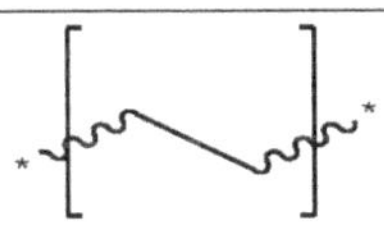 Polyethylene	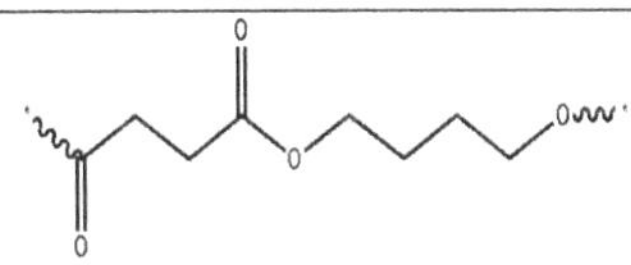 Polybutylene succinate
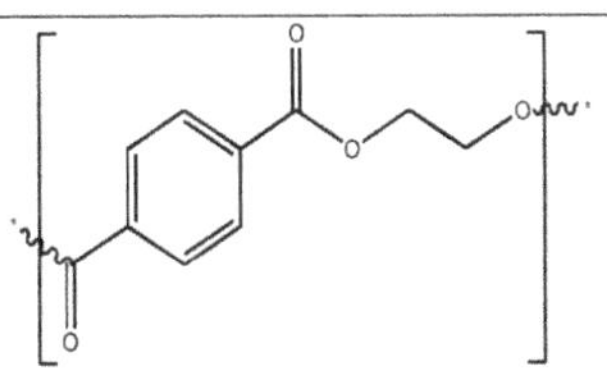 Polyethyleneterepthalate	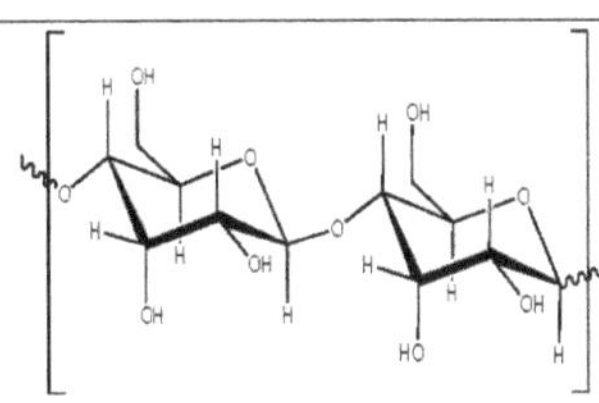 Cellulose
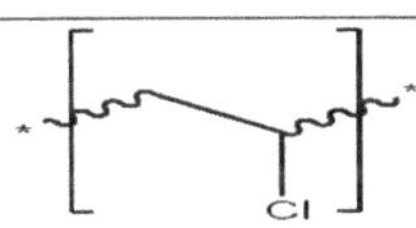 Polyvinylchloride	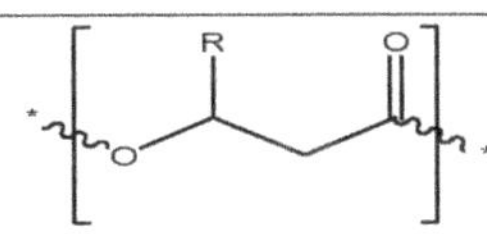 Polyhydroxyalkanoates
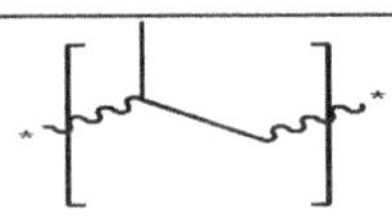 Polypropylene	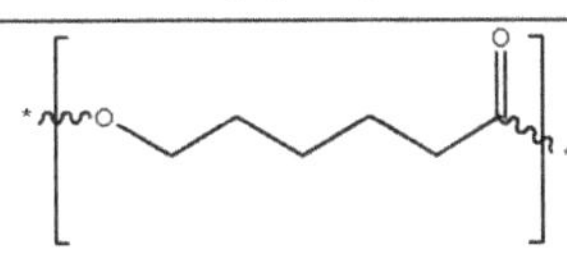 Polycaprolactone

7.4 Phytoremediation

Phytoremediation has a wide scope due to its applicability to various types of contaminated soil. In this process, plants are used for the absorption of MNP from soil. A study by Li et al. (2020) observed that wheat and lettuce plants can absorb

2.0-μm polymethylmethacrylate microbeads and 200-nm polystyrene nanobeads. Although the phytoremediation process only transfers pollutants from soil to plants do not degrade them. Phytoremediation is an ecologically friendly and cost-effective technique to remove pollutants from soil. The efficiency of the removal of pollutants can be improved by introducing genetic factors for the hyperaccumulation of MNPs (Fasani et al., 2018). Some studies found that MP can only reach the root of plants but their further transportation to the aerial part of plants is also verified by some studies by SEM (scanning electron microscope). The rhizosphere can also change the movement of MNP by changing soil properties, microbial population, etc. Similarly, root exudates released from plant roots such as enzymes, amino acids, carbohydrates, etc also influence the fate of MNPs in the soil. These root exudates change the distribution and dispersion of MNP and modify their properties by forming aggregation. The amino acids released in the form of root exudates provide different charges to MNP which determine their fate to get remediated by plants. They can also reduce the mobilization of MNP and restrict their movement by immobilization. It is also verified by some studies that NP can be remediated more efficiently by plants compared to MP. Plant species are important factors in determining the success of the phytoremediation technique (Gong et al., 2023). Some plant species provide no evidence of uptake of MNPs while they can accumulate it. The degree of accumulation also varies from species to species. The response of plants to MNPs also depends on the type of species. MNPs can cause stress in plants and plants have different ways to deal with stress. Sometimes the toxicity caused by MNPs can cause growth inhibition, blocking pores and low biomass, oxidative stress in plants. The effect of MNPs can also lower the remediation efficiency of plants.

7.5 Biotechnology

Along with mitigation and preventive measures to reduce MNP pollution, it is necessary to increase efforts by providing scientific techniques. Biotechnology can improve microbe's efficiency in degrading MNPs. Gene editing is a method in genetic engineering that allows desired gene expression in plants and microbes. This method helps in the incorporation of genes that are associated with MNP degradation (Feng et al., 2023). It is necessary to find out suitable gene for this technique. The research has proved it through various experiments like the deletion of the wspF gene in *Pseudomonas aeruginosa* enhancing its potential for capture of MP in its biofilm. However, genetically modified microbes perform well in laboratory settings but sometimes they fail to give satisfactory results in field conditions. This approach results in the scavenging of MP from a place and helps in remediation. Mutation in specific genes to increase its ability for the formation of particular types of enzymes which can degrade microplastic is also an approach to use biotechnological approach for mitigation of microplastic. E coli strain modified by genetic engineering can also help in MP degradation by producing enzymes manganese peroxidase (Anand et al., 2023).

7.6 Policy and Regulations

In the last few decades, the legislature has also continuously evolved to control MNP pollution. The first plastic policy came in Germany in 1991 to bring down plastic consumption. Then many other countries banned plastic bags in their countries and made provisions to impose fines on manufacturing and consumption of plastic bags. Serious prohibitions are seen in Kenya where even carrying a plastic bag can result in imprisonment. In some countries complete ban on plastic bags below a certain thickness was implemented. This approach has given positive outcomes in some countries whereas in some countries increase in demand for plastic with more thickness has increased after imposing the ban (Lam et al., 2018). After the ban, some manufacturers shifted their production towards more sustainable fabric bags, and some relocated to other countries where rules are more flexible. Labeling of plastic by type of plastic has also helped in preventing plastic mixing during recycling. Providing an alternative to plastic bags also helps in changing the behavior of end users. Another approach that gave positive results in some places is the campaign. Most plastic is used in the packaging of products therefore change in end-user behavior toward plastic consumption and recycling can be an effective strategy to reduce MNP pollution. Some countries like the United States, Canada and Australia have passed laws to stop the manufacturing and sale of microbeads. These microbeads are used in cleaning and cosmetic products. Imposing a ban on microbead production has led to a search for alternatives and now many countries have started using abrasives such as silica, cellulose microcrystal, etc to replace the microbeads in cosmetics (Deme et al., 2022). The lack of quantitative information on MNP pollution restricts the implementation of laws. There are some challenges associated with the enforcement of rules. The public is often not aware of policies, the impact of the consumption of plastic products, and alternatives available in the market for plastic (Seay & Ternes, 2022). Other than this poor governance in the implementation of the ban cause damage by plastic pollution. The plastic sector needs standards for MP in wastewater and quality standards for the production of plastic to manage plastic pollution. Trade policies on plastic can also help in reducing the risk associated with plastic and can create opportunities for economic expansion.

Even with a wide distribution of MNPs in the terrestrial ecosystem, very little quantitative information is available. Frequent monitoring of movement and interaction of MNPs is required to know the associated risk. The knowledge and actual behavior of MPs and NPs in the environment are still in their infancy. There is a need to collect more information about the exposure of MNPs in soil.

8. CONCLUSION AND FUTURE CHALLENGES

Plastic removal from the lives of people is difficult because its demands are very high. If it is released into the environment, it produces micro and nanoplastic on exposure to environmental factors. MNPs are difficult to remove from the environment due to their small size. However, the focus on effective management of plastic pollution is continuously increasing. A lack of regulations on microplastic and insufficient information on abundance, degradation rate and composition, interaction, and behavior of MNPs in the terrestrial environment is an obstacle to mitigating pollution caused by MNPs. There are various impacts of MNPs have been recognized so far by researchers. They affect soil properties, accumulate in plants and can enter the food chain. They can also cause health-related issues in humans like cardiovascular disorders, and gastrointestinal problems. Therefore, plastic waste management is a necessity to save the environment from further damage. Implementation of policies, reducing the waste dispersion, reducing consumption, use of remediation technologies and raising awareness can help to mitigate pollution caused by MNPs.

REFERENCES

Aeschlimann, M., Li, G., Kanji, Z. A., & Mitrano, D. M. (2022). Potential impacts of atmospheric microplastics and nanoplastics on cloud formation processes. *Nature Geoscience*, 15(12), 967–975. DOI: 10.1038/s41561-022-01051-9

Anand, U., Dey, S., Bontempi, E., Ducoli, S., Vethaak, A. D., Dey, A., & Federici, S. (2023). Biotechnological method to remove microplastic: A review. *Environmental Chemistry Letters*, 21(3), 1787–1810. DOI: 10.1007/s10311-022-01552-4

Auta, H. S., Emenike, C. U., & Fauziah, S. H. (2017). Screening of Bacillus strains isolated from mangrove ecosystems in Peninsular Malaysia for microplastic degradation. *Environmental Pollution*, 231, 1552–1559. DOI: 10.1016/j.envpol.2017.09.043

Banerjee, T., & Srivastava, R. K. (2012). Plastics waste management and resource recovery in India. *International Journal of Environment and Waste Management*, 10(1), 90–111. DOI: 10.1504/IJEWM.2012.048153

Bermúdez, J. R., & Swarzenski, P. W. (2021). A microplastic size classification scheme aligned with universal plankton survey methods. *MethodsX*, 8, 101516. DOI: 10.1016/j.mex.2021.101516

Blackburn, K., & Green, D. (2022). The potential effects of microplastics on human health: What is known and what is unknown. *Ambio*, 51(3), 518–530. DOI: 10.1007/s13280-021-01589-9

Borrelle, S. B., Ringma, J., Law, K. L., Monnahan, C. C., Lebreton, L., McGivern, A., & Rochman, C. M. (2020). Predicted growth in plastic waste exceeds efforts to mitigate plastic pollution. *Science*, 369(6510), 1515–1518. DOI: 10.1126/science.aba3656

Bostan, N., Ilyas, N., Akhtar, N., Mehmood, S., Saman, R. U., Sayyed, R. Z., & Pandiaraj, S. (2023). Toxicity assessment of microplastic (MPs); a threat to the ecosystem. *Environmental Research*, 116523, 116523. Advance online publication. DOI: 10.1016/j.envres.2023.116523

Burke, D. J., Zhu, S., Pablico-Lansigan, M. P., Hewins, C. R., & Samia, A. C. S. (2014). Titanium oxide nanoparticle effects on composition of soil microbial communities and plant performance. *Biology and Fertility of Soils*, 50(7), 1169–1173. DOI: 10.1007/s00374-014-0938-3

Campanale, C., Galafassi, S., Savino, I., Massarelli, C., Ancona, V., Volta, P., & Uricchio, V. F. (2022). Microplastics pollution in the terrestrial environments: Poorly known diffuse sources and implications for plants. *The Science of the Total Environment*, 805, 150431. DOI: 10.1016/j.scitotenv.2021.150431

Chamas, A., Moon, H., Zheng, J., Qiu, Y., Tabassum, T., Jang, J. H., & Suh, S. (2020). Degradation rates of plastics in the environment. *ACS Sustainable Chemistry & Engineering*, 8(9), 3494–3511. DOI: 10.1021/acssuschemeng.9b06635

Chen, H., Wang, Y., Sun, X., Peng, Y., & Xiao, L. (2020). Mixing effect of polylactic acid microplastic and straw residue on soil property and ecological function. *Chemosphere*, 243, 125271. DOI: 10.1016/j.chemosphere.2019.125271

Cox, K. D., Covernton, G. A., Davies, H. L., Dower, J. F., Juanes, F., & Dudas, S. E. (2019). Human consumption of microplastics. *Environmental Science & Technology*, 53(12), 7068–7074. DOI: 10.1021/acs.est.9b01517

De Souza Machado, A. A., Lau, C. W., Kloas, W., Bergmann, J., Bachelier, J. B., Faltin, E., & Rillig, M. C. (2019). Microplastics can change soil properties and affect plant performance. *Environmental Science & Technology*, 53(10), 6044–6052. DOI: 10.1021/acs.est.9b01339

Deme, G. G., Ewusi-Mensah, D., Olagbaju, O. A., Okeke, E. S., Okoye, C. O., Odii, E. C., & Sanganyado, E. (2022). Macro problems from microplastics: Toward a sustainable policy framework for managing microplastic waste in Africa. *The Science of the Total Environment*, 804, 150170. DOI: 10.1016/j.scitotenv.2021.150170

Feng, W., Deng, Y., Cao, Y., Liu, J., Han, Y., Liu, J., & Giesy, J. P. (2023). Biotechnology remediation and environmental behavior of microplastics in soils: A review. *Reviews of Environmental Contamination and Toxicology*, 261(1), 13. DOI: 10.1007/s44169-023-00038-w

Gao, B., Yao, H., Li, Y., & Zhu, Y. (2021). Microplastic addition alters the microbial community structure and stimulates soil carbon dioxide emissions in vegetable-growing soil. *Environmental Toxicology and Chemistry*, 40(2), 352–365. DOI: 10.1002/etc.4916

Gong, X., Shi, G., Zou, D., Wu, Z., Qin, P., Yang, Y., & Zhou, Y. (2023). Micro-and nano-plastics pollution and its potential remediation pathway by phytoremediation. *Planta*, 257(2), 35. DOI: 10.1007/s00425-023-04069-4

Haldar, S., Muralidaran, Y., Míguez, D., Mulla, S. I., & Mishra, P. (2023). Ecotoxicity of nano-plastics and its implication on human metabolism: Current and future perspective. *The Science of the Total Environment*, 861, 160571. DOI: 10.1016/j.scitotenv.2022.160571

Hartmann, N. B., Rist, S., Bodin, J., Jensen, L. H., Schmidt, S. N., Mayer, P., & Baun, A. (2017). Microplastics as vectors for environmental contaminants: Exploring sorption, desorption, and transfer to biota. *Integrated Environmental Assessment and Management*, 13(3), 488–493. DOI: 10.1002/ieam.1904

He, D., Zhang, Y., & Gao, W. (2021). Micro (nano) plastic contaminations from soils to plants: Human food risks. *Current Opinion in Food Science*, 41, 116–121. DOI: 10.1016/j.cofs.2021.04.001

Hou, J., Xu, X., Yu, H., Xi, B., & Tan, W. (2021). Comparing the long-term responses of soil microbial structures and diversities to polyethylene microplastics in different aggregate fractions. *Environment International*, 149, 106398. DOI: 10.1016/j.envint.2021.106398

Huang, D., Tao, J., Cheng, M., Deng, R., Chen, S., Yin, L., & Li, R. (2021). Microplastics and nanoplastics in the environment: Macroscopic transport and effects on creatures. *Journal of Hazardous Materials*, 407, 124399. DOI: 10.1016/j.jhazmat.2020.124399

Iqbal, S., Xu, J., Allen, S. D., Khan, S., Nadir, S., Arif, M. S., & Yasmeen, T. (2020). Unraveling consequences of soil micro-and nano-plastic pollution on soil-plant system: Implications for nitrogen (N) cycling and soil microbial activity. *Chemosphere*, 260, 127578. DOI: 10.1016/j.chemosphere.2020.127578

Khalid, N., Aqeel, M., Noman, A., & Rizvi, Z. F. (2023). Impact of plastic mulching as a major source of microplastics in agroecosystems. *Journal of Hazardous Materials*, 445, 130455. DOI: 10.1016/j.jhazmat.2022.130455

Lam, C. S., Ramanathan, S., Carbery, M., Gray, K., Vanka, K. S., Maurin, C., & Palanisami, T. (2018). A comprehensive analysis of plastics and microplastic legislation worldwide. *Water, Air, and Soil Pollution*, 229(11), 1–19. DOI: 10.1007/s11270 018-4002-z

Li, H. H., Chen, X. W., Zhai, F. H., Li, Y. T., Zhao, H. M., Mo, C. H., & Li, H. (2024). Arbuscular Mycorrhizal Fungus Alleviates Charged Nanoplastic Stress in Host Plants via Enhanced Defense-Related Gene Expressions and Hyphal Capture. *Environmental Science & Technology*, 58(14), 6258–6273. DOI: 10.1021/acs.est.3c07850

Lima, J. Z., Cassaro, R., Ogura, A. P., & Vianna, M. M. G. R. (2023). A systematic review of the effects of microplastics and nanoplastics on the soil-plant system. *Sustainable Production and Consumption*, 38, 266–282. DOI: 10.1016/j.spc.2023.04.010

Lin, D., Yang, G., Dou, P., Qian, S., Zhao, L., Yang, Y., & Fanin, N. (2020). Microplastics negatively affect soil fauna but stimulate microbial activity: insights from a field-based microplastic addition experiment. *Proceedings of the Royal Society B, 287*(1934), 20201268. https://doi.org/doi:10.1098/rspb.2020.1268

Liu, Y., Shao, H., Liu, J., Cao, R., Shang, E., Liu, S., & Li, Y. (2021). Transport and transformation of microplastics and nanoplastics in the soil environment: A critical review. *Soil Use and Management*, 37(2), 224–242. DOI: 10.1111/sum.12709

Mitrano, D. M., & Wohlleben, W. (2020). Microplastic regulation should be more precise to incentivize both innovation and environmental safety. *Nature Communications*, 11(1), 5324. DOI: 10.1038/s41467-020-19069-1

Mohajerani, A., & Karabatak, B. (2020). Microplastics and pollutants in biosolids have contaminated agricultural soils: An analytical study and a proposal to cease the use of biosolids in farmlands and utilise them in sustainable bricks. *Waste Management (New York, N.Y.)*, 107, 252–265. DOI: 10.1016/j.wasman.2020.04.021

Ng, E. L., Lwanga, E. H., Eldridge, S. M., Johnston, P., Hu, H. W., & Geissen, V. (2018). An overview of microplastic and nanoplastic pollution in agroecosystems. *The Science of the Total Environment*, 627, 1377–1388. DOI: 10.1016/j.scitotenv.2018.01.341

Paço, A., Duarte, K., da Costa, J. P., Santos, P. S., Pereira, R., Pereira, M. E., & Rocha-Santos, T. A. (2017). Biodegradation of polyethylene microplastics by the marine fungus Zalerion maritimum. *The Science of the Total Environment*, 586, 10–15. DOI: 10.1016/j.scitotenv.2017.02.017

Parkinson, S. J., Tungsirisurp, S., Joshi, C., Richmond, B. L., Gifford, M. L., Sikder, A., Lynch, I., O'Reilly, R. K., & Napier, R. M. (2022). Polymer nanoparticles pass the plant interface. *Nature Communications*, 13(1), 7385. DOI: 10.1038/s41467-022-35066-y

Pignattelli, S., Broccoli, A., & Renzi, M. (2020). Physiological responses of garden cress (L. sativum) to different types of microplastics. *The Science of the Total Environment*, 727, 138609. DOI: 10.1016/j.scitotenv.2020.138609

Rai, P. K., Lee, J., Brown, R. J., & Kim, K. H. (2021). Micro-and nano-plastic pollution: Behavior, microbial ecology, and remediation technologies. *Journal of Cleaner Production*, 291, 125240. DOI: 10.1016/j.jclepro.2020.125240

Ramasamy, B. S. S., & Palanisamy, S. (2021). A review on occurrence, characteristics, toxicology and treatment of nanoplastic waste in the environment. *Environmental Science and Pollution Research International*, 28(32), 43258–43273. DOI: 10.1007/s11356-021-14883-6

Rehm, R., Zeyer, T., Schmidt, A., & Fiener, P. (2021). Soil erosion as transport pathway of microplastic from agriculture soils to aquatic ecosystems. *The Science of the Total Environment*, 795, 148774. DOI: 10.1016/j.scitotenv.2021.148774

Sajjad, M., Huang, Q., Khan, S., Khan, M. A., Liu, Y., Wang, J., & Guo, G. (2022). Microplastics in the soil environment: A critical review. *Environmental Technology & Innovation*, 27, 102408. DOI: 10.1016/j.eti.2022.102408

Sarkar, B., Dissanayake, P. D., Bolan, N. S., Dar, J. Y., Kumar, M., Haque, M. N., & Ok, Y. S. (2022). Challenges and opportunities in sustainable management of microplastics and nanoplastics in the environment. *Environmental Research*, 207, 112179. DOI: 10.1016/j.envres.2021.112179

Seay, J., & Ternes, M. E. (2022). A review of current challenges and legal advances in the global management of plastic waste. *Clean Technologies and Environmental Policy*, 24(3), 731–738. DOI: 10.1007/s10098-022-02289-y

Shamsuyeva, M., & Endres, H. J. (2021). Plastics in the context of the circular economy and sustainable plastics recycling: comprehensive review on research development, standardisation and market. *Composites Part C: Open Access,* p 100168. doi:DOI: 10.1016/j.jcomc.2021.100168

Sheng, Y., Liu, Y., Wang, K., Cizdziel, J. V., Wu, Y., & Zhou, Y. (2021). Ecotoxicological effects of micronized car tire wear particles and their heavy metals on the earthworm (Eisenia fetida) in soil. *The Science of the Total Environment*, 793, 148613. DOI: 10.1016/j.scitotenv.2021.148613

Sun, X. D., Yuan, X. Z., Jia, Y., Feng, L. J., Zhu, F. P., Dong, S. S., & Xing, B. (2020). Differentially charged nanoplastics demonstrate distinct accumulation in Arabidopsis thaliana. *Nature Nanotechnology*, 15(9), 755–760. DOI: 10.1038/s41565-020-0707-4

Surendran, U., Jayakumar, M., Raja, P., Gopinath, G., & Chellam, P. V. (2023). Microplastics in terrestrial ecosystem: Sources and migration in soil environment. *Chemosphere*, 318, 137946. DOI: 10.1016/j.chemosphere.2023.137946

Syberg, K., Nielsen, M. B., Oturai, N. B., Clausen, L. P. W., Ramos, T. M., & Hansen, S. F. (2022). Circular economy and reduction of micro (nano) plastics contamination. *Journal of Hazardous Materials Advances*, 5, 100044. DOI: 10.1016/j.hazadv.2022.100044

Taylor, S. E., Pearce, C. I., Sanguinet, K. A., Hu, D., Chrisler, W. B., Kim, Y.-M., Wang, Z., & Flury, M. (2020). Polystyrene nano- and microplastic accumulation at Arabidopsis and wheat root cap cells, but no evidence for uptake into roots. *Environmental Science. Nano*, 7(7), 1942–1953. DOI: 10.1039/D0EN00309C

Wahl, A., Le Juge, C., Davranche, M., El Hadri, H., Grassl, B., Reynaud, S., & Gigault, J. (2021). Nanoplastic occurrence in a soil amended with plastic debris. *Chemosphere*, 262, 127784. DOI: 10.1016/j.chemosphere.2020.127784

Wang, W., Zhang, J., Qiu, Z., Cui, Z., Li, N., Li, X., & Zhao, C. (2022). Effects of polyethylene microplastics on cell membranes: A combined study of experiments and molecular dynamics simulations. *Journal of Hazardous Materials*, 429, 128323. DOI: 10.1016/j.jhazmat.2022.128323

Wright, S. L., Ulke, J., Font, A., Chan, K. L. A., & Kelly, F. J. (2020). Atmospheric microplastic deposition in an urban environment and an evaluation of transport. *Environment International*, 136, 105411. DOI: 10.1016/j.envint.2019.105411

Wu, X., Lyu, X., Li, Z., Gao, B., Zeng, X., Wu, J., & Sun, Y. (2020). Transport of polystyrene nanoplastics in natural soils: Effect of soil properties, ionic strength and cation type. *The Science of the Total Environment*, 707, 136065. DOI: 10.1016/j.scitotenv.2019.136065

Yang, Z., Lü, F., Zhang, H., Wang, W., Shao, L., Ye, J., & He, P. (2021). Is incineration the terminator of plastics and microplastics? *Journal of Hazardous Materials*, 401, 123429. DOI: 10.1016/j.jhazmat.2020.123429

Yu, Y., & Flury, M. (2021). Current understanding of subsurface transport of micro-and nanoplastics in soil. *Vadose Zone Journal*, 20(2), e20108. DOI: 10.1002/vzj2.20108

Chapter 7
Wastewater Treatment Using Nanotechnology: A Sustainable Approach

Om Akolkar
MIT Art, Design, and Technology University, India

Arya Kawale
MIT Art, Design, and Technology University, India

Susanta Das
https://orcid.org/0000-0002-9314-3988
Ajeenkya D.Y. Patil University, India

ABSTRACT

Economy & growth of society depend on access to clean, safe water. However, rapid population increase, industrialization, & urbanization have led to generation of wastewater that pollutes (contaminates) water leading to several health issues. Millions die annually after consuming pathogen infected water. Many limitations restrict the use of various methods for treating wastewater which includes among others - chemical application, formation of disinfection by products (DBPs), time taken and cost incurred. Nanotechnology has achieved significant successes in different areas including water treatment. Nanomaterials are characterized by high surface area to volume ratio, high reactivity/sensitivity/adsorption capacity/ease of functionalization, hence suitable for use in treating waste water. This chapter aims to provide an overview of techniques for wastewater treatment & management based on nanotechnology Moreover, the fate of nanomaterials in wastewater treatment has also been reported and the hazardous effects of their use.

DOI: 10.4018/979-8-3693-9163-1.ch007

1. INTRODUCTION

Water is a pure source on earth and its availability in a pure state is essential for both humans and other living beings, for without water the concept of life is unimaginable. The supply of fresh water is greatly impacted by two major factors: the world's population expansion and climate change. Overall, 80% of diseases worldwide are directly or indirectly attributable to poor water quality. In addition, 19% of global deaths are due to infections caused by harmful micro-organisms in water (Diallo et al., 2008). The risk of water problems is particularly high in developing countries. This is exacerbated by the fact that analyzing the microbial and chemical composition of water can be both expensive and technically challenging. Contaminated water is not only harmful to humans but also has adverse effects on wildlife and the environment (Cheriyamundath & Vavilala, 2021; United Nations Sustainable Development Goals, 2023). Clean water is therefore crucial, not only for human health but also for the well-being of ecosystems and biodiversity. Although the construction of piped water supplies to households has increased worldwide over the last two decades, a large proportion of the population still does not have access to clean drinking water. Around 780 million people worldwide are affected by this inequality, and the trend is rising (Cooley et al., 2014). In 2010, 690 million people had access to water that was only partially treated (WHO UNICEF, 2012). Although conserving water is crucial to lowering water usage, it is insufficient on its own, particularly in light of the world's expanding population. Other solutions are needed to effectively address this challenge. In addition, the supply of clean and safe water is further hampered by poor sanitation practices. Ensuring access to clean water for all therefore requires a comprehensive approach that includes not only conservation but also improvements in sanitation infrastructure and management. Therefore, new innovative and cost-effective ways of treating wastewater are crucial (Madhura et al, 2019). Only 3% of the world's water is freshwater, and this is decreasing. Of this 3%, around 70% is in the form of ice, and the remaining 30% is groundwater, which can be difficult to access (Bodzek et al., 2020). A world population of 9 billion people is predicted for 2050, and by 2075, 75% of the world's population will not have enough freshwater (Tlili & Alkanhal, 2019). This again emphasizes the importance of novel and scalable methods for wastewater treatment.

In conventional methods of wastewater treatment, chemicals and physical processes are used to purify the water. These include the addition of chemicals such as chlorine and ammonia to clump contaminants, the use of membranes to filter out impurities, and the use of varying levels of oxygen to break down harmful substances. However, these methods require a lot of chemicals and energy as well as complex machinery and infrastructure. In addition, they do not always succeed in removing all toxins, phosphorus, nitrogen and heavy metals from water. Despite their drawbacks, they do help to reduce pollution to some extent, but they can be costly and time-consuming (Kumar et al., 2014; Yamamura et al., 2011).

Nanotechnology can be used to solve serious environmental problems such as solid waste management and air and water pollution (Taran et al., 2021). The extremely small impurities, which range in size from 1 to 100 nanometers, are the hardest to remove from wastewater. That's why using nanotechnology-based methods is a good idea because they're specifically designed to remove these super-small pollutants. Nanotechnologies are excellent for water purification because they're tiny and have amazing properties that make them very effective in removing pollutants. Nanotechnology research and development is accelerating worldwide, not only for water purification but also for other areas such as improving energy storage, medicine, clothing manufacturing and keeping food fresh. In the USA, 6 billion dollars are invested annually in nanotechnology testing and progress (Diallo et al., 2008). However, nanoparticles can come from natural sources or be produced in a laboratory. Scientists are very interested in producing new types of nanoparticles with special properties to better purify industrial wastewater. Nanoparticles come in different sizes, shapes and properties. These factors, such as their dissolution capacity, their shape, the size of their surface and whether they carry an electrical charge, influence their chemical, biological and physical behavior (Cheriyamundath & Vavilala, 2021). NF, RO, MF and UF are all processes that use tiny particles to purify water. Nanofiltration is the most commonly used because it can remove salts, minerals, germs and other substances dissolved in water. This technology is also very well suited to getting rid of harmful germs such as viruses, protozoa and bacteria. This is important to prevent the spread of diseases from contaminated water, which can affect both humans and animals (Tlili & Alkanhal, 2019).

Urban and industrial wastewater is regarded as a significant pollutant carrier and needs to be appropriately handled and inclined. The need to control organic/inorganic pollutants and complex compounds present in wastewater has engaged the awareness of researchers. The use of productive technologies for the treatment of these compounds is one of the most important ways to improve water quality. The purpose of this chapter is to summarize the application of NMs in water treatment, including adsorption, recovery of analytes, membrane modification, photocatalysis, disinfection of pathogens, and removal of metal ions. The properties, advantages and disadvantages of the various NMs and their role in the water treatment process have also been examined and presented schematically so that they can be understood by a wide range of readers.

2. CURRENT WASTEWATER TREATMENT TECHNOLOGIES

The removal of pollution, pathogens, organic matter and toxic chemicals, as well as the need to follow guidelines and legislation for the safe discharge of wastewater, are some of the main reasons for wastewater treatment. The type and mix of waste that ends up in a wastewater treatment plant depends on the sources. The first step in achieving adequate treatment is to characterize the wastewater to determine its specific constituents. Parameters such as pH, biological oxygen demand (BOD), chemical oxygen demand (COD), alkalinity, hardness, mineral composition, anionic composition and non-ionic composition are important. The right technologies can then be used for treatment. Treatment can be physical, chemical or biological, but usually a combination of all three methods is used (Riffat, 2012).

2.1. Limitations of Current Treatment Steps

Despite the success of the methods outlined above, their large-scale application is often associated with some problems. For example, chlorination, which is used to eliminate pathogens, leaves an undesirable taste and odor in the water, which can often only be prevented by further dichlorination. An alternative to chlorine is ozonation, which also effectively inactivates pathogens (Epelle et al., 2022a, 2022b). Some methods of eliminating germs in water, such as the use of ozone or ultraviolet light, can be expensive as they consume a lot of energy. Other methods, such as ion exchange, aren't particularly suitable for eliminating all germs. However, there are some really good methods, such as nanofiltration, reverse osmosis, microfiltration and ultrafiltration (Liu et al., 2021; Joo & Tansel, 2015). These methods are great for removing tiny contaminants, but sometimes they can clog up the equipment. We're always looking for new ways to better purify water, and nanotechnology seems to be a promising option (Amin et al., 2014).

3. NANOTECHNOLOGY

Nanomaterials, that have a huge floor area as compared to their size, can purify water thoroughly. In an assessment of traditional methods, nanotechnology solutions are efficient and do not require doubtlessly harmful chemicals. They might therefore play a crucial function in offering secure water for all (Ajith et al., 2020; Varghese, 2004). In developing countries, water treatment faces problems such as poor maintenance, unreliable supply, pollution and the use of chlorine, which can produce harmful substances. However, chlorine-free water filtration methods based on nanotechnology offer advantages. These methods purify water without the use of chlorine, thus reducing health risks. Nanotechnology helps to filter the water more effectively, making it safer to drink. Simply put, nanotechnology methods provide a safer and better way to treat water in places where it's difficult to ensure clean drinking water. When irradiated with UV light, a substance called titanium dioxide can perform two important tasks: break down organic pollutants and kill microorganisms (Gleick, 2003; Mintz et al., 1995). This process is called photocatalysis. So when titanium dioxide is irradiated with UV light, it purifies the water particularly effectively by breaking down pollutants and eliminating germs (Li et al., 2008; Danion et al., 2004) Nanomembranes coated with titanium dioxide are attracting a great deal of attention in water treatment. These nanomembranes can also be coated with antibacterial and photocatalytic material by using the atomic layer deposition technique. The creation of water filtration systems in underdeveloped countries could benefit from the use of these nanomembranes. Toxic metal ions such as chromium (VI), cadmium (II), lead (II), mercury (II), chromium (III), cobalt (II), nickel (II), copper (II), arsenic (III), arsenic (V) and silver (I) can be removed from water using nanoparticles such as magnetic nanoparticles and carbon nanotubes as well as iron zeolite. These metal ions cause serious damage to human health (Vouk, 1979; Nowack, 2008).

Due to their large surface reactivity and surface area, nanoparticles can transform materials more quickly, forming fewer harmful by-products than granular forms. Many of the current water quality problems could be solved or greatly reduced by the use of nanosorbents, carbon nanotubes, pharmacologically active nanomaterials, nanostructured catalytic membranes, nanowires, magnetic nanoparticles, granules, flakes and supramolecular metal particle assemblies with large surface areas and characteristic length scales of 8-9 nm, including clusters, two-dimensional nanoparticles and colloids. Compared to a microbial cell, which has a diameter of about 1.0 m, nanoparticles are characterized by their extremely small particle sizes (1-100 nm) as adaptable tools for water remediation (1000 nm). Therefore, the groundwater flow can transport nanoparticles efficiently (Theron et al., 2008). The implementation of an in is made easier by the long durations of time that these nanoparticles may stay floating in water.

4. NANOTECHNOLOGY IN WASTEWATER MANAGEMENT

Various nanomaterials have been investigated for applications in wastewater treatment. These include NPs, metal nanoparticles, carbon-based nanomaterials (such as carbon nanotubes), graphene, activated carbon, zeolites, SAMMS, biopolymers, and many other materials. Each of these nanomaterials has specific properties and functions that can meet specific needs in wastewater treatment, making them promising candidates for solving these problems (Baruah et al., 2019). Nanotechnology plays an important role in wastewater treatment through various methods. The main processes include adsorption and biosorption, where nanomaterials effectively capture microorganisms in the water. In addition, nanofiltration removes pollutants based on size, while photocatalysis uses nanomaterials to break down pollutants in light. Nanotechnology also helps in disinfection and disease control, ensuring the safety of water. Additionally, monitoring and analysis technologies can accurately identify and track contaminants, thus improving wastewater treatment.

4.1. Adsorption and Biosorption Processes

Adsorption and Biosorption are both an exothermic and superficial phenomenon. Molecules or ions, whether liquid or gaseous, must be transferred from a phase to a solid surface. This process leads to the formation of a monolayer on the surface by physicochemical or chemical interactions under certain conditions. (Gusain et al., 2020; Crawford & Quinn, 2017). Biosorption is a type of adsorption in which certain biological materials, such as bacteria, algae or fungi, act as adsorbents due to their binding properties. Adsorption processes can capture heavy metals from even very aqueous chemicals using metabolic drugs. Pathways involving ATP or physicochemical uptake mechanisms do not require ATP consumption. The biosorption process mainly includes microprecipitation, ion exchange and cell surface complexation (Derco & Varana, 2018; Seung et al., 2006; Yoon et al., 2006; Fard et al., 2011).

Yang et al. The biosorption of Cr (VI) from synthetic wastewater using algal aerobic granular sludge was investigated. They found that Cr (VI) biosorption depends on the pH value, and the maximum Cr (VI) biosorption capacity of algal aerobic granular sludge at pH 2 was 51.0 mg g.1. Removal is usually biosorption combined with bioreduction. Desorption with NaHCO3 can recover 64% to 73% of the adsorbed Cr, mostly in the form of Cr (III). Compared with aerobic granular sludge, algal aerobic granular sludge has excellent biosorption capacity and improved granular stability (Yang et al., 2020). Ding et al. An alginate-immobilized AAM biosorbent was developed to remove Th ions, especially Th (IV) ions, from wastewater. The AAM biosorbent showed good biosorption performance with an adsorption capacity of 303.95 mg. g at pH 6 and 40 °C. Moreover, rapid recovery of Th (IV) can be achieved in less than 100 minutes using biosorbent (Ding et al., 2019).

4.1.1. Nanostructured Carbon Adsorbents

In different areas of science and innovation, carbon science has been examined for decades. Nanostructures of carbon are known to have diverse low-dimension allotropes of carbon such as actuated carbon, CNTs, and the C60 family of bucky-balls, graphite, and graphene (Nasir et al., 2018). Carbon nanostructures are broadly utilized as nano adsorbents for wastewater treatment owing to their inexhaustible accessibility, cost-effectiveness, tall chemical and warm stabilities, tall dynamic surface regions, amazing adsorption capacities, and ecologically inviting nature (Meng et al., 2010).

CNTs are without a doubt round and hollow atoms made up of carbon iotas organized in a hexagonal design. They can be SWCNTs, made by rolling up a single sheet of graphene, or MWCNTs, made by rolling up numerous layers of graphene (Li et al., 2017). Researchers are profoundly curious about CNTs due to their momentous properties, counting mechanical adaptability, tall particular surface ranges, and huge pore volumes. These one-of-a-kind characteristics make CNTs amazingly important in different applications, especially in wastewater treatment. By leveraging these properties, CNTs can workably adsorb contaminants from water, contributing to the refinement and remediation of wastewater assets (Zhou et al., 2018 & Verma et al., 2020). CNTs display prevalent execution as adsorbents compared to other materials due to a few key components. These incorporate their tunable surface chemistry, which permits surface alterations to upgrade adsorption properties. Moreover, their chemically dormant nature, empty structure, tall particular surface region, light mass thickness, and tall porosity contribute to their uncommon adsorption capabilities. Moreover, CNTs have a solid interaction with toxins, assist improving their viability in wastewater treatment applications (Moradi et al., 2013). All these properties make them fabulous to be utilized in wastewater treatment (Yadav et al., 2017). Overwhelming metals and particles in water pose critical dangers to both the environment and human well-being. Yadav and Srivastava conducted an investigation of the adsorption and desorption of Mn^{7+} particles by CNTs. They found that CNTs proficiently adsorbed Mn^{7+} particles, and UV-visible spectrophotometric examination showed a lessening in Mn^{7+} concentration from 150 ppm to 3 ppm when treated with CNTs. In their ponder, laboratory-grade KMnO4 was utilized as a source of Mn^{7+} particles. This inquiry highlights the potential of CNTs for the expulsion of overwhelming metal particles from water, contributing to endeavors aimed at moving forward water quality and shielding human well-being and the environment (Zhao et al., 2021).

Demand for pharmaceutical items has developed a parcel as a result of population development and broad sickness. Anti-microbials are one of the foremost broadly expended pharmaceutical items and in this way, their release within the encompassing environment has moreover expanded. Such items, when expended by people or other living life forms, cause genuine well-being issues. The arranged MWCNTs illustrated an estimated extend of 10.8 emu/g, attractive distinguishableness with an attractive minute of 10.8 emu/g, and a tall adsorption capacity (qm = 494.91 mg/g at 308 K) for TC. In group tests, MWCNTs showed over 80% productivity in adsorptive TC removal across a pH run of 2–10, with TC concentrations underneath 80 mg/L. This effectiveness was decided through adsorption energy, adsorption isotherms, and the examination of the beginning pH's impact on TC adsorption by MWCNTs. The adsorption of TC onto attractive MWCNTs was well portrayed by the Langmuir isotherm and pseudo-second-order energetic adsorption demonstrated. This proposes that the adsorption handle essentially includes chemical intuitive and monolayer adsorption on the surface of the MWCNTs (Lu et al., 2016).

4.1.2. Nanomaterials for Metal Adsorption:

Right now, metal-based nanomaterials as adsorbents are catching the eyes of analysts (Sadegh et al., 2017). Their oxides such as Fe_3O_4, TiO_2, MnO_2, MgO, ZnO and CdO are broadly utilized to evacuate overwhelming metals, particles, and colors from wastewater (Sayed, 2020). Nanometal oxides are regularly considered more viable adsorbents than enacted carbon, particularly when the evacuation of overwhelming metals and radioactive metals is the essential concern. Moreover, their little measure and expansive surface zone permit a diminished intraparticle dissemination separation, which can be compressed without changing their surface range. The sorption handle is represented by the complexation between broken-down metals and the oxygen display in metal oxides (Qu et al., 2013; He et al., 2021).

Wastewater from oil refineries is contaminated with a variety of metals and particles, including Cu^{2+} and Ca^{2+}. He et al. set up reusable nano adsorbents based on Fe_3O_4/GO-COOH by magnetizing and carboxylating GO to remove metals and particles. The nano adsorbents appeared 78.4% and 51% expulsion of Ca^{2+} and Cu^{2+} individually, at 60 min. The nano adsorbent held tall recuperation rates (82.1% for Ca^{2+} and 91.8% for Cu^{2+}) and evacuation rates (72.3% for Ca^{2+} and 49.33% for Cu^{2+}) after five adsorption-desorption cycles.

Peralta et al., (2021) created attractive nano adsorbents based on silica and assessed its potential for the disposal of natural poisons such as contaminants of rising concern (CECs), polyaromatic hydrocarbons (PAHCs), and aliphatic hydrocarbons (AHCs). The planning preparation included the covering of attractive press oxide NPs with a crossover shell comprising of silica and 3-(trimethoxysilyl) propyl-octadecyl dimethyl-ammonium chloride (3-TPODAC) as a structure-directing operator. The arranged attractive MSNPs were encouraged and altered by implies of trimethoxyphenylsilane, to induce nano adsorbents functionalized with phenyl MSNPs-ph.

Both materials experienced characterization and were subjected to group sorption tests with single and blended contaminants. The discoveries showed that MSNPs-ph displayed more prominent effectiveness in adsorbing PAHCs and AHCs compared to the other materials. The nearness of phenyl and 3-TPODAC moieties on the mesostructured silica frameworks was found to be a key calculation in getting tall PAHC take-up from fluid media. The arranged MSNPs-ph was advance assessed for its adsorption capacity against CECs counting carbamazepine, diclofenac, and ibuprofen. Whereas MSNPs for the most part displayed superior adsorption capacities for CECs, MSNPs-ph appeared especially tall uptakes of ibuprofen and diclofenac. An adsorbent based on an S-NaOH from horticultural biomass was created by Arshadi et al. (2014) and enhanced with NZVIPs. The manufactured framework was subjected to the examination of adsorption/reduction of inorganic toxins such as Pb^{2+} particles. The NZVIPs appeared great scattering (ca.10–70 nm) over the surface of the particular squander. The manufactured framework had a double and unconstrained adsorption profile for Pb^{2+} (Pandey et al., 2017).

4.1.3. Nanoadsorbents Using Polymers

During the last numerous years, sizeable studies have been done on polymer nanocomposites for the development of environmental sustainability and wastewater treatment. It offers a high surface area for instant decontamination, stepped forward processability, extraordinary stability, stepped forward processability, cost-effectiveness, and selectivity to get rid of special pollutants in wastewater (Berber, 2020; Abdi, 2020). Nanocellulosics, which are derived from cellulose, on the advantages of being safe, ubiquitous, splendid adsorbents, and feature ease of surface amendment, rendering them appropriate for wastewater remediation. These days, lignin-derived nanoparticles have been taught and have proven to be effective at treating wastewater and water. They have been observed effective in the catalytic degradation of dyes, and nitroarenes, and the removal of heavy metals (Moharrami & Motamedi, 2020).

Abdi and Abedini evolved an MOF adsorbent composed of ZIF integrated into polymeric nanocomposite beads, especially PES, for the green removal of MG from wastewater. The effect of numerous parameters consisting of pH, adsorbent dosage, and initial MG awareness changed into investigated for the developed PES ZIF eight/ZIF sixty-seven and compared with PES beads alone. The findings indicated that PES beads alone had very little capacity for adsorption, while PES-ZIF-8/ZIF-sixty-seven showed a remarkable capacity for adsorption of ~ 613 mg/g and an exceptional performance for MG removal of ~ 89%. The fragrant-fragrant interaction (π-π stacking) between the imidazole ligand in the ZIF framework and the earrings of the MG dye is responsible for the adsorption mechanism (Nithya et al., 2021). Chen and coworkers fabricated bi-functionalized β-CD and PEI magnetic nano adsorbents (β-CD/PEI- Fe3O4) to seize MO and Pb^{2+} simultaneously from wastewater with spatially separated sorption websites. Inside the developed system, the nitrogen-bearing groups of PEI, which deliver a high-quality price, and the cavities of β-CD have been believed to be by and large answerable for the removal of MO. This elimination took place via a combination of electrostatic attraction and host-guest inclusion mechanisms. For the effective uptake of Pb^{2+} from wastewater, the edges of β-CD containing oxygen-bearing groups and free amino moieties present at the PEI functioned as active sites (Ali et al., 2020).

4.1.4. Zeolites

There are well-defined structures of voids and conduits of distinct sizes in zeolites, which are easily accessible through the well-defined pores of their normal framework. Synthetically prepared zeolites can be used as magnetically modified, bio-zeolite, etc. zeolitic nanomaterials have attracted attention in environmental applications due to their stability in water, their low cost of production, high surface area, selectivity, and compatibility with the natural environment. Several investigations have verified that metal cations are removed from wastewater. Due to its ability to regenerate and reuse, zeolite is a good candidate for wastewater treatment and is found to possess numerous research applications in industries (Zhao., 2010; Torabian., 2010). The zeolite is composed of Na_2O/Al_2O_3. It is obtained from natural halloysite minerals, which are used as source material for the adsorption of NH^{4+} from wastewater. The maximum capacity of the prepared NaA zeolite was found to be ~ 44%. There is a potential application for the elimination of NH^{4+} pollutants from wastewater (Pandey et al., 2015).

Bandpi and colleagues created an adsorbent based on NZ coated with Fe_3O_4 CNZ nanoparticles for CEX expulsion from a fluid arrangement. The greatest CEX evacuation productivity for CNZ and NZ was found to be 93% and 28%, respectively (Gugushe et al., 2019).

4.2. Nanofilters

Water filtration is the method of stripping or bringing down the concentration of particulate matter, such as suspended particles, microorganisms as well as other hindering natural and chemical contaminants from contaminated or contaminated water to form secure and clean water for drinking, pharmaceutical and restorative applications (Charcosset, 2012). Film technology has picked up impressive consideration in later a long time and the foremost vital progression in layer innovation is the NF layer. As the title itself suggests, NF films have MWCO for uncharged particles within the nanometer extent (Nagy, 2019). NF films are relatively recent and are the foremost favored strategy for drinking water and wastewater treatment (Timmer, 2001). NF may be a pressure-driven film preparation that lies between UF and RO, with a weight between 5–20 bars and a pore estimate between 0.5 and 2.0 nm characterized by a tall dismissal of divalent or higher-valent ions, a moo dismissal of monovalent particles, and tall flux and vitality utilization compared to RO, and a tall dismissal compared to UF (Mulyanti & Susanto, 2018; Abdel-Fatah, 2018; Bruggen et al., 2013). NF may be a generally modern progression in layer innovation that can be utilized for both watery and non-aqueous applications. It is considered one of the foremost critical and commonly utilized methods in wastewater treatment due to its special filtration component and the accessibility of a wide extend of films suited for distinctive purposes. NF is suitable for filtering out most natural and inorganic pollutants, including a significant number of harmful organisms from wastewater (Benfer et al., 2001; Mostafavi et al., 2009; Parham et al., 2013). NF layers are profoundly adaptable, cost-effective, and simple to create. The two most commonly utilized sorts of NF films are polymeric layers and ceramic layers. The polymeric layers show a brief lifetime due to their second-rate chemical resistance and a high fouling rate (Han et al., 2013). On the other hand, ceramic layers have higher mechanical, chemical, and warm stabilities. Mostafavi et al. (2009) manufactured an NF based on CNTs for the expulsion of the MS2 infection from the water. The MS2 infection may belong to the same family of bacterial illnesses that infect Escherichia coli and other members of the Enterobacteriaceae family. The NF film was first characterized to evaluate its porosity and surface morphology. Along these lines, it was assessed for its effectiveness in expelling MS2 infection. The result illustrated tall evacuation effectiveness at 8–11 weight bars (Saravanan et al., 2021). Additionally, Parham et al. created a channel based on a CNT ceramic composite for yeast filtration. The manufactured channel illustrated tall (98%) filtration effectiveness for yeast, and nearly a 100% removal efficiency for overwhelming metal particles from water (Margan & Haghighi, 2018).

4.3. Disinfection and Pathological Control

WHO highlights a critical issue: in developing countries, 80% of diseases stem from consuming water contaminated by various pathogens such as bacteria, viruses, prions, fungi, protozoa, and rickettsia. This contamination poses a significant threat to human health and results in millions of deaths annually worldwide. The statistics underscore the urgent need to improve water quality and sanitation infrastructure to safeguard public health in vulnerable regions (Rikta, 2019). Disinfection involves reducing the microbial count on surfaces or within materials to a level considered safe or acceptable. This reduction can be achieved through various physical or chemical methods, such as heat, irradiation, or the use of disinfectants like chlorine or alcohol.

The aim is to eliminate or deactivate harmful microorganisms, thus minimizing the risk of infection or contamination. Indeed, traditional disinfection techniques used in wastewater treatment encompass a range of methods. These include (Asselt, 2005):

1. Free chlorine: Chlorine gas or sodium hypochlorite is commonly used to disinfect wastewater by killing or inactivating pathogens.
2. Hypochlorite: Sodium hypochlorite, also known as bleach, is an effective disinfectant used to treat wastewater.
3. Chlorine dioxide: Another chlorine-based disinfectant, chlorine dioxide, is utilized to kill bacteria, viruses, and other harmful microorganisms in wastewater.
4. Chloramines: Chloramines are formed by combining chlorine with ammonia and are employed as disinfectants in wastewater treatment.
5. Ozone: Ozone gas is a powerful oxidizing agent that is used to disinfect wastewater by destroying organic contaminants and pathogens.
6. Reverse osmosis: While primarily a filtration method, reverse osmosis can also contribute to disinfection by removing contaminants, including microorganisms, from wastewater.
7. Peracetic acid: Peracetic acid is a strong oxidizing agent and disinfectant used in wastewater treatment to kill bacteria and other pathogens.

These traditional disinfection techniques play crucial roles in ensuring that wastewater is treated to meet regulatory standards and is safe for discharge or reuse (Qin, 2019). Certainly, traditional disinfection techniques like those you mentioned have limitations such as high energy consumption, the requirement for costly equipment, and the formation of potentially harmful DBPs. These factors not only contribute to operational costs but also raise environmental and health concerns (Liu, 2017; Zhang, 2016). Consequently, there's a pressing need to develop alternative disinfection methods that are efficient, sustainable, cost-effective, and less burdensome in terms of time and resources. Such methods could include innovative technologies like AOPs, UV irradiation, photocatalysis, electrochemical disinfection, and plasma treatment. These approaches offer promising avenues for improving disinfection efficiency while minimizing energy consumption, equipment costs, and the formation of undesirable byproducts. Moreover, they align with the broader goals of sustainability and environmental stewardship in wastewater treatment practices (Guidetti et al., 2018).

Nanomaterials are invested with tip-top usefulness for the inactivation of pathogens in water, such as huge surface zones and particular reactivity, which cannot be accomplished using ordinary strategies (Xiu et al., 2012). The inactivation component of these nanomaterials incorporates surface-based electrostatic interaction and photochemical responses, which actuate the era of ROS, the disturbance of cell dividers, and focus on conveyance of cleaning operators (Wang et al., 2015). Due to their momentous surface properties and reactivity, nanomaterials make it moderately simple to restrain pathogens in water (Mauter et al., 2018).

5. CONCLUSION

The quick increment in population combined with industrialization, urbanization and huge agricultural activities has led to an elevated demand for clean and safe water across the world. Currently, there are different ways of cleaning and purifying water. However, many of these methods use chemicals thus they are energy-demanding as well as operationally complicated needing specialized engineering and infrastructure. There is a need for innovative approaches to wastewater treatment to ensure the efficient removal of contaminants from it. Nanotechnology could be used here which refers broadly to enabling treatment systems based on nanomaterials or those involving nanoparticles themselves being used directly such as adsorbents, photocatalysts etcetera which help remove pollutants present within waste waters Much creativity is required if we want new techniques capable of efficiently eliminating any type of contaminant found in dirty streams This can only come about through novel ways being designed specifically made up around this area while considering all factors related towards sustainable development including costs effectiveness among others, In addition, nanotechnologies provide several avenues towards achieving sustainable development goals especially when it comes to materials with nanostructures having unique features like high surface area/volume ratio sensitivity/reactivity rates adsorptive capacities ease functionalizing. Due to their little measure, nanomaterials have the potential to enter the bodies of people and other oceanic living beings, posturing dangers of poisonous quality. The degree of poisonous quality can change depending on variables such as pH, concentration, and term of presentation. Whereas researchers have made critical advances in investigating nanotechnology, there's still much to reveal. The utilization of nanotechnology holds incredible guarantees for wastewater administration, but it requires concerted exertion from the logical community and government bodies to guarantee its secure and compelling usage.

REFERENCES

Abdel-Fatah, M. A. (2018). Nanofiltration systems and applications in wastewater treatment: Review article. *Ain Shams Engineering Journal*, 9(4), 3077–3092. DOI: 10.1016/j.asej.2018.08.001

Abdi, J., & Abedini, H. (2020). MOF-based polymeric nanocomposite beads as an efficient adsorbent for wastewater treatment in batch and continuous systems: Modelling and experiment. *Chemical Engineering Journal*, 400, 125862. DOI: 10.1016/j.cej.2020.125862

Ajith, M. P., Priyadarshini, E., & Rajamani, P. (2020). Effective and selective removal of heavy metals from industrial effluents using sustainable Si–CD conjugate based column chromatography. *Bioresource Technology*, 314, 314. DOI: 10.1016/j.biortech.2020.123786

Ali, M. E., Hoque, M. E., Hossain, S. K. S., & Biswas, M. C. (2020). Nanoadsorbents for wastewater treatment: Next generation biotechnological solution. *International Journal of Environmental Science and Technology*, 17(9), 4095–4132. DOI: 10.1007/s13762-020-02755-4

Amin, M. T., & Alabama Manzoor, U. (2014). A review of removal of pollutants from water/wastewater using different types of nanomaterials. *Advances in Materials Science and Engineering*, 825910. Advance online publication. DOI: 10.1155/2014/825910

Arshadi, M., Soleymanzadeh, M., & Salvacion, J., & SalimiVahid, F. (2014). Nanoscale Zero-Valent Iron (NZVI) supported on singular waste for Pb (II) removal from aqueous solution: Kinetics, thermodynamic and mechanism. *Journal of Colloid and Interface Science*, 426, 241–251. DOI: 10.1016/j.jcis.2014.04.014

Asselt, V. A., & Te Giffel, M. C. (2005). Pathogen resistance, and adaptation to disinfectants and sanitisers. In *Understanding Pathogen Behaviour* (pp. 484–506). Elsevier., DOI: 10.1533/9781845690229.3.484

Baruah, A., Chaudhary, V., Malik, R., & Tomer, V. K. (2019). Nanotechnology Based Solutions for Wastewater Treatment. In Nanotechnology in Water and Wastewater Treatment: Theory and Applications (pp. 337–368). Elsevier. doi:DOI: 10.1016/B978-0-12-813902-8.00017-4

Benfer, S., Popp, U., Richter, H., Siewert, C., & Tomandl, G. (2001). Development and characterization of ceramic nanofiltration membranes. *Separation and Purification Technology*, 22(1-2), 231–237. DOI: 10.1016/S1383-5866(00)00133-7

Berber, M.R. (2020). Current Advances of Polymer Composites for Water Treatment and Desalination. *J. Chem., 2020 (Article ID 7608423*), 19 pages. doi:DOI: 10.1155/2020/7608423

Bodzek, M., Konieczny, K., & Kwiecińska-Mydlak, A. (2020). Nanotechnology in water and wastewater treatment. Graphene–the nanomaterial for the next generation of semipermeable membranes. *Critical Reviews in Environmental Science and Technology*, 50(15), 1515–1579. DOI: 10.1080/10643389.2019.1664258

Bruggen, B., Hoek, E. M., & Tarabara, V. V. (2013). *Nanofiltration. Encyclopedia of Membrane Science and Technology*. John Wiley & Sons.

Charcosset, C. (2012). Some perspectives. Membrane Processes in Biotechnology and Pharmaceutics. Elsevier.

Cheriyamundath, S., & Vavilala, S. L. (2021). Nanotechnology-based wastewater treatment. *Water and Environment Journal : the Journal / the Chartered Institution of Water and Environmental Management*, 35(1), 123–132. DOI: 10.1111/wej.12610

Cooley, H., Ajami, N., Ha, M. L., Srinivasan, V., Morrison, J., Donnelly, K., & Christian-Smith, J. (2014). Global water governance in the twenty-first century. The World's Water. Springer.

Crawford, C. B., & Quinn, B. (2017). The interactions of microplastics and chemical pollutants. In *Microplastic Pollutants* (pp. 131–157). Elsevier., DOI: 10.1016/B978-0-12-809406-8.00006-2

Danion, A., Disdier, J., Guillard, C., Abdelmalek, F., & Jaffrezic-Renault, N. (2004). Characterization and study of a single-TiO_2-coated optical fiber reactor. *Applied Catalysis B: Environmental*, 52(3), 213–223. DOI: 10.1016/j.apcatb.2004.04.005

Derco, J., & Vrana, B. (2018). Introductory Chapter: Biosorption. In *Biosorption*. IntechOpen: London, UK.

Diallo, M., Duncan, J., Savage, N., Street, A., & Sustich, R. (2008). Nanotechnology Applications for Clean Water: Solutions for Improving Water Quality, 0815519737.

Ding, H., Luo, X., Zhang, X., & Yang, H. (2019). Alginate-immobilized Aspergillus niger: Characterization and biosorption removal of thorium ions from radioactive wastewater. *Colloids and Surfaces. A, Physicochemical and Engineering Aspects*, 562, 186–195. DOI: 10.1016/j.colsurfa.2018.11.032

El-Sayed, M. E. (2020). Nanoadsorbents for water and wastewater remediation. *The Science of the Total Environment*, 739, 139903. DOI: 10.1016/j.scitotenv.2020.139903

Epelle, E. I., Emmerson, A., Nekrasova, M., Macfarlane, A., Cusack, M., Burns, A., Mackay, W., & Yaseen, M. (2022b). Microbial Inactivation: Gaseous or Aqueous Ozonation? *Industrial & Engineering Chemistry Research*, 61(27), 9600–9610. DOI: 10.1021/acs.iecr.2c01551

Epelle, E. I., Macfarlane, A., Cusack, M., Burns, A., Amaeze, N., Richardson, K., Mackay, W., Rateb, M. E., & Yaseen, M. (2022a). Stabilisation of Ozone in Water for Microbial Disinfection. *Environments (Basel, Switzerland)*, 9(4), 45. DOI: 10.3390/environments9040045

Fard, R. F., Azimi, A., & Bidhendi, G. N. (2011). Batch kinetics and isotherms for biosorption of cadmium onto biosolids. *Desalination and Water Treatment*, 28, 69–74. DOI: 10.5004/dwt.2011.2203

Gleick, P. H. (2003). Global freshwater resources: Soft-path solutions for the 21st century. *Science*, 302(5650), 1524–1528. DOI: 10.1126/science.1089967

Gugushe, A. S., Mpupaa, A., & Nomngongoabc, P. N. (2019). Ultrasound-assisted magnetic solid phase extraction of lead and thallium in complex environmental samples using magnetic multi-walled carbon nanotubes/zeolite nanocomposite. *Microchemical Journal*, 149, 103960. DOI: 10.1016/j.microc.2019.05.060

Guidetti, G., Giuri, D., Zanna, N., Calvaresi, M., Montalti, M., & Tomasini, C. (2018). Biocompatible and Light-Penetrating Hydrogels for Water Decontamination. *ACS Omega*, 3(7), 8122–8128. DOI: 10.1021/acsomega.8b01037

Gusain, R., Kumar, N., & Ray, S. S. (2020). Recent Advances in Carbon Nanomaterial-Based Adsorbents for Water Purification. *Coordination Chemistry Reviews*, 405, 213111. DOI: 10.1016/j.ccr.2019.213111

Han, Y., Xu, Z., & Gao, C. (2013). Ultrathin Graphene Nanofiltration Membrane for Water Purification. *Advanced Functional Materials*, 23(29), 3693–3700. DOI: 10.1002/adfm.201202601

He, L., Wang, L., Zhu, H., Wang, Z., Zhang, L., Yang, L., Dai, Y., Mo, H., Zhang, J., & Shen, J. (2021). A reusable Fe3O4/GO-COOH nano adsorbent for Ca^{2+} and Cu^{2+} removal from oilfield wastewater. *Chemical Engineering Research & Design*, 166, 248–258. DOI: 10.1016/j.cherd.2020.12.019

Joo, S. H., & Tansel, B. (2015). Novel technologies for reverse osmosis concentrate treatment: A review. *Journal of Environmental Management*, 150, 322–335. DOI: 10.1016/j.jenvman.2014.10.027

Kumar, S., Ahlawat, W., Bhanjana, G., Heydarifard, S., Nazhad, M. M., & Dilbaghi, N. (2014). Nanotechnology-Based Water Treatment Strategies. *Journal of Nanoscience and Nanotechnology*, 14(2), 1838–1858. DOI: 10.1166/jnn.2014.9050

Li, Q., Mahendra, S., Lyon, D. Y., Brunet, L., Liga, M. V., Li, D., & Alvarez, P. J. (2008). Antimicrobial nanomaterials for water disinfection and microbial control: Potential applications and implications. *Water Research*, 42(18), 4591–4602. DOI: 10.1016/j.watres.2008.08.015

Li, Y., Zhou, B., Zheng, G., Liu, X., Li, T., Yan, C., Cheng, C., Dai, K., Liu, C., Shen, C., & Guo, Z. (2017). Continuously prepared highly conductive and stretchable SWNT/MWNT synergistically composited electrospun thermoplastic polyurethane yarns for wearable sensing. *Journal of Materials Chemistry. C, Materials for Optical and Electronic Devices*, 6(9), 2258–2269. DOI: 10.1039/C7TC04959E

Liu, Y., Zeng, G., Zhong, H., Wang, Z., Liu, Z., Cheng, M., Liu, G., Yang, X., & Liu, S. (2017). Effect of rhamnolipid solubilization on hexadecane bioavailability: Enhancement or reduction? *Journal of Hazardous Materials*, 322, 394–401. DOI: 10.1016/j.jhazmat.2016.10.025

Liu, Z. Q., Huang, C., Li, J.-Y., Yang, J., Qu, B., Yang, S.-Q., Cui, Y.-H., Yan, Y., Sun, S., & Wu, X. (2021). Activated carbon catalytic ozonation of reverse osmosis concentrate after coagulation pretreatment from coal gasification wastewater reclamation for zero liquid discharge. *Journal of Cleaner Production*, 286, 124951. DOI: 10.1016/j.jclepro.2020.124951

Lu, H., Wang, J., Stoller, M., Wang, T., Bao, Y., & Hao, H. (2016). An Overview of Nanomaterials for Water and Wastewater Treatment. *Advances in Materials Science and Engineering*, 2016, 1–10. DOI: 10.1155/2016/4964828

Madhura, L., Singh, S., Kanchi, S., Sabela, M., & Bisetty, K., & Inamuddin. (2019). Nanotechnology-based water quality management for wastewater treatment. *Environmental Chemistry Letters*, 17(1), 65–121. DOI: 10.1007/s10311-018-0778-8

Margan, P., & Haghighi, M. (2018). Sono-coprecipitation synthesis and physicochemical characterization of CdO-ZnO nano photo-catalyst for removal of acid orange 7 from wastewater. *Ultrasonics Sonochemistry*, 40, 323–332. DOI: 10.1016/j.ultsonch.2017.07.003

Mauter, M. S., Zucker, I., Perreault, F., Werber, J. R., Kim, J. H., & Elimelech, M. (2018). The role of nanotechnology in tackling global water challenges. *Nature Sustainability*, 1(4), 166–175. DOI: 10.1038/s41893-018-0046-8

Meng, H., Xue, M., Xia, T., Zhao, Y. L., Tamanoi, F., Stoddart, J. F., Zink, J., & Nel, A. E. (2010). Autonomous in vitro anticancer drug release from mesoporous silica nanoparticles by pH-sensitive nano valves. *Journal of the American Chemical Society*, 132(36), 12690–12697. DOI: 10.1021/ja104501a

Mintz, E. D., Reiff, F. M., & Tauxe, R. V. (1995). Safe water treatment and storage in the home: A practical new strategy to prevent waterborne disease. *Journal of the American Medical Association*, 273(12), 948–953. DOI: 10.1001/jama.1995.03520360062040

Moharrami, P., & Motamedi, E. (2020). Application of cellulose nanocrystals prepared from agricultural wastes for the synthesis of starch-based hydrogel nanocomposites: Efficient and selective nano adsorbent for removal of cationic dyes from water. *Bioresource Technology*, 313, 123661. DOI: 10.1016/j.biortech.2020.123661

Moradi, O., Fakhri, A., Adami, S., & Adami, S. (2013). Isotherm, thermodynamic, kinetics, and adsorption mechanism studies of Ethidium bromide by single-walled carbon nanotube and carboxylate group functionalised single-walled carbon nanotube. *Journal of Colloid and Interface Science*, 395, 224–229. DOI: 10.1016/j.jcis.2012.11.013

Mostafavi, S., Mehrnia, M., & Rashidi, A. (2009). Preparation of nanofilter from carbon nanotubes for application in virus removal from water. *Desalination*, 238(1-3), 271–280. DOI: 10.1016/j.desal.2008.02.018

Mulyanti, R., & Susanto, H. (2018). Wastewater treatment by nanofiltration membranes. *IOP Conference Series. Earth and Environmental Science*, 142, 012017. DOI: 10.1088/1755-1315/142/1/012017

Nagy, E. (2019). *Nanofiltration. Basic Equations of Mass Transport Through a Membrane Layer*. Elsevier.

Nasir, S., Hussein, M. Z., Zainal, Z., & Yusof, N. A. (2018). Carbon-Based Nanomaterials/Allotropes: A Glimpse of Their Synthesis, Properties and Some Applications. *Materials (Basel)*, 11(2), 295. DOI: 10.3390/ma11020295

Nithya, P. (2021). Alginate coated layered double hydroxide/reduced graphene oxide nanocomposites for removal of toxic As (V) from wastewater. *Phys. E: Low-Dimens.Physica E, Low-Dimensional Systems and Nanostructures*, 127, 114527. DOI: 10.1016/j.physe.2020.114527

Nowack, B. (2008). Pollution prevention and treatment using nanotechnology. In H. Krug (Ed.), *Nanotechnology general, Nanotechnology, Volume 2: environmental aspects* (1–15). Wiley-VCH. doi:DOI: 10.1002/9783527628155.nanotech010

Pandey, N., Shukla, S. K., & Singh, N. B. (2017). Water purification by polymer nanocomposites: An overview. *Nanocomposites*, 3(2), 47–66. DOI: 10.1080/20550324.2017.1329983

Pandey, P. K., Sharma, S. K., & Sambi, S. S. (2015). Removal of lead (II) from wastewater on zeolite-NaX. *Journal of Environmental Chemical Engineering*, 3(4), 2604–2610. DOI: 10.1016/j.jece.2015.09.008

Parham, H., Bates, S., Xia, Y., & Zhu, Y. (2013). A highly efficient and versatile carbon nanotube/ceramic composite filter. *Carbon*, 54, 215–223. DOI: 10.1016/j.carbon.2012.11.032

Peralta, M. E., Mártire, D. O., Moreno, M. S., Parolo, M. E., & Carlos, L. (2021). Versatile nano adsorbents based on magnetic mesostructured silica nanoparticles with tailored surface properties for organic pollutants removal. *Journal of Environmental Chemical Engineering*, 9(1), 104841. DOI: 10.1016/j.jece.2020.104841

Qin, L., Zeng, Z., Zeng, G., Lai, C., Duan, A., Xiao, R., Huang, D., Fu, Y., Yi, H., Li, B., Liu, X., Liu, S., Zhang, M., & Jiang, D. (2019). Cooperative catalytic performance of bimetallic Ni-Au nanocatalyst for highly efficient hydrogenation of nitroaromatics and corresponding mechanism insight. *Applied Catalysis B: Environmental*, 259, 118035. DOI: 10.1016/j.apcatb.2019.118035

Qu, X., Alvarez, P. J., & Li, Q. (2013). Applications of nanotechnology in water and wastewater treatment. *Water Research*, 47(12), 3931–3946. DOI: 10.1016/j.watres.2012.09.058

Riffat, R. (2012). *Fundamentals of Wastewater Treatment and Engineering*. CRC Press., DOI: 10.1201/b12746

Rikta, S. Y. (2019). Application of Nanoparticles for Disinfection and Microbial Control of Water and Wastewater. In *Nanotechnology in Water and Wastewater Treatment* (pp. 159–176). Elsevier., DOI: 10.1016/B978-0-12-813902-8.00009-5

Sadegh, H., Ali, G. A. M., Gupta, V. K., Makhlouf, A. S. H., Shahryari-Ghoshekandi, R., Nadagouda, M. N., Sillanpää, M., & Megiel, E. (2017). The role of nanomaterials as effective adsorbents and their applications in wastewater treatment. *Journal of Nanostructure in Chemistry*, 7(1), 1–14. DOI: 10.1007/s40097-017-0219-4

Saravanan, R., Gracia, F., & Stephen, A. (2017). Basic Principles, Mechanism, and Challenges of Photocatalysis. In M.M. Khan, D. Pradhan, Y. Sohn, Y. (Eds.), *Nanocomposites for Visible Light-Induced Photocatalysis*. Book series: Springer Series on Polymer and Composite Materials. Springer.

Seung, J. K., Jae, H. C., Tae, Y. K., & Sung, Y. C. (2006). Biosorption of heavy metals and cyanide complexes on biomass. In New Developments and Application in Chemical Reaction Engineering; Studies in Surface Science and Catalysis (pp. 141–144). Elsevier.

Taran, M., Safaei, M., Karimi, N., & Almasi, A. (2021). Benefits and application of nanotechnology in environmental science: An overview. *Biointerface Research in Applied Chemistry*, 11, 7860–7870.

Theron, J., Walker, J. A., & Cloete, T. E. (2008). Nanotechnology and water treatment: Applications and emerging opportunities. *Critical Reviews in Microbiology*, 34(1), 43–69. DOI: 10.1080/10408410701710442

Timmer, J. M. K. (2001). *Properties of nanofiltration membranes: model development and industrial application*. [PhD Thesis 1 (Research TU/e / Graduation TU/e), Chemical Engineering and Chemistry]. Technische Universiteit Eindhoven. doi:DOI: 10.6100/IR545659

Tlili, I., & Alkanhal, T. A. (2019). Nanotechnology for water purification: Electrospun nanofibrous membrane in water and wastewater treatment. *Journal of Water Reuse and Desalination*, 9(3), 232–248. DOI: 10.2166/wrd.2019.057

Torabian, A., Kazemian, H., Seifi, L., Bidhendi, G. N., Azimi, A. A., & Ghadiri, S. K. (2010). Removal of Petroleum Aromatic Hydrocarbons by Surfactant-modified Natural Zeolite: The Effect of Surfactant. *Clean (Weinheim)*, 38(1), 77–83. DOI: 10.1002/clen.200900157

United Nations Sustainable Development Goals. Available online: https://sdgs.un.org/goals (accessed on 14 September 2023).

Varghese, A. (2004). A comparative risk approach to assessing point-of-use water treatment systems in developing countries. In I. Linkov & A.B. Ramada (Eds.) *Proceedings of Comparative risk assessment and environmental decision making*. NATO Science Series IV Earth and Environment Science NAIV, *38*, 99–112.

Verma, B., & Balomajumder, C. (2020). Surface modification of one-dimensional Carbon Nanotubes A review for the management of heavy metals in wastewater. *Environmental Technology & Innovation*, 17, 100596. DOI: 10.1016/j.eti.2019.100596

Vouk, V. B. (1979). *Handbook on the Toxicology of Metals*. Elsevier-North-Holland Biomedical Press.

Wang, H., Zhou, Y., Jiang, X., Sun, B., Zhu, Y., Wang, H., Su, Y., & He, Y. (2015). Simultaneous Capture, Detection, and Inactivation of Bacteria as Enabled by a Surface-Enhanced Raman Scattering Multifunctional Chip. *Angewandte Chemie International Edition*, 54(17), 5132–5136. DOI: 10.1002/anie.201412294

WHO UNICEF. (2012). *Progress on Drinking Water and Sanitation: 2012 Update; WHO*. UNICEF.

Xiu, Z.-M., Zhang, Q.-B., Puppala, H. L., Colvin, V. L., & Alvarez, P. J. J. (2012). Negligible Particle-Specific Antibacterial Activity of Silver Nanoparticles. *Nano Letters*, 12(8), 4271–4275. DOI: 10.1021/nl301934w

Yadav, D. K., & Srivastava, S. (2017). Carbon nanotubes as adsorbent to remove heavy metal ion (Mn^{+7}) in wastewater treatment. *Materials Today: Proceedings*, 4(2), 4089–4094. DOI: 10.1016/j.matpr.2017.02.312

Yamamura, H., Hoshino, M., & Kubota, N. (2011). Membrane Filtration. *Principle and Applications in Water Treatment.*, 67, 3.

Yang, X., Zhao, Z., Yu, Y., Shimizu, K., Zhang, Z., Lei, Z., & Lee, D. J. (2020). Enhanced biosorption of Cr(VI) from synthetic wastewater using algal-bacterial aerobic granular sludge: Batch experiments, kinetics and mechanisms. *Separation and Purification Technology*, 251, 117323. DOI: 10.1016/j.seppur.2020.117323

Yoon, J., Cao, X., Zhou, Q., & Ma, L. Q. (2006). Accumulation of Pb, Cu, and Zn in native plants growing on a contaminated Florida site. *The Science of the Total Environment*, 368(2-3), 456–464. DOI: 10.1016/j.scitotenv.2006.01.016

Zhang, C., Lai, C., Zeng, G., Huang, D., Yang, C., Wang, Y., Zhou, Y., & Cheng, M. (2016). Efficacy of carbonaceous nanocomposites for sorbing ionizable antibiotic sulfamethazine from aqueous solution. *Water Research*, 95, 103–112. DOI: 10.1016/j.watres.2016.03.014

Zhao, W., Tian, Y., Chu, X., Cui, L., Zhang, H., Li, M., & Zhao, P. (2021). Preparation and characteristics of a magnetic carbon nanotube adsorbent: Its efficient adsorption and recoverable performances. *Separation and Purification Technology*, 257, 117917. DOI: 10.1016/j.seppur.2020.117917

Zhao, Y., Zhang, B., Zhang, X., Wang, J., Liu, J., & Chen, R. (2010). Preparation of highly ordered cubic NaA zeolite from halloysite mineral for adsorption of ammonium ions. *Journal of Hazardous Materials*, 178(1-3), 658–664. DOI: 10.1016/j.jhazmat.2010.01.136

Zhou, B., Li, Y., Zheng, G., Dai, K., Liu, C., Ma, Y., Zhang, J., Wang, N., Shen, C., & Guo, Z. (2018). Continuously fabricated transparent conductive polycarbonate/carbon nanotube nanocomposite films for switchable thermochromic applications. *Journal of Materials Chemistry. C, Materials for Optical and Electronic Devices*, 6(31), 8360–8371. DOI: 10.1039/C8TC01779D

KEY TERMS AND DEFINITIONS

Adsorption: Adsorption: It includes the attachment of pollutants to the outer layer of a solid substance (adsorbent), like activated carbon or specific nanomaterials because of the forces of attraction.

Biosorption: Biosorption is a form of adsorption wherein living or dead biological substances are utilized as the adsorbent and which include; certain types of bacteria, algae or fungi. These biological materials can attach pollutants on their outer membranes or within their structures and help in their elimination from the wastewater.

Carbon: Carbon in wastewater treatment is the term for activated carbon or other carbonaceous materials used as adsorbents that are utilized in the removal of organic contaminants, odors, as well as some dissolved substances from water via processes of adsorption.

Disinfection: Disinfection in the scenario of wastewater treatment entails the elimination of pathogenic microorganisms in water including bacteria, viruses and parasites to safe discharge or recycling level. The types of disinfections are chemical disinfectants such as chlorine, UV radiation, ozonation as well as the advanced oxidation processes (AOPs). Some methods are more effective than others and these define the type of method to be used depending on the efficiency, cost, and environmental friendliness.

Metals: The heavy elements of lead and mercury from the heavy metal wastewater can be toxic and destroy all water sources. The techniques of treatment comprise adsorption and precipitation that are applied to separate the contaminants from the water before it is disposed of or used again.

Nanofilters: Nanofilters are the new generation of filters to purify water with the help of bits as small as nanosized ones. These filters employ such aspects as nanotubes, nano-particles or nano-fibers that may trap the properties including; heavy metals, pathogens or organic matter in contaminated water thereby promoting the standard of drinking water or water that has undergone through tertiary treatment.

Nanotechnology: Nanotechnology deals with objects that are at least one dimension less than 100 nm, to create special properties. It presents fresh possibilities in the field of wastewater treatment for the proper withdrawal of pollutants and water purification.

Sensors: In wastewater treatment, devices called sensors identify and gauge selected physical, chemical or biological parameters to keep watch over and regulate operations in an efficient and environmentally friendly manner.

Wastewater treatment: Water reclamation is the process of cleaning the sewage or industrial wastewater by removing impurities and unwanted substances so that it is safe to be disposed of or used again. The primary goal of this approach is to prevent the detrimental consequences that water pollution has on both human populations and ecosystems.

Zeolites: Zeolites are porous minerals used in wastewater treatment to absorb ions and pollutants that in effect clear the water by using them as catalysts, effectively trapping contaminants in their crystal's structural configuration.

Chapter 8
Application of Nanotechnology in Sustainable Food Packaging

Aditya Shrivastav
Ajeenkya D.Y. Patil University, India

Yadnyesh Khapekar
Ajeenkya D.Y. Patil University, India

Devanshi Soni
Ajeenkya D.Y. Patil University, India

Sunil Sankathala
Ajeenkya D.Y. Patil University, India

Susanta Das
https://orcid.org/0000-0002-9314-3988
Ajeenkya D.Y. Patil University, India

ABSTRACT

This thorough analysis examines how nanotechnology is transforming a number of industries, with a primary focus on packaging materials. Through the improvement of biodegradable biopolymer-based materials functional performance, nanotechnology has completely transformed the packaging sector. These developments support food safety and sustainability in addition to enhancing packaging qualities. The study also addresses the use of nanotechnology in water filtration systems to lessen

DOI: 10.4018/979-8-3693-9163-1.ch008

the use of plastic bottles, addressing issues related to plastic pollution that affect the environment and human health. It also draws attention to the difficulties and possibilities associated with applying nanotechnology and biodegradable polymers to reduce the amount of plastic trash generated and boost the effectiveness of environmental cleanup procedures. The paper also explores the advantages and disadvantages of nanoplastic particles, highlighting the need for more investigation to fully comprehend their effects on the environment.

INTRODUCTION

Nanotechnology is a growing field which involves materials at small scale but has applications at large scale in various sectors such as health industry (Modi et al., 2021), medicinal industry (Haleem et al., 2023), biotechnology (Contera et al., 2020), waste water treatment (Werkneh & Rene, 2018), pollution control (Ibrahim et al., 2016), agriculture (Elizabath, 2019), cosmetics (Gupta et al., 2022), textile (Shah et al., 2022), semiconductor devices (Hossain et al., 2023), construction (Shariati et al., 2017), and the food packaging industry (Galstyan et al., 2018; Samal, 2017) to name of few.

In food industry, NPs are specially focused on packaging of food for sustainable environment, as plastic packaging of food are hazardous for environment creating waste due to which it affects the living being (Barage et al., 2022). NPs such as $TiO_{2,}$ Zn, SiO_2 are used as preservatives, additives, flavouring agents, and microbial sensors (Samal, (2017)). Recent studies have shown that NPs have been used to improve food safety by increasing shelf life and developing antimicrobial packaging materials (De Sousa et al., 2013; Galstyan et al., 2018; Samal, 2017). If we see the history of food packaging, thousands of years ago leaves, shells, and animal skins were used (Samal, 2017; Barage et al., 2022). As technology got advanced many new manufacturing techniques got discovered which led to development of ceramic glass containers, metal packaging (which was developed in 17th Century by French chef Nicolas Appert) (Samal, 2017; Barage et al., 2022). But they have limitations as they were not sustainable for solid foods, affecting texture and taste of food and these packages were non-biodegradable which affected environment by polluting the water bodies creating blockage in Food Chain and were quite expensive (Samal, 2017). And due to all reasons now food industry is focusing on biodegradable food packaging, which are not expensive and reusable (De Sousa et al., 2023; Galstyan et al., 2018; Samal, 2017; Barage et al., 2022).

Many studies were conducted to solve this issue, and researchers found that nanotechnology has the potential to solve this issue by making the packaging materials stronger, more durable, and sustainable for environment and improving food security, flavour, taste, and nutrients (Primozic et al., 2021). But still there is a growing concern as it has various toxic effects on environment and health of all living beings (Samal, 2017).

In this chapter, we are going to discuss in detail about types of packaging through various nanomaterials and their applications and how these nanomaterials are used for sustainable food packaging and also considering their toxicological effects on human health and environment.

2. NANOPARTICLES IN FOOD PACKAGING & IT'S CLASSIFICATION

Food delivers essential nutrients to the human body, but it is highly perishable and susceptible to bacterial infections. Researchers are employing nanomaterials in polymers to develop packaging material that has better strength and endures longer while protecting food from contamination and degradation. Packaging categories include enhanced, dynamic, functional and bio based. Advanced packaging utilises carbon nanomaterials to build and upgrade polymers by making them more flexible, stronger, and stable to shifts in temperature, pH level and humidity. There are also catalogue packaging being made of active substances such as antimicrobial agents and oxygen absorbers for the preservation of the food's quality. Smart packaging uses technology components such as nano sensors that evaluate the status of foods without having to modify them. Bio-based packaging is a form of packaging material that is made out of materials developed from proteins, fats, and polysaccharides instead of the conventional inorganic plastics (Chausali et al., 2022).

Smart and interactive packaging is advanced technique which not only detects the changer in environment and product history but display quality of the food products which are packed. It can also monitor food freshness and even alert the consumer when the food is no longer safe to be ingested – a key issue attributed to food expiry. Some of the smart indicators are naturally occurring pigments like anthocyanins, alizarin, and beta lain that help in detecting the changes a midst. Active packaging systems communicate with the food in one way or indirectly to improved safety and quality of food. Some of these antibacterial qualities are useful in their capability to avoid microbial invasion, prolong food shelf life, protect articles from ultraviolct light and inhibit spoilage. Smart packaging features of the active and intelligent packaging type incorporate sensors that evaluate the quality and freshness of food with possible antimicrobial and antioxidant properties (De Sousa et al., 2023).

Nanomaterials can also be in the form of nanostructured or nanomodified materials and they hold the promise of developing functional nanomaterials with improved beneficial properties; active nanomaterials with antimicrobial, antioxidant, and

UV protective properties. Other areas which also benefit from these advancements include intelligent gas nano sensors and environmentally friendly developments of biodegradable, biocompatible, and sustainable packaging (Ashfaq et al., 2022; Samal, 2017; Primozic et al., 2021). Table 1 gives further elaboration about various food packaging systems containing nanomaterials including packaging type, NPs used, polymer base, applications, and advantages.

Table 1. Various types of packaging

Type of packaging	Nanoparticles used	Polymer matrix	Applications	Advantages	References
Improved packaging	Clay nanoparticles	EVOH copolymer	Improved oxygen and water vapor barrier properties, reduced permeability	Improved shelf life, reduced food spoilage	(Primozic et al., 2021)
	Nano clay	Polypropylene	Improved oxygen and water vapor barrier properties, reduced permeability	Improved shelf life, reduced food spoilage	(Primozic et al., 2021), (Samal, 2017), (Ashfaq et al., 2022)
	Cellulose nanomaterials	Polymers (e.g. PLA, PBAT)	Improved oxygen barrier properties, reduced water vapor permeability	Biodegradable, renewable, and sustainable	(Ashfaq et al., 2022)
Active packaging	Ag nanoparticles	Gelatine film	Antimicrobial properties, reduced bacterial growth	Extended shelf life, improved food safety	(Samal, 2017)
	ZnO nanoparticles	Starch-poly vinyl alcohol-based film	pH sensitivity, potential for use in smart food packaging	Real-time monitoring of food spoilage	(Primozic et al., 2021)
Intelligent/ smart packaging	Soluble soybean polysaccharide film with curcumin and SiO_2	Soluble soybean polysaccharide film	Indicates shrimp spoilage through colour changes	Real-time monitoring of food spoilage, improved food safety	(Ashfaq et al., 2022)
	Starch-poly vinyl alcohol-based film with zinc oxide nanoparticles and phytochemicals	Starch-poly vinyl alcohol-based film	pH sensitivity, potential for use in smart food packaging	Real-time monitoring of food spoilage, improved food safety	(Ashfaq et al., 2022)

3. NANOMATERIALS USED IN FOOD PACKAGING

Nanotechnology, both naturally occurring and intentionally created by humans, is revolutionizing the food packaging industry by providing enhanced, smart, and active packaging options (Chaudhary et al., 2020). These types of materials have exceptional mechanical, barrier and heat resistance properties, making them ideal for improving food packaging materials (Chaudhary et al., 2020; Barage et al., 2022).

The food packaging industry is currently being revolutionized by nanotechnology by producing various types of NPs specially made from metal such as Au, Ag, Zn, TiO_2, SiO_2, MgO and ZnO which are also known as metal NPs (M-NPs) (Barage et al., 2022). Metal like Ag and Au have high plasmon resonance which helps to detect environmental molecules which ensure food quality and prevent spoilage agents (Kumar et al., 2021). Table 2 list various types of other nanomaterials with their properties, advantages, disadvantages and applications in food packaging.

Table 2. Lists of nanomaterials along with their properties

Nanomaterials	Properties	Application in Food Packaging	Benefits	Disadvantages	References
AgNPs (Silver Nanoparticles)	Antimicrobial, antifungal, antiviral	Active packaging, extended shelf life	Improved shelf life, reduced food spoilage	Risk of migration into food, imbalance in biomass production	(Ashfaq et al., 2022), (Chadha et al., 2022), (Barage et al., 2022), (Peidaei et al., 2021), (Hoseinnejad et al., 2017), (Kumar et al., 2021), (Chaudhary et al., 2020)
TiO_2-NPs (Titanium Dioxide Nanoparticles)	Thermo-stability, inertia, photocatalytic	Whitener, photocatalyst, antimicrobial	Improved shelf life, UV light filter, inert and non-toxic	Migration into lipid matrix, rancidity	(Ashfaq et al., 2022), (Chadha et al., 2022), (Peidaei et al., 2021), (Hoseinnejad et al., 2017), (Kumar et al., 2021), (Chaudhary et al., 2020)
ZnO-NPs (Zinc Oxide Nanoparticles)	Antimicrobial, affordable, non-toxic	Cosmetics, medical devices, food packaging	Improved shelf life, reduced food spoilage, cost-effective	Visible on skin as opaque white layer, non-eco-friendly byproducts	(Ashfaq et al., 2022), (Chadha et al., 2022), (Barage et al., 2022), (Hoseinnejad et al., 2017), (Kumar et al., 2021), (Chaudhary et al., 2020)

continued on following page

Table 2. Continued

CuO-NPs (Copper Oxide Nanoparticles)	Antimicrobial, high stability	Textiles, coatings, food packaging	-	-	(Chadha et al., 2022), (Barage et al., 2022), (Hoseinnejad et al., 2017), (Kumar et al., 2021)
Se-NPs (Selenium Nanoparticles)	Antimicrobial, antiviral, antioxidant	Novel antibacterial polymeric coating substance	-	-	(Hoseinnejad et al., 2017)
Au-NPs (Gold Nanoparticles)	Antimicrobial, high potential for standardization	Electronics, cosmetics, food packaging	Strong antibacterial activity, bioactive films	Expensive, irreversible agglomeration, loss of red colour	(Chadha et al., 2022), (Hoseinnejad et al., 2017), (Kumar et al., 2021)
Nano clay	Improved physical properties, barrier resistance	Enhanced mechanical strength, barrier properties	Extend shelf life of meat products and reduce moisture	Low specific gravity and high aspect ratio, less compatible	(Ashfaq et al., 2022), (Chadha et al., 2022), (Barage et al., 2022), (Kumar et al., 2021), (Chaudhary et al., 2020)
Nano-Starch	Enhanced mechanical and barrier properties	Flexible food packaging, filler	Inexpensive, widely available, increase shelf life	Cannot be used directly, must be added with other nanomaterials	(Chaudhary et al., 2020)
Nano-Scaled Cellulose	High axial stiffness, biodegradable	Reinforce-ment fibber, polymer composites	Excellent barrier properties	-	(Chaudhary et al., 2020)
CNTs (Carbon Nanotubes)	Enhanced mechanical strength, antimicrobial	Intelligent sensors, packaging applications	Mechanical and antimicrobial properties	-	(Ashfaq et al., 2022), (Chadha et al., 2022), (Chaudhary et al., 2020)
Nano-Silica	Hydrophobic, self-cleaning	Food packaging, improved flowability	Disinfectant photocatalytic agent, UV blocker, water retention	Hazardous during production, application, and disposal, high cost	(Chaudhary et al., 2020)
Iron Oxide Nanoparticles	Magnetic, enzyme immobilization, food analysis	Oxygen scavenging, ethylene detection	Low toxicity, hydroxyl radical formation	-	(Chadha et al., 2022)

4. ENVIRONMENTAL SAFETY AND TOXICOLOGICAL CONSIDERATIONS

Nanomaterials are rising in the food packing industry due to its ability to offer a healthier, safer, and better option. However, there are still safety and capability fitness implications that come with the concerns of nanofoods. Due to the enhancement of interaction with cells, tissues and organs through size, transfer of nanomaterials from packaging to food poses potential toxicological effects. In order to respond to these concerns, specific objective criteria are required to assess the relative safety of nano-food products. They should be able to identify and measure nanomaterials in food matrices, define their distribution and migration and offer credible analytical instruments. Several techniques have been suggested for the evaluation of engineered NPs in the substrates and they include the use of dynamic light scattering and mass spectroscopy. Nano-structured surfaces can also be used as sensor systems also NPs in solution are effective at identifying dangerous compounds in foods. Raman Nano Chip is a food safety detection system to collect test samples with nanorods and assess the migration of NPs from Food Contact Materials into Food. Novel technologies need to be established and refined in order to accurately measure and identify nanomaterials in food compounds to adequately protect the consumer and safeguard nano-food products. EPA may have to place certain limitations concerning particles as small as NPs until all the risks are conclusively managed (Chaudhary et al., 2020). In the contemporary food industry, Edible packaging is also becoming popular because it is environmentally friendly, improves meals quality, and safety. It encompasses materials that may be fed on with foods thus minimise on the conventional packages that are feared to have the effects on the environment. It can release active substances such as antimicrobials, vitamins, and other additives leading to an increased capability and reduced wastage. Public communication can be in healthy to be eaten packaging material, which is agar a polysaccharide derived from seaweed, combined with the biopolymers such as starch, proteins, and lipids. The technique of creating packaging in shape to be eaten includes numerous steps, which includes casting, dipping, spraying, and spreading. The substances may be extruded, temporized, or dried to create the final product. Edible packaging may be used for diverse foods, together with end result, veggies, meats, and dairy merchandise. Edible packaging is environmentally pleasant, in contrast to conventional materials like plastics, metals, and glass. It might not harm the environment or human fitness if improperly disposed of. Consumer demand for sustainable and eco-friendly merchandise drives using edible packaging. However, demanding situations which include high manufacturing charges, limited shelf lifestyles, and worries approximately safety and efficacy of safe to eat packaging are still to be addressed as these conditions are quite challenging. Currently, eating packaging is a trend that has been thought to transform meals packaging and consumption in the society than affecting the world environment and people's health significantly (Chadha et al., 2022).

Although nanomaterials have many benefits there are concerns about their potential effects on human life and environmental sustainability. When nanomaterials are used in meals packaging or sensing programs, they could migrate into the human body through inhalation, pores and skin penetration, or ingestion (Primozic et al., 2021; Chausali et al., 2022). This can happen while nanocomponents leach from packaging or sensing factors into food, or while packaging is stored in landfills and releases nanomaterials into the surroundings. Some researchers have established that NPs can have poisonous consequences on organic structures, however the toxicity relies upon the kind and size of the NPs (Primozic et al., 2021; Chausali et al., 2022). Exposure to NPs can purpose the formation of reactive oxygen species, protein denaturation, and mitochondrial ailment. A major state of affairs is the migration of nanomaterials from packaging into food, which can arise while we devour food that has nanomaterials. The houses of the gastrointestinal tract, which includes pH and the presence of digestive enzymes, can have an effect on the absorption of nanomaterials and alternate their houses. Some steel and metal oxide NPs are known for its physique however they are still capable of inducing pro-inflammatory reactions and oxidative stress. Carbon nanotubes can accumulate in dwelling organisms and can produce reactive oxygen species, but are toxic in nature. There is various research done on the migration of a number of metallic oxides such as TiO_2, ZnO and SiO_2 through the packaging material into the food and no high value of migration was found (Primozic et al., 2021; Chausali et al., 2022).

However, it is advocated to have a further barrier to separate food from nanocomposite materials. The migration of biopolymers, such as PLA, additionally desires to be studied and taken into consideration. Some biodegradable substances have been proven to be toxic, and powerful guidelines and regulations are had to ensure the secure use of NPs inside the meals industry. There are some guidelines by the America FDA, FSANZ, and EFSA on the usage of NPs in meals and packaging. However, more study is needed to address the challenges and data constraints of nanomaterials in the food sector (Primozic et al., 2021). The shift closer to herbal merchandise within the meals industry has caused the exploration of biodegradable and natural polymer-primarily based packaging with revolutionary technology like nanotechnology and biotechnology. However, worries about ability toxicity of nanomaterials rise up from their migration into food objects. Although some nanomaterials such as nano clays are considered safe, there are others such as Organo changed clay and carbon nanotubes that have been found to be toxic (Chausali et al., 2022). Although nanocellulose is generally biodegradable and non-toxic, it is potentially dangerous during inhalation. Clay NPs exhibited enhancement in mechanical power and possessed an antimicrobial property, nevertheless, they were found to migrate into the meals which should be taken into account (Chausali et al., 2022). Regulatory companies like the FDA and EFSA have mounted recommendations for NPs migration from packaging materials. Factors like temperature can have an effect on migration rates. Metal oxides like ZnO, TiO_2, and SiO_2 can have toxicity, even as steel alloys and carbon nanotubes can alter protein molecules (Chausali et al., 2022). To make sure safe commercial packages, the physicochemical residences and dose of nanomaterials ought to be evaluated and quantified. Factors like pH, osmotic awareness, absorption, metabolism, excretion, and toxicity have to also be considered (Chausali et al., 2022).

Nanotechnology has the capacity to revolutionize food packaging and coating, but there are nonetheless worries approximately its protection. The protection of nanotechnology in food packaging is closely tied to the substances used and the additives introduced to them. Countries have mounted requirements and legal guidelines to adjust using these materials, consisting of the European Union, America, China, and Brazil. However, with the arrival of nanotechnology, new legal guidelines and policies are had to make certain the safe production, use, and trade of those substances. Nanostructures introduced to biopolymer matrices can trade the residences of the packaging fabric, affecting its strength, flexibility, and capacity to protect meals from spoilage. Some nanostructures can even upload new houses, which include antimicrobial activity. There are not any uniform rules on the usage of nanostructures in meals packaging, as distinct international locations develop their very own lists of accredited and prohibited nanostructures. In America, elements such as aluminium and Ag are allowed in meals packaging while the European Union has legalised titanium nitride and SiO_2 (De Sousa et al., 2023). This is because as nanotechnology in food packaging continues to surge, each country has had to come up with their unique tests and regulations to guarantee the safe use of the nanotechnologies. According to the European Commission, migration checking out is the process of evaluating the release of substances from an item or an article into food or a meals surrogate. For this reason, more floor location to extent ratios, which can stimulate toxicity within the frame if they translocate into food. The effects of NPs can only be attributed to the type of NPs, the concentration and exposure time, and the susceptibility of the individual. These organic NPs such as proteins, lipids, starch, and chitosan are deemed to be non-toxic because they can be digested, bioavailable and possess dietary properties. While Ag NPs can favour the habitation of packing materials substances, they can also trigger toxic reactions including neurotoxicity, genotoxicity, and organ localization such as liver, kidney, testes & brain (Chadha et al., 2022). Other NPs, for instance, nano clay, could also enter into food based on factors such as temperature, direct contact to meals, and interface with the polymer matrix. Hence, it becomes crucial to evaluate the migration, toxicity and PEL of NPs before employing these in food packaging materials. Consumers need for meals products has led to research for improved effective and reliable meals packaging. Use of Nanotechnology in food packaging therefore helps satisfy a wish for enhancing the bodily, mechanical, and functional properties of the food packaging; however, need to have into account possible risks implicated by way of NPs migration and toxicity when practising them in food packaging agents (Ashfaq et al., 2022). Nano-merchandise added within the food packing cloth can migrate to food products and cause damage to men thru bodily intake, inhalation and pores and skin contact. These increase as their sizes locate big floor areas thereby producing numerous sickness and other health issues. This investigation exposes that Ag NPs may penetrate meals products below sure situations; nonetheless, their penetration is far beneath the fashionable restrict stipulated by the EU authorities. However, ZnO NPs in general is non-poisonous, but when NPs in it is carcinogenic to the people. Chemical vapour deposition is also one of the techniques used to develop NPs it has some drawbacks such as it requires specialised equipment, and it is toxic in nature. Carbon nanotubes were found to be potential for food packaging however a few of them were found to be toxic to some micro-organisms and are possibly carcinogenic. It is also being used to produce clear films that are non-toxic, cheaper, and resistant to light; but this substance has been proved to cause lung tumours in some instances. Starch is an abundant renewable biodegradable and non-toxic polysaccharide and is being investigated for use in food packaging. As much as they benefit the process of meals packaging, their harm and impacts on human health should not be neglected (Chadha et al., 2022).

When ingested, inhaled or even during packaging and handling processes of foods containing NPs, these particles are dangerous to health. They can be a reason to damage the liver, and lead to cardiovascular sicknesses, lung tumors, and blood clotting. The migration of NPs from packaging to merchandise relies upon factors which includes polymer complexes, particle size, solubility, and awareness. Interaction with the frame can lead to risky consequences, which include carcinogenicity and genotoxicity. Nano-TiO_2 and nano-Ag have high toxicity, while large-sized ZnO is much less toxic. Inhaling nano-TiO_2 can increase lung tumor danger, at the same time as ZnO can inhibit cell proliferation and viability, leading to oxidative strain. TiO_2-NPs can distort intestinal tissue shape, affecting nutrient absorption. The regulatory system plays an essential position in assessing the dangers associated with M-NPs in the metal region (Kumar et al., 2021). The increasing use of NPs in food packaging has raised issues about their ability, toxicity and potential to harm living beings. The FDA has set up a venture force to identify the dangerous use of NPs to living beings. Ag NPs have been found to be poisonous and can motive continual illnesses like hypersensitive reactions, allergies, and cancer. The migration of NPs from food packaging to meals itself is some other situation, as factors like temperature, pH, and food type can have an effect on the migration system. To cope with these worries, regulatory businesses are growing suggestions for the secure use of NPs in meals packaging, comparing the risks and growing standardized processes to investigate their impact on human beings and the environment. Identifying and characterizing NPs in metals is vital due to their capability dangers to consumers. The dissolution of steel-primarily based NPs can launch poisonous ions which could have interaction with organic systems and cause damage. The agglomeration of NPs also can have an effect on their toxicity, with large agglomerates potentially inflicting greater harm than smaller ones. Conducting in vivo and in vitro studies is important to ensure the safety of natural nanomaterials. Testing the migration of NPs into meals materials is critical for the secure use of metallic based totally substances in food packaging (Adeyemi & Fawole, 2023).

5. APPLICATIONS OF NANOMATERIALS IN FOOD PACKAGING

The food industry faces the mission of preserving meals fresh and secure from infection. NPs can be used to increase food packaging substances with houses together with blocking off CO_2, oxygen, and water vapor, chemical and thermal balance, recyclable, and biodegradable. Some of these substances can reduce the packaging weight, cost, and waste technology in addition. Nanocomposites primarily based totally on biopolymers and NPs are anticipated to increase product shelf lifestyles, specifically throughout emergencies or navy storage. Another area of research is the intelligent packaging or the use of nano sensors to identify when meals are spoiled. These sensors can encounter changes in humidity, freshness, oxygen, CO_2, temperature, microbes, and chemical spoilage. Still, there are concerns with concern to the high cost of this technology and issues to do with the law. Some of the many agencies and universities are choosing for more intelligent packaging material and the US market it is predicted to be worth at $54 billion (Basavegowda et al., 2019). The meat export enterprise had grown as much as $15 billion by way of 2020 (Basavegowda et al., 2019). Recently, scientists have evolved a nanocomposite-based smart packaging method that can hit upon food spoilage and release preservatives to extend shelf life, assembly the developing call for energetic packaging that increases food product freshness (Basavegowda et al., 2019).

Active loaded packaging materials have been developed to improve the pleasant, protection, and shelf lifestyles of various meals products. Seafood, because of its excessive nutrient and moisture content, is exceedingly perishable. Meat, like seafood, is also susceptible to microbial and chemical spoilage. Researchers have evolved active-loaded packaging substances to inhibit the growth of micro-organism and yeast on meat merchandise, which includes incorporating Trachyspermum Ammi oil nano emulsions into alginate suitable for eating coatings. Nutrient-rich food such as Cheese, is likewise vulnerable to microbial and chemical spoilage (Khezerlou et al., 2021). Researchers have advanced active-loaded packaging substances to increase the shelf lifestyles of cheese through inhibiting the boom of spoilage microorganisms (Khezerlou et al., 2021). For instance, thyme oil-loaded nanoliposomes into chitosan coatings extended the shelf lifestyles of Karish cheese (Khezerlou et al., 2021). Bread, like wheat bread, has also been extended via incorporating carvacrol nano emulsions into carboxymethylcellulose films (Khezerlou et al., 2021). Fruits and greens, which can be enormously perishable because of their high-water content material and susceptibility to fungal boom, have additionally been prolonged through the use of lively-loaded packaging materials (Khezerlou et al., 2021).

6. CONCLUSION

It has been revealed that nanotechnology, a field that focuses on materials measured at the nano-level, has numerous applications in the food industry. In this state-of-art sector, the synthesis and fabrication of NPs that can act as a flavouring agent, preservative, additive and as an antimicrobial sensor are created. These NPs enhance food safety by prolonging shelf-life and preventing spoilage of foods as well as recognising potential lethal pathogens. Although the use of packaging in food products has its roots deep in history, modern advancements in nanotechnology are revolutionising this sector. The current generation NPs are ideal for food packaging due to desirable mechanical, functional and physical characteristics. Among these materials are metal NPs, such as ZnO, Ag and gold which are being used in manufacturing active packaging to sense molecules in the surrounding air and keep food fresh. Furthermore, new types of packaging materials that are edible and biodegradable in nature are being developed by using several types of nanomaterials. However, there are still some issues on exposure and effects of nanomaterials on human beings and the environment. When present in food packaging, through ingestion, dermal or via inhalation NPs can influence human body. Nevertheless, it is equally important to have a look at the many benefits that come with the application of nanotechnology in food packaging. This may use NPs for packaging materials that are resistant to water vapour and carbon dioxide and oxygen and possible chemical and thermal stability. These materials may also be biodegradable and recyclable. Another amazing innovation is the intelligent packaging that deploys nano sensors to detect food spoilage. Therefore, the advancement of mechanical, functional, and physical properties of the nanotechnology backing could overhaul the food packaging industry and enhance the concept of sustainability of the food packaging industry and food safety.

REFERENCES

Adeyemi, J. O., & Fawole, O. A. (2023). Metal-Based Nanoparticles in Food Packaging and Coating Technologies: A Review. *Biomolecules*, 13(7), 1092. DOI: 10.3390/biom13071092

Ashfaq, A., Khursheed, N., Fatima, S., Anjum, Z., & Younis, K. (2022). Application of nanotechnology in food packaging: Pros and Cons. *Journal of Agriculture and Food Research*, 7, 100270. DOI: 10.1016/j.jafr.2022.100270

Barage, S., Lakkakula, J., Sharma, A., Roy, A., Alghamdi, S., Almehmadi, M., Hossain, M. J., Allahyani, M., & Abdulaziz, O. (2022). Nanomaterial in Food Packaging: A Comprehensive Review. *Journal of Nanomaterials*, 2022(1), 1–12. DOI: 10.1155/2022/6053922

Basavegowda, N., Mandal, T. K., & Baek, K. H. (2019). Bimetallic and Trimetallic Nanoparticles for Active Food Packaging Applications: A Review. *Food and Bioprocess Technology*, 13(1), 30–44. DOI: 10.1007/s11947-019-02370-3

Chadha, U., Bhardwaj, P., Selvaraj, S. K., Arasu, K., Praveena, S., Pavan, A., Khanna, M., Singh, P., Singh, S., Chakravorty, A., Badoni, B., Banavoth, M., Sonar, P., & Paramasivam, V. (2022). Current Trends and Future Perspectives of Nanomaterials in Food Packaging Application. *Journal of Nanomaterials*, 2022(1), 1–32. DOI: 10.1155/2022/2745416

Chaudhary, P., Fatima, F., & Kumar, A. (2020). Relevance of Nanomaterials in Food Packaging and its Advanced Future Prospects. *Journal of Inorganic and Organometallic Polymers and Materials*, 30(12), 5180–5192. DOI: 10.1007/s10904-020-01674-8

Chausali, N., Saxena, J., & Prasad, R. (2022). Recent trends in nanotechnology applications of bio-based packaging. *Journal of Agriculture and Food Research*, 7, 100257. DOI: 10.1016/j.jafr.2021.100257

Contera, S., De La Serna, J. B., & Tetley, T. D. (2020). Biotechnology, nanotechnology, and medicine. *Emerging Topics in Life Sciences*, 4(6), 551–554. DOI: 10.1042/ETLS20200350

De Sousa, M. S., Schlogl, A. E., Estanislau, F. R., Souza, V. G. L., Coimbra, J. S. D. R., & Santos, I. J. B. (2023). Nanotechnology in Packaging for Food Industry: Past, Present, and Future. *Coatings*, 13(8), 1411. DOI: 10.3390/coatings13081411

Elizabath, A. (2019). Application of Nanotechnology in Agriculture. *International Journal of Pure & Applied Bioscience*, 7(2), 131–139. DOI: 10.18782/2320-7051.6493

Galstyan, V., Bhandari, M., Sberveglieri, V., Sberveglieri, G., & Comini, E. (2018). Metal Oxide Nanostructures in Food Applications: Quality Control and Packaging. *Chemosensors (Basel, Switzerland)*, 6(2), 16. DOI: 10.3390/chemosensors6020016

Gupta, V., Mohapatra, S., Mishra, H., Farooq, U., Kumar, K., Ansari, M., Aldawsari, M., Alalaiwe, A., Mirza, M., & Iqbal, Z. (2022). Nanotechnology in Cosmetics and Cosmeceuticals-A Review of Latest Advancements. *Gels (Basel, Switzerland)*, 8(3), 173. DOI: 10.3390/gels8030173

Haleem, A., Javaid, M., Singh, R. P., Rab, S., & Suman, R. (2023). Applications of nanotechnology in medical field: A brief review. *Global Health Journal (Amsterdam, Netherlands)*, 7(2), 70–77. DOI: 10.1016/j.glohj.2023.02.008

Hoseinnejad, M., Jafari, S. M., & Katouzian, I. (2017). Inorganic and metal nanoparticles and their antimicrobial activity in food packaging applications. *Critical Reviews in Microbiology*, 44(2), 161–181. DOI: 10.1080/1040841X.2017.1332001

Hossain, N., Mobarak, M. H., Mimona, M. A., Islam, M. A., Hossain, A., Zohura, F. T., & Chowdhury, M. A. (2023). Advances and significances of nanoparticles in semiconductor applications – A review. *Results in Engineering*, 19, 101347. DOI: 10.1016/j.rineng.2023.101347

Ibrahim, R. K., Hayyan, M., AlSaadi, M. A., Hayyan, A., & Ibrahim, S. (2016). Environmental application of nanotechnology: Air, soil, and water. *Environmental Science and Pollution Research International*, 23(14), 13754–13788. DOI: 10.1007/s11356-016-6457-z

Khezerlou, A., Tavassoli, M., Sani, M. A., Mohammadi, K., Ehsani, A., & McClements, D. J. (2021). Application of Nanotechnology to Improve the Performance of Biodegradable Biopolymer-Based Packaging Materials. *Polymers*, 13(24), 4399. DOI: 10.3390/polym13244399

Kumar, A., Choudhary, A., Kaur, H., Mehta, S., & Husen, A. (2021). Metal-based nanoparticles, sensors, and their multifaceted application in food packaging. *Journal of Nanobiotechnology*, 19(1), 256. Advance online publication. DOI: 10.1186/s12951-021-00996-0

Modi, S., Prajapati, R., Inwati, G. K., Deepa, N., Tirth, V., Yadav, V. K., Yadav, K. K., Islam, S., Gupta, P., Kim, D., & Jeon, B. (2021). Recent Trends in Fascinating Applications of Nanotechnology in Allied Health Sciences. *Crystals*, 12(1), 39. DOI: 10.3390/cryst12010039

Peidaei, F., Ahari, H., Anvar, S. A., & Ataee, M. (2021). Nanotechnology in Food Packaging and Storage: A Review. *DOAJ (DOAJ: Directory of Open Access Journals)*.https://doi.org/doi:10.22059/ijvm.2021.310466.1005130

Primozic, M., Knez, E., & Leitgeb, M. (2021). (Bio)Nanotechnology in Food Science Food Packaging. *Nanomaterials (Basel, Switzerland)*, 11(2), 292. DOI: 10.3390/nano11020292

Samal, D. (2017). Use of Nanotechnology in the Food Industry: A review. *International Journal of Environment, Agriculture and Biotechnology*, 2(4), 2270–2278. DOI: 10.22161/ijeab/2.4.90

Shah, M. A., Pirzada, B. M., Price, G., Shibiru, A. L., & Qurashi, A. (2022). Applications of nanotechnology in smart textile industry: A critical review. *Journal of Advanced Research*, 38, 55–75. DOI: 10.1016/j.jare.2022.01.008

Shariati, S., Abedi, M., Saedi, A., Yazdani-Chamzini, A., Tamosaitiene, J., Saparauskas, J., & Stupak, S. (2017). Critical factors of the application of nanotechnology in construction industry by using anp technique under fuzzy intuitionistic environment. *Journal of Civil Engineering and Management*, 23(7), 914–925. DOI: 10.3846/13923730.2017.1343202

Werkneh, A. A., & Rene, E. R. (2018). Applications of Nanotechnology and Biotechnology for Sustainable Water and Wastewater Treatment. In *Energy, environment, and sustainability* (pp. 405–430). https://doi.org/doi:10.1007/978-981-13-3259-3_19

KEY WORDS AND DEFINITIONS

Active Packaging: This means packaging that interferes with the food in some or other manner to utilize, preserve or even add value, to the food item for its intended use.

Antimicrobial: Capable of lowering the microbial load and was believed to be utilized for the purpose of preserving foods for several days.

Antioxidant: They are chemical being used in preservation to avoid that which is spoilt through oxidation of the food products.

Bio-Based Packaging: Reusable product containers made from biological sources as a substitution of conventional plastics.

Biodegradable: Or biodegradable meaning, packaging materials that can be disintegrated through natural decay processes.

Biomolecules: Polymeric materials that come from living organisms and which find application in packaging because they are biodegradable and eco-friendly.

Biopolymer: The synthetically produced chemical that is derived from renewable materials, and which substitute ordinary plastics or polysaccharides used in food packaging.

Coating Technologies: Processes employed to deposit one layer of material over a substrate base, employed in food packaging for reinforcing the barrier functionality and enhancing usefulness of the product.

Food Industry: The sector included food production and processing and food distribution.

Food Packaging: The act of putting food into a certain type of material and similarly also the covering or pack material.

Food Safety: Cleaning, washing, and organizing methods and techniques that involve avoiding the spread of dangerous microbes.

Metal-Based Nanoparticles: Metals – including food packaging nanoparticles in foods to fight microbes and those with antioxidant properties.

Nanomaterials: Nanomaterials with specific properties by virtue of size, applied in food packing on accounts of their antibacterial, antioxidizing and barrier characteristics.

Nanoparticles: The nanoparticles appearing as small solids smaller than 100 nm and larger than 1 nm they have been employed in food packaging.

Nanotechnology: It means a capability to change them on a molecular and the atomic level in order to produce substances that have special purposes.

Packaging Materials: The packaging materials required to package and safeguard food stuffs during transit, storage, and marketing.

Quality Control: There is the procedure of making sure that food products conform to specific standards in quality and safety.

Sensors: Smart sensors that change their operations in Realtime, used in food packaging for evaluating the quality and safety of foods to consume.

Shelf Life: The time duration for which the food product is still safe and acceptable for human consumption.

Sustainability: The capability to sustain a characteristic or performance level in a system or a process and is used to characterize food packaging that has a low ecological footprint.

Chapter 9
Economic Impacts of Micro and Nano Plastic Pollution

Mohit Yadav
https://orcid.org/0000-0002-9341-2527
O.P. Jindal Global University, India

Ashutosh Pandey
https://orcid.org/0000-0002-8255-8459
FORE School of Management, New Delhi, India

Xuan-Hoa Nghiem
https://orcid.org/0000-0003-2292-0257
International School, Vietnam National University, Hanoi, Vietnam

Rashika Singh
Gautam Buddha University, India

ABSTRACT

This research examines the economic implications of micro- and nano plastic pollution. The study assesses the substantial costs associated with environmental cleanup, healthcare, and biodiversity loss. Economic impacts on sectors such as fisheries, tourism, and agriculture are analyzed. The paper highlights the urgent need for effective policies, technological advancements, and international cooperation to mitigate these costs. By quantifying the economic consequences, this research emphasizes the imperative for immediate action to address plastic pollution and its detrimental effects on economies and societies worldwide.

DOI: 10.4018/979-8-3693-9163-1.ch009

INTRODUCTION

Plastic is a material that was earlier extolled for its versatility and durability but now is an invasive and insidious pollutant. Its fragmentation into the microscopic and nanoscopic range presents problems of magnanimous proportions to ecosystems, human health, and economies worldwide. The magnitude of economic implications resulting from micro- and nano plastic pollution is massive, covering such areas as environmental cleanups, healthcare, and loss of biodiversity. This chapter considers the economic dimensions of this vexing global problem by assessing the cost and potential benefits of alleviating the issue (da Costa, 2018).

The Ubiquity and Persistence of Plastic

Plastic is a synthetic polymer obtained from fossil fuels that have transformed industries related to packaging, construction, healthcare, and electronics, among others (Sarkar et al., 2022). The low cost, durability, and versatility of the material have definitely factored in its wide application. However, the same features that make plastics so useful also make them environmentally persistent. For instance, once discharged into the environment, it will take hundred years for plastic products to decay (Osuna-Laveaga et al., 2023).

The Emergence of Micro and Nano Plastics

Increasingly worrying is plastic fragmentation into micro- and nano plastics. These small particles, mostly invisible to the naked eye, can be formed by several methods, such as processing of larger plastic pieces, industrial processes, or even through using personal care products containing plastic microbeads (Osuna-Laveaga et al., 2023).

Microplastics and nano-plastics are ubiquitous in all environments, including oceans, rivers, and even soil and the air we breathe. Because of their small size, they can infiltrate ecosystems and organisms at a cellular level with far-reaching and potentially devastating consequences.

The Economic Costs of Being Ignorant

Economic costs of micro- and nano plastic pollution are high, yet largely underrated. The ceteris paribus of inaction can lead to a multi-faceted range of effects, which can be described as follows:

- **Degradation of the environment:** The pollution from plastics badly upsets ecologies, deteriorating the habitat and thereby causing damage to the flora and fauna. These, in turn, can lead to enormous economic costs borne by industries that significantly depend on healthy ecology namely, fisheries, tourism, and agriculture (da Costa, 2018).
- **Human health risks**: Emerging evidence is linking micro- and nano plastic exposure to health effects such as respiratory problems, reproductive disorders, and cancer. Treatment of such health effects is economically costly (Sarkar et al., 2022).
- **Loss of biodiversity:** Plastic pollution is partially responsible for the decline in biodiversity, which has some very key ecosystem services like pollination, cleaning of water, and climate regulation. These services have very huge economic implications once lost (Jiang et al., 2020).
- **Infrastructure Damage**: Large plastic debris can cause blockages in waterways and infringements on structures, raising the costs of public works maintenance (Seuront et al., 2022).
- **Damage to Business:** Among sectors worst hit by the economic impacts of plastic pollution include fisheries, aquaculture, and tourism.

Why We Must Act

Fully understanding the economic costs of micro- and nano plastic pollution is essential in view of the enormity of the problem. Such knowledge forms a base for the development of policies, incentives for innovation, and mobilization of resources for mitigation (Jiang et al., 2020).

Quantifying the economic impacts of plastic pollution helps underline the urgency of action and, based on that, informs decision-making at all levels. The chapter therefore aims to present an overview of current knowledge about the economic costs due to micro- and nano plastic pollution and to explore the potential economic benefits that could be reaped from its reduction.

THE ECONOMIC COSTS OF MICRO- AND NANO PLASTIC POLLUTION

Environmental Cleanup Costs

The accumulation of micro- and nano plastics comes at a high cost in terms of cleanup and remediation. Among others, some direct costs include:

- **Waste management and collection:** The infrastructure, workforce, and technology required to collect and deliver plastic wastes from different sources, such as landfills, waterways, and beaches.
- **Recycling and Processing:** Cleaning and processing plastic wastes to produce re-usable materials requires investment in sorting and cleaning facilities. However, it is most often the case that the poor value of most recycled plastics and contamination of the waste streams by nature and/or by human activities dampen the economic feasibility of plastic recycling (Rai et al., 2021).
- **Incineration and Disposal:** The incineration or landfilling of plastic wastes involves costs connected with facility construction and operation, monitoring environmental performance.

The indirect costs associated with cleaning up the environment are harder to quantify but no less important:

- **Ecosystem impairment**: Plastic pollution alters ecosystems, leading to losses of habitats, biodiversity, and ecosystem function. These changes in turn can lead to economic losses for the many businesses sectors dependent upon healthy functioning ecosystems, including fisheries, tourism, and agriculture (Rai et al., 2021).
- **Contamination of the water resources:** Micro- and nano plastics penetrate into the water bodies, posing a danger to aquatic life and human health. Restoring polluted water sources and providing safe drinking water supplies is extremely expensive.
- **Pollution in the soil:** Plastic pollution within the soils compromises the soils' quality, yield of crops, and food safety. This might cause economic loss for the farming sector.

Healthcare Costs

The potential human health impacts of micro- and nano plastic pollution are growing in concern. This is what is being investigated, and emerging evidence points to links to various health issues, including:

- **Respiratory problems:** By inhalation, the microplastics lead to respiratory ailments like asthma, bronchitis, allergic reactions (da Costa, et al., 2016).
- **Digestive problems:** On ingestion, MPs cause problems in the gastrointestinal tract, inflammatory bowel disease, and liver damage.

- **Reproductive and Development Effects:** Microplastics are responsible for reproductive disorders, birth defects, and reduced fetal growth.
- **Endocrine Disruptors:** There are some plastics that comprise endocrine disruptors that can cause hormone imbalance and consequent health conditions.

The Economic Costs Associated with Health Impacts Are:

- **Direct Medical Costs**: The diagnosis, treatment, and hospitalization costs that result from plastic-related illnesses
- **Indirect Costs:** Losses in productivity that result from sickness, disability, and premature death.
- **Long-term care costs**: Expenses associated with chronic conditions and disabilities.

The Loss of Biodiversity and Ecosystem Services

Biodiversity and degradation of ecosystem services from plastic pollution entail economic costs incurred. A fit state in ecosystems delivers various benefits as described below:

- **Provisioning services:** food, water, raw materials
- **Regulating services**: climate regulation, flood control, water purification
- **Cultural services:** recreation, tourism, spiritual values
- **Supporting services:** nutrient cycling, soil formation, and pollination.

Their loss can have severe impacts on various sectors, even though the economic value of these services is often underestimated. For instance, plastic-induced declines of fish populations can mean economic hardship for communities dependent on fisheries.

ECONOMIC SECTOR AND INDUSTRY IMPACTS

FIsheries and Aquaculture

Fisheries and aquaculture are the most vulnerable sectors to be hit in cases of micro- and nano plastic pollution.

- **Marine ecosystem disruption:** Plastic is affecting marine ecosystems and, as a result, food chains and fish populations. This further leads to lower volumes of catch at higher costs of fishing, essentially by way of increased fuel use and time spent in pursuit of fish.

- **Economic losses for fishing communities:** Diminished fish stock levels translate to reduced earned incomes for fishermen and thus the level of their livelihoods and local economies.
- **Aquaculture challenges**: Plastic pollution may cause contamination of the sites for aquaculture and result in fish diseases, growth retardation, and mortalities. This results in the economic losses of the businesses in aquaculture (da Costa, et al., 2016).
- **Market impacts**: Perceptions about plastic contamination of seafood may result in reduced consumer confidence and a lower demand, consequently further impacting the profitability of the industry.

Tourism and Recreation

The tourism and recreation industry is both directly and indirectly severely impacted by plastic pollution.

- **Visual pollution:** Plastic litter on beaches, in coastal waters, and natural landscapes represents a serious diminution of the visual amenity of destinations, which impacts on visitor numbers and expenditure (Alprol et al., 2021).
- **Economic losses for tourism-dependent regions**: If tourism slows down, unemployment, reduced tax revenues, and decreased investment in local communities could result.
- **Impact on marine tourism:** Diving, snorkeling, and whale watching are all hampered by plastic pollution, often taking away much of the excitement of seeing these ocean activities (Cárdenas-Alcaide et al., 2022).

Damage to infrastructure: Plastic waste has the ability to block irrigation systems, destroy boats, and also cause problems for other marine infrastructures. This raises the maintenance costs of companies connected with tourism.

Agriculture

There are numerous influences of micro- and nano plastic contamination on farming.

- **Soil contamination:** The plastic pollution in the land may hinder crop growth, affect the fertility of the soil, and increase risks of soil erosion.

- **Economic loss to farmers:** Reduced yield, higher cost of productions, and possible damage to farm machinery can cause huge economic loss to the farmers.
- **Food safety risk:** Plastic contamination in agricultural produce is harmful to human health. It may cause product withdrawal and reputational damage.
- **Effect on livestock:** When microplastics are ingested by livestock, the result can be health complications, poor weight gain, and a decrease in milk production.

Other Affected Industries

- **Packaging industry**: The increasing concern over the reduction in plastic packaging and replacement with sustainable alternative materials is likely to change the face of the industry and bring an economic impact on the businesses.
- **Textile industry:** Issues related to the release of microplastics during washing of synthetic clothing are increasingly being highlighted. There is a pressing need for pressure on the industry for innovative, more sustainable production pathways and materials.
- **Automotive industry:** Plastics find extensive applications in automotive parts. The recycling and disposal of end-of-life vehicles present an environmental issue and an economic challenge.

This section reviews how the impacts will affect specific sectors, underlining that the impacts of micro and nano plastic pollution are very wide-ranging and extend beyond pure environmental impacts.

ECONOMIC BENEFITS OF REDUCING MICRO- AND NANO PLASTIC POLLUTION

Cost-Benefit Analysis

Very relevant economic benefits could be reaped from reducing micro- and nano plastic pollution and could far outweigh financial costs for mitigation measures. It is therefore very relevant to conduct a full cost-benefit analysis for defining the most efficient and effective interventions.

- **Quantifying costs and benefits:** The CBA shall be based on estimation of the economic value of the reduction of plastic pollution in terms of costs saved in environmental cleanup, cost in health issues, and cost in biodiversity loss. It also quantifies the different forms of mitigation, such as research and development, policy, and infrastructure investments.

- **Identifying cost-effective interventions:** Comparing costs and benefits across options allows the prioritization of those with the highest return on investing resources to get the biggest bang for the buck in maximizing the overall economic impact.

Policy and Regulatory Frameworks

Effective policies and regulations are the necessary drivers of plastic pollution reduction. This can generate economic benefits by:

- **Creating new markets:** Incentivizing policies for recycling, reusing, and developing alternative materials can drive innovation and create new business.
- **Generating revenue:** This can be done through imposing taxes or fees on plastic products to establish a fund for environmental programs while at the same time incentivizing the reduction of pollution.
- **Protection of Public Health**: Restrictions on the plastic content in specific products, like personal-care items, will contribute to a decrease in the exposure to microplastics and bring down health costs.
- **Level playing field:** Homogenous regulation can prevent unfair competition and incentivize businesses to undertake sustainable business behavior. Technological Innovations Investment in research and development of new technologies that reduce plastic and enhance recycling can have huge economic dividends:
- **New industries created:** Developing advanced technologies for recycling, plastic alternatives, and waste handling can create new jobs and industries (Cárdenas-Alcaide et al., 2022).
- **Resource efficiency:** Technologies that improve plastic recycling and recovery rates can lead to virgin material savings—reducing costs and environmental impacts.
- **Circular economy models:** Innovative business models that emphasize product re-use, repair, and recycle can generate revenue and create value chains that are sustainable.

Circular Economy and Sustainable Business Models

A shift towards a circular economy—maintaining the use of resources for as long as possible—can generate tremendous economic advantages, which include the following:

- **Waste reduction:** Designing products for durability, reusing, and recyclability can diminish generated waste and related costs to businesses.
- **New revenue streams:** Businesses that provide repair, refurbishment, and recycling open new revenue streams from captured value.
- **Improved brand image:** Adopting sustainable practices enhances brand image and helps to attract environmentally conscious consumers.

International Cooperation Opportunities

Addressing micro- and nano plastic pollution requires a global and unified solution. The international collaboration is instrumental in the sharing of knowledge and resources for developing effective solutions and creates a level playing field amongst nations.

Key Themes for International Cooperation

- **Research and Development:**
  - Joint research efforts about sources, pathways, and impacts of micro and nano plastics.
  - Collaboration on innovative technologies for plastics' reduction, recycling, and remediation.
  - Sharing data and methodologies to speed up scientific advancement.
- **Policy and Regulation Harmonization**:
  - Establishment of common standards and regulations in the production, use, and disposal of plastics.
  - Making the sharing among countries best practices and policies.
  - Dealing with the problems posed by transboundary plastic pollution.
- **Capacity Building**:
  - The provision of technical assistance and training to developing countries to enhance and increase their capacity in handling plastic waste.
  - Infrastructure support in developing waste management and recycling.
  - Public awareness and education on plastic pollution.
- **Financial Cooperation**:

- Establishment of international funds to advance plastic pollution reduction research, development and implementation programs.
- Financial incentives for developing countries to adopt sustainable practices.
- Mobilizing investment in the private sector toward plastic pollution solutions.

- **Global Monitoring and Assessment:**
  - Standardization of monitoring and assessment techniques for micro- and nano plastics.
  - Supporting the establishment of a global database of trends and impacts.

International Platforms for Cooperation

- **United Nations Environment Program:** The leading global environmental authority, which can provide the needed coordination at the international level to combat plastic pollution.
- **Global Environment Facility:** It is a facility that offers financial support to developing countries in efforts toward global environmental problems and therefore is relevant to plastic pollution.
- **Intergovernmental Oceanographic Commission:** Deals with marine pollution and can help combat plastic pollution in oceans.
- **Regional Sea Conventions:** These promote cooperation among coastal states to protect the marine environment and, therefore, from plastic pollution.
- **Public-Private Partnerships:** Collaborations among the government, business, and civil society can more speedily add to reductions.

It is only with international partnerships that countries could work jointly on developing and implementing robust solutions to the global issue of micro and nano plastic pollution. It would require a shared commitment to sustainability, innovation, and equity (Soares et al., 2020).

Case Studies

Case of The Great Pacific Garbage Patch

The Great Pacific Garbage Patch acts as a very souring pointer to just how destructive plastic pollution can be. This enormous accumulation of marine debris, a good amount of which is microplastic, poses severe threats to marine ecosystems and the economies of the world. Contrary to common perception, the patch does not appear as a solid island of trash but is rather dynamic and spread over a really huge area of the ocean. This causes harm to marine life, breaks ecosystems, and entails serious losses in fisheries and tourism. To sort out the situation, a multi-faceted approach is required that incorporates plastic consumption, effective waste management, innovation in technologies for cleaning up, and building international cooperation. The GPGP stands as a stark reminder of the need for urgent global action against plastic pollution and protection of our oceans.

The Case of The Aral Sea:

Over the past several decades, the Aral Sea, formerly the world's fourth-biggest inland water body, has been facing a catastrophic decline due to human activities. The massive water diversion for irrigation started in the 1960s and led to gradual shrinkage of the sea, bringing about an acute ecological and socio-economic crisis. Due to the disappearance of the Aral Sea, regional climate conditions changed dramatically: salinity and desertification increased; and temperature extremes became sharper. This meant the demise of what had been a prosperous fishing industry, consequently plunging local communities into economic misery. Moreover, the seabed—exposed now—has given rise to toxic dust storms that have brought health risks to residents. The tragedy of the Aral Sea serves as a grave warning to the irreversibility of unsustainably managed water and what that means for ecological balance in human wellbeing.

The Case of The Fishing Industry in the Philippines

The Philippines simply cannot afford to be oblivious to the increasing plastic pollution, which is posing serious challenges to its fishing sector. With this as a primary source of livelihood and protein for millions of people, this sector's health means much. Plastic pollution is therefore causing the disruption of marine ecosystems, such as decline in fish populations, entanglement, and damage of fishing gear, which have seen fishers obtain reduced catch and consequently spend more in operations, reducing their income. This has economically hurt fishing-dependent coastal communities as fishers struggle to make ends meet. Moreover, the degradation of marine habitats due to plastic pollution has broader implications for the economy through tourism and other ocean-based industries. The problem of plastic pollution in Philippine waters needs foremost attention if sustainable fisheries and human well-being within coastal communities have to be addressed.

The Case of Tourism in Bali, Indonesia:

Known worldwide for its natural beauty and cultural richness, Bali has forever been a brand destination of tourism. This industry makes the backbones of the economy of this island by providing employment and financial revenue (Iqbal et al., 2020). The effect of tourism in such a large inflow brought with it huge amounts of environmental problems, the plastic waste on the beaches and the marine ecosystems have tarnished the entire face of a beautiful island and frayed off probable visitors. This drop in tourist arrival directly impacts the local economy through businesses, transporters, and hospitality sectors. Currently, due to the vast amounts of waste and lax regulations that enable its accumulation, Bali needs environmental preservation and waste management for long-term tourism sustainability. Sustainable tourism practices will help safeguard Balinese natural assets and economic vitality.

IMPLICATIONS FOR POLICYMAKERS AND BUSINESSES

Policymakers:

This means that policymakers at all levels will mainly have to contend with the economic impediments arising from micro- and nano plastic pollution. It will need effectual policies for stimulating a change in behavior, provoking innovation, and creating a level playing field among businesses (Walker & Fequet, 2023).

- **TCM:** The tight regulatory framework is necessary to bring out all-comprehending legislation that takes in the entire life cycle of plastics, beginning with production and ending with waste. In essence, these are extended producer responsibility, deposit return schemes, and prohibition on harmful plastic products.
- **Economic Instruments**: These include taxes, subsidies, and other market-based mechanisms that are instrumental in encouraging source reduction and pollution prevention (Syberg et al., 2022).
- **Legal and Regulatory Frameworks:** There has to be strengthening of legal and regulatory frameworks at all government levels for the management of plastic wastes across the value chain (Iqbal et al., 2020).
- **Investment in Infrastructure:** One must invest in infrastructure to handle waste management, such as waste collection, sorting, and recycling facilities, to prevent plastic leakage into the environment (Mitrano et al., 2021).
- **Research and Development Support:** These innovative plastic alternatives and modern recycling technologies should be financially supported, and an

enabling environment should be provided by the government for accelerating solutions.

- **International Cooperation**: Cooperation among countries on the issue of cross-border plastic pollution and sharing best practice is key to global impact.

Business

Businesses have a very significant role in the reduction of plastic pollution and mitigation of its economic repercussions. Redirection towards sustainable practices is henceforth not only an environmental imperative but also has substantial economic value attached to it.

- **Product Design and Innovation:** Designing products with less plastic, from recycled materials, made for recyclability, or designed for re-use improves brand image while saving money.
- **Supply Chain Management:** Mapping and streamlining supply chains to decrease plastic packaging and waste could realize cost savings and efficiency in supply chain management.
- **Circular Economy Adoption:** Engaging in circular economy practices by including repair, re-use, and recycling into business models opens new revenue channels (Mofijur, et al., 2021).
- **Investment in Technology:** Adopt state-of-the-art technologies for plastics recycling, waste management, and product design that can help innovate the sector and enhance competitiveness (Jaiswal et al., 2022).
- **Transparency and Accountability:** Engage stakeholders through the dissemination of information about environmental initiatives and performance. This shall enhance trust and lure environmentally conscious consumers. A collaborative effort of the policymakers and businesses is needed to ensure that the environment where economic growth happens is at par with protection of the environment.

Future Research Directions

Develop an economic valuation of the ecosystem services and the human health impacts associated with plastic pollution. This will depend on improved data collection and modeling capabilities through standardized measurement protocols and better analytical tools (Soares et al., 2020). It could also include studying in what ways the business is done in a circular economy and studying some new technologies for the feasibility of arresting plastic pollution. Knowledge gaps in human exposure pathways, health risk assessment, and international cooperation will be important in the development of comprehensive strategies for mitigating the economic consequences of micro- and nano plastic pollution.

CONCLUSIONS AND RECOMMENDATIONS

Conclusions

Economic implications of micro- and nano plastic pollution are such that they are deep and extensive. This chapter has been able to depict the huge costs pertaining to environmental cleanups, health, and biodiversity loss. Along with this, specific sector analysis has shown great economic impacts to fisheries, aquaculture, tourism, agriculture, and other industries.

Clearly, plastic pollution comes at a high economic cost. Cleanup, remediation, and healthcare costs are high, while the degradation of ecosystem services and biodiversity adds to the tangible economic cost. Added to this is the damage to key productive sectors that underscores the issue.

Recommendations

A multi-faceted approach will help in reducing the economic burden of micro- and nano plastic pollution. The recommendations include:

- **Comprehensive Policy Framework:** There is an imperative need for a robust policy framework that governs the overall plastic production, uses, and disposal. This has to include extended producer responsibility, deposit return schemes, or bans on harmful plastic products (Mofijur, et al., 2021).
- **Investing in Research and Development: The** urgency of the situation demands necessary investment in research into innovative technologies for plastic reduction and recycling or remediation.
- **Circular Economy Transition:** Building a model with an emphasis on reuse, repair, and recycling of plastics can significantly reduce plastic wastes while creating economic opportunities.
- **Strong Waste Management Infrastructure:** Waste collection facilities, sorting, and recycling devices are essential to preventing plastic leakage into the environment (Jaiswal et al., 2022).
- **Public Awareness and Education:** Raising general public awareness about the impact of plastic pollution so that through behavioral change and demand, only sustainable products result.

- **International Cooperation: It** may prove necessary to help generate international cooperation around issues such as transboundary plastic pollution, good practice sharing, and ensuring the possibility of actions at the global level.
- **Economic Incentives:** Introducing economic instruments through taxation, subsidies, and market-based instruments gives incentives that deter pollution and reward behavior that is socially more desirable.

Following these recommendations, governments, businesses, and individuals can bring significant action to reduce the economic costs associated with plastic pollution and create opportunities for sustainable growth and development.

REFERENCES

Alprol, A. E., Gaballah, M. S., & Hassaan, M. A. (2021). Micro and Nanoplastics analysis: Focus on their classification, sources, and impacts in marine environment. *Regional Studies in Marine Science*, 42, 101625. DOI: 10.1016/j.rsma.2021.101625

Cárdenas-Alcaide, M. F., Godínez-Alemán, J. A., González-González, R. B., Iqbal, H. M., & Parra-Saldívar, R. (2022). Environmental impact and mitigation of micro (nano) plastics pollution using green catalytic tools and green analytical methods. *Green Analytical Chemistry*, 3, 100031. DOI: 10.1016/j.greeac.2022.100031

da Costa, J. P. (2018). Micro-and nanoplastics in the environment: Research and policymaking. *Current Opinion in Environmental Science & Health*, 1, 12–16. DOI: 10.1016/j.coesh.2017.11.002

da Costa, J. P., Santos, P. S., Duarte, A. C., & Rocha-Santos, T. (2016). (Nano) plastics in the environment–sources, fates and effects. *The Science of the Total Environment*, 566, 15–26. DOI: 10.1016/j.scitotenv.2016.05.041

Iqbal, S., Xu, J., Allen, S. D., Khan, S., Nadir, S., Arif, M. S., & Yasmeen, T. (2020). Unraveling consequences of soil micro-and nano-plastic pollution on soil-plant system: Implications for nitrogen (N) cycling and soil microbial activity. *Chemosphere*, 260, 127578. DOI: 10.1016/j.chemosphere.2020.127578

Jaiswal, K. K., Dutta, S., Banerjee, I., Pohrmen, C. B., Singh, R. K., Das, H. T., Dubey, S., & Kumar, V. (2022). Impact of aquatic microplastics and nanoplastics pollution on ecological systems and sustainable remediation strategies of biodegradation and photodegradation. *The Science of the Total Environment*, 806, 151358. DOI: 10.1016/j.scitotenv.2021.151358

Jiang, B., Kauffman, A. E., Li, L., McFee, W., Cai, B., Weinstein, J., Lead, J. R., Chatterjee, S., Scott, G. I., & Xiao, S. (2020). Health impacts of environmental contamination of micro-and nanoplastics: A review. *Environmental Health and Preventive Medicine*, 25(1), 1–15. DOI: 10.1186/s12199-020-00870-9

Mitrano, D. M., Wick, P., & Nowack, B. (2021). Placing nanoplastics in the context of global plastic pollution. *Nature Nanotechnology*, 16(5), 491–500. DOI: 10.1038/s41565-021-00888-2

Mofijur, M., Ahmed, S. F., Rahman, S. A., Siddiki, S. Y. A., Islam, A. S., Shahabuddin, M., & Show, P. L. (2021). Source, distribution and emerging threat of micro-and nanoplastics to marine organism and human health: Socio-economic impact and management strategies. *Environmental Research*, 195, 110857. DOI: 10.1016/j.envres.2021.110857

Osuna-Laveaga, D. R., Ojeda-Castillo, V., Flores-Payán, V., Gutiérrez-Becerra, A., & Moreno-Medrano, E. D. (2023). Micro-and nanoplastics current status: Legislation, gaps, limitations and socio-economic prospects for future. *Frontiers in Environmental Science*, 11, 1241939. DOI: 10.3389/fenvs.2023.1241939

Rai, P. K., Lee, J., Brown, R. J., & Kim, K. H. (2021). Micro-and nano-plastic pollution: Behavior, microbial ecology, and remediation technologies. *Journal of Cleaner Production*, 291, 125240. DOI: 10.1016/j.jclepro.2020.125240

Sarkar, B., Dissanayake, P. D., Bolan, N. S., Dar, J. Y., Kumar, M., Haque, M. N., Mukhopadhyay, R., Ramanayaka, S., Biswas, J. K., Tsang, D. C. W., Rinklebe, J., & Ok, Y. S. (2022). Challenges and opportunities in sustainable management of microplastics and nanoplastics in the environment. *Environmental Research*, 207, 112179. DOI: 10.1016/j.envres.2021.112179

Seuront, L., Zardi, G. I., Uguen, M., Bouchet, V. M., Delaeter, C., Henry, S., Spilmont, N., & Nicastro, K. R. (2022). A whale of a plastic tale: A plea for interdisciplinary studies to tackle micro-and nanoplastic pollution in the marine realm. *The Science of the Total Environment*, 846, 157187. DOI: 10.1016/j.scitotenv.2022.157187

Soares, J., Miguel, I., Venâncio, C., Lopes, I., & Oliveira, M. (2020). Perspectives on Micro (Nano) Plastics in the marine environment: Biological and societal considerations. *Water (Basel)*, 12(11), 3208. DOI: 10.3390/w12113208

Syberg, K., Nielsen, M. B., Oturai, N. B., Clausen, L. P. W., Ramos, T. M., & Hansen, S. F. (2022). Circular economy and reduction of micro (nano) plastics contamination. *Journal of Hazardous Materials Advances*, 5, 100044. DOI: 10.1016/j.hazadv.2022.100044

Walker, T. R., & Fequet, L. (2023). Current trends of unsustainable plastic production and micro (nano) plastic pollution. *Trends in Analytical Chemistry*, 160, 116984. DOI: 10.1016/j.trac.2023.116984

Chapter 10

Scaling Legal Framework for Plastic Pollution and Advancing Cutting Edge Water Governance:

Reducing and Eliminating Marine Pollution in Alignment With SDG 14 (Life Below Water)

Bhupinder Singh
https://orcid.org/0009-0006-4779-2553
Sharda University, India

Christian Kaunert
https://orcid.org/0000-0002-4493-2235
Dublin City University, Ireland

Gursahib Singh
K.R. Mangalam World School, Greater Noida, India

ABSTRACT

The growing danger of plastic pollution and its catastrophic effects on marine ecosystems highlight the pressing need for creative approaches to water governance as well as a thorough legal framework. Marine pollution comprises a mixture of chemicals and debris, primarily originating from terrestrial sources and finding its way into the ocean through runoff or wind dispersion. This contamination leads to adverse effects on the environment, the well-being of various organisms and the global economic structures. The significant factors such as- littering, storm winds

DOI: 10.4018/979-8-3693-9163-1.ch010

and inadequate waste management collectively contribute to the buildup of this refuse. This chapter explores the complex interactions between state-of-the-art governance models and legal tools, with an emphasis on protecting marine life and reducing plastic pollution. In line with the objectives of Sustainable Development Goal 14 (SDG 14) Life below Water.

1. INTRODUCTION AND BACKGROUND

The persistence of plastic waste poses a significant environmental challenge, as these items can take several centuries to break down (Ge et al., 2024). Also, the typical forms of marine debris encompass a range of plastic items such as shopping bags, beverage bottles, cigarette butts, bottle caps, food wrappers and fishing gear (Shukla et al., 2024). Marine debris refers to all human-made objects, predominantly plastic, that find their way into the ocean (Singh & Kaunert, 2024). The long-term environmental impact of plastics is made worse by their tenacity as they take hundreds of years to disintegrate (Wang et al., 2023). In order to protect the health and biodiversity of our world, addressing these concerns demands rapid global action (Singh et al., 2024). This action includes strengthening international collaboration, implementing methods to limit plastic consumption, and improving waste management (ma et al., 2023). When plastics enter the marine environment, the effects of marine litter extend beyond ecological degradation and environmental pollution (Gaur et al., 2023) (Meegoda & Hettiarachchi, 2023). They also have an impact on human health, leisure, and livelihoods (Singh & Kaunert, 2024).

The social and economic effects of policies must be considered while developing them, particularly with regard to vulnerable populations (Mojiri et al., 2024). Selecting and developing cost-effective methods that target high-impact solutions that can be adopted nationally can be made easier with an awareness of the elements that contribute to marine plastic pollution such as behavior and access to services (Singh, 2024) (Gong et al., 2023). This emphasizes the necessity of carefully weighing the facts behind a variety of potential policy interventions before deciding on the best course of action (Geremia et al., 2023) (Pittura et al., 2023). Fighting plastic pollution is essential, especially for marine habitats because it protects biodiversity and ecosystems from permanent harm (Singh et al., 2024). With stopping the spread of microplastics via the food chain, plastic contamination may be stopped, protecting human health (Singh & Kaunert, 2024)

Figure 1. Landscapes of introduction split sections (Original)

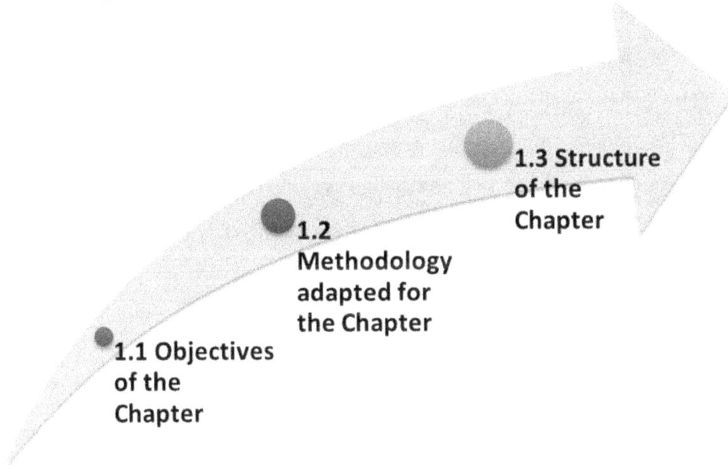

1.1 Objectives of the Chapter

This chapter has the following objectives to-

- examine and compile the national and international legislative frameworks that already exist to address the issue of plastic pollution in marine areas (marfella et al., 2024).
- observe the weaknesses and difficulties in the present laws and governance frameworks that could prevent marine pollution from being effectively reduced or eliminated (Valiyaveettil Salimkumar et al., 2024).
- provide creative legal and regulatory solutions to the problem of marine pollution, with an emphasis on lowering plastic waste and enhancing water governance in line with SDG 14 (Singh, 2024).
- determine how current and prospective legal frameworks affect marine ecosystems and how they safeguard life below the surface of the ocean (Fraissinet et al., 2024).
- provide advice on practical ways to improve water governance and lessen marine pollution to states, international organizations, and policymakers, with a focus on teamwork to meet SDG 14 objectives (Yu et al., 2024).

Figure 2. Objectives of the chapter (Original)

1.2 Methodology Adapted for the Chapter

The extensive analysis of international accords, treaties and Indian Legal Provisions, this chapter aims to evaluate the effectiveness of current legal frameworks in tackling the ubiquitous problem of marine plastic debris. It investigates methods that can be scaled up to mitigate and eventually eradicate marine pollution by integrates policy assessments, legal framework, and water governance tactics via an interdisciplinary approach to suggest practical for addressing this worldwide water pollution issue. This chapter employs exploratory and doctrinal methodologies, incorporating a wide range of literature sources such as books, articles, book chapters, commentaries, dictionaries, encyclopedias, journal articles, monographs, newspaper pieces, and websites. To evaluate the environmental impacts and resource use of plastic products throughout their entire life cycle, the study considers life cycle assessment and circularity as crucial concepts. The literature review points out the importance of innovative approaches for reducing, reusing, recycling, and recovering plastics, along with exploring eco-friendly alternatives. It suggests that legal frameworks should be established to empower and educate individuals and communities to reduce plastic waste and adopt alternatives. To mitigate the harmful effects of plastic pollution on ecosystems, it's essential to prioritize collaboration on plastic waste management, enact regulatory reforms, and increase public awareness.

1.3 Structure of the Chapter

This chapter deeply dives into the Scaling Legal Framework for Plastic Pollution and Advancing Cutting Edge Water Governance: Reducing and Eliminating Marine Pollution in Alignment with SDG 14 (Life below Water). Section 2 expresses the Plastic Pollution and Advancing Cutting Edge Water Governance. Section 3 elaborates the International Legal Frameworks on Marine and Plastics Pollution. Section 4 lays down the Biodiversity and Species-Oriented Legal Instruments. Section 5 specifies the Governance of Marine Plastics. Section 6 highlights the Case Studies: Countries have Proven Success in Embark upon Plastic and Marine Pollution. Finally, Section 7 Conclude the Chapter with Future Scope.

Figure 3. Flow of this chapter (Original)

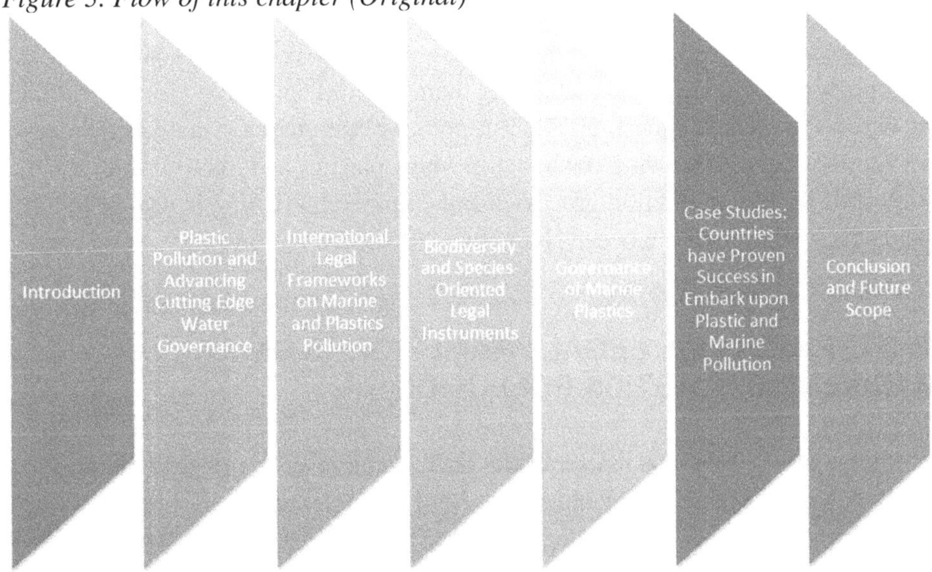

2. PLASTIC POLLUTION AND ADVANCING CUTTING EDGE WATER GOVERNANCE

Plastic, a synthetic organic polymer derived from petroleum, possesses versatile properties suitable for various applications such as- packaging, construction, household items, sports equipment, vehicles, electronics, and agriculture (Singh, 2024) (Liu et al., 2024). Annual plastic production exceeds millions of tons with half of it dedicated to the manufacturing of disposable items like shopping bags, cups, and straws (Lv et al., 2024) (Singh & Kaunert, 2024). Inadequate disposal of plastic waste can have detrimental effects on the environment and biodiversity (Chen et al., 2024) (Singh, 2024). Land-based activities such as urban and stormwater runoff, sewage overflows, littering, incorrect waste disposal, industrial operations, tire abrasion, construction and illegal dumping are the main causes of plastic debris that ends up in the ocean (Varshney, 2024). Since they have been around for a little more than a century, plastics made from fossil fuels have had a revolutionary effect on our way of life. They made it easier to develop life-saving tools and put cutting-edge technology into action that changed the path of history (Sandil & Zaray, 2024).

The broad adoption of a disposable attitude exposed the negative aspects of plastic, including the serious harm it presents to our environment and way of life, despite the material's many benefits (Singh, 2023). Littering is a major contributor to the plastic waste seen in the waters. This happens when buy or use throwaway plastic products including- food wrappers, bags, bottles, razors and so on and when don't properly dispose of them, allowing them to wind up in rivers and eventually the ocean (Singh & Kaunert, 2024). But not all plastic garbage seen in the ocean is a result of people throwing trash into it. Many plastics and microplastics are the product of flawed production processes and commercial fishing is responsible for the plastic pollution in the ocean (Naz et al., 2024).

3. INTERNATIONAL LEGAL FRAMEWORKS ON MARINE AND PLASTICS POLLUTION

Marine plastic litter and microplastics are causing severe damage to marine ecosystems. Marine species are facing risks from entanglement, ingestion of plastics and their chemicals, and habitat degradation (Mustapha et al., 2024). Microplastics, smaller than 5 mm in size, pose further threats as they can enter the food chain and impact human health. Plastics' non-biodegradable nature and the ability for plastic waste to travel vast distances contribute to this problem. The Great Pacific Garbage Patch, a massive floating collection of plastic waste far from any coastline, spans over 1.6 million square kilometers and contains roughly 80,000 tons of plastic (Singh, 2024). Because they are inexpensive, safe, long-lasting, and easily moldable into a wide variety of shapes, plastics have become more and more common in a wide range of applications. But one of the most pressing environmental issues of day is the rise in single-use and throwaway plastics, which has caused a sharp rise in the amount of plastic garbage that ends up in the seas. More than eight million tons of plastic debris are discarded into the oceans each year (Arif et al., 2024).

A growing amount of plastic is being consumed worldwide; according to some studies, the oceans may contain more plastic by weight by 2050 than fish (Singh, 2023). Millions of tons of plastic debris still end up in the seas every year, despite the several treaties and laws intended to combat marine plastic pollution. This raises questions about how successful these laws are (Sun et al., 2024). The ongoing pollution issue raises concerns about the adequacy of the current regulations and whether their inadequate execution is to blame. Legislation that is both explicit and enforceable and vigorously enforced are essential to maintaining the rule of law, which is a fundamental aspect of ocean governance (Singh, 2024). This case brings to light a serious threat to the rule of law. This chapter looks at how to combat marine plastics and microplastic contamination by bolstering the rule of law. It examines the present legal framework that governs marine plastics and microplastics, noting any gaps in the law and evaluating potential solutions (Deng et al., 2024).

Figure 4. International legal frameworks on marine and plastics pollution (Original)

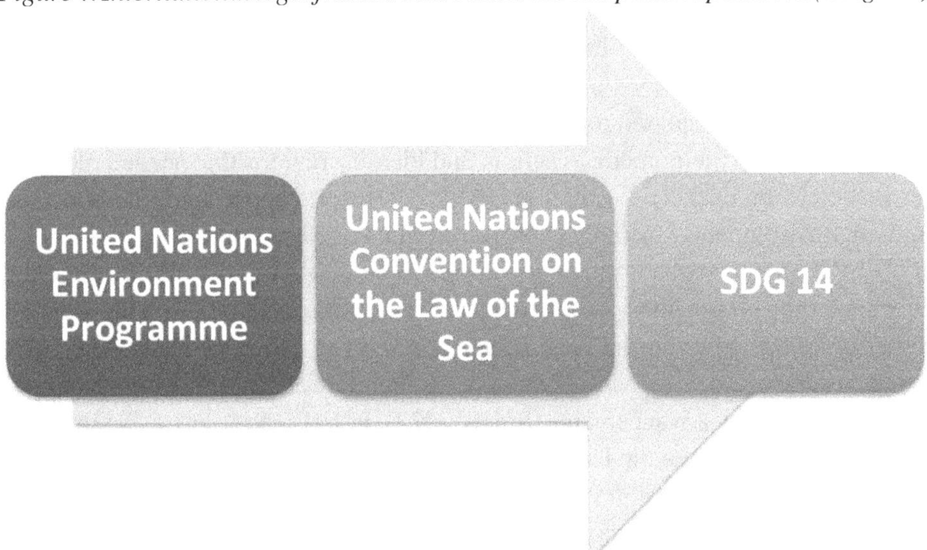

3.1 UNEP (United Nations Environment Program): Recognizes Urgency of Tackling Plastic Pollution

The extensive and swiftly escalating amount of marine litter and plastic pollution poses a significant threat to the health of all the world's oceans and seas. Plastics, including microplastics have become omnipresent, serving as a prominent symbol of the Anthropocene, the current geological epoch and are gradually becoming integrated into the Earth's fossil record. Plastics have even led to the emergence of a novel marine microbial habitat known as the "plastisphere." Through a number of significant resolutions, the UN Environment Assembly has continuously tackled the growing problem of marine litter, which includes plastics and microplastics (Kharraz et al., 2024). A number of resolutions have been adopted on the subject of marine plastic debris and microplastics, including UNEP/EA.1/Res.6 (2014), UNEP/EA.2/Res.11 (2016), UNEP/EA.3/Res.7 (2017), UNEP/EA.4/Res.6 (2019) and UNEP/EA.4/Res.9 (2019) on Addressing Single-use Plastic Products Pollution (Singh, 2024).

"Marine Plastic Debris and Microplastics Global Lessons and Research to Inspire Action and Guide Policy Change" is a study that the United Nations Environment Program (UNEP) produced in 2016 in response to UNEP/EA.1/Res.6 (Singh & Kaunert, 2024). The study's main objectives were to pinpoint the main causes of marine plastic waste and microplastics, offer precautionary ways to stop them from building up, advocate immediate action, and identify regions that needed more investigation (Singh, 2024). Resolution UNEP/EA.4/Res.6 paragraph 2 at the United Nations Environment Assembly's fourth meeting assigned the Executive Director of UNEP the mission of advancing scientific and technical understanding of marine litter, including plastic litter and microplastics (Rajendran et al., 2024). This required gathering information on the origins, spread, and risks associated with trash, as well as its detrimental impacts on ecosystems, human health, and environmentally friendly technology advancements (Singh, 2023). In light of the extensive research that has been done since the UNEP report from 2016, this evaluation offers an update that builds on the findings of that study and incorporates new discoveries. The purpose is to provide background information for the United Nations Environment Assembly's fifth session (UNEA-5.2) (Ojha & Skariyachan, 2024). A Scientific Advisory Committee on Marine Litter and Microplastics (SAC) was created by UNEP's Executive Director in 2019 to offer advice and recommendations for this evaluation. SAC members provided expertise through scientific knowledge, statistics, experiences, opinions and evaluations. They were nominated by United Nations Member States, specialized agencies and accredited key groups and stakeholders (Singh, 2023). The SAC suggested that UNEP create the assessment using data that has been published, with an emphasis on the causes and sources of marine litter, the kinds and amounts of plastics in waste streams that enter the ocean, the pathways within various ecosystems, the hazards, and impacts, monitoring initiatives and technological and solution examples (Duncan et al., 2024).

3.2 UNCLOS (United Nations Convention on the Law of the Sea): Provides Legal Framework for Ocean Conservation and Sustainable Use

Parties are required under the United Nations Convention on the Law of the Sea to take all reasonable steps in accordance with the Convention to avoid, minimize and regulate pollution of the marine environment from all sources (Sharma & Singh, 2022). The Convention covers pollution from a range of sources, such as air pollution, vessel-related pollution, dumping, land-based activities, and seabed operations (Lazar et al., 2024). It expressly forbids dumping without the coastal state's previous consent in the exclusive economic zone, territorial sea, or on the continental shelf (Liu et al., 2024). States must pass the necessary laws and regulations and, if necessary, coordinate regional policy. According to the Convention, pollution is the introduction of materials or energy, either directly or indirectly, into the marine environment that has negative consequences on human health, the environment, or marine life and resources (Bao et al., 2024). It can also impede legitimate marine activities like fishing and interfere with other uses of the sea. It also includes the deterioration of the quality of the saltwater and the decrease in facilities. As a result, these national policies seek to safeguard the natural advantages that the seas offer, while promoting economic growth (Singh, 2022).

3.3 SDG 14: Aims to Prevent and Significantly Reduce Marine Pollution

The preservation and sustainable use of seas, oceans, and marine resources are the main objectives of Goal 14 (Adler at al., 2024). Seas and oceans are essential to human survival and the continuation of life on Earth. Three-quarters of the Earth's surface is made up of the ocean, which holds 97% of the planet's water and, by volume, makes up 99 percent of all living space (Singh, 2024). When it comes to supplying vital natural resources like food, medicine, biofuels, and other goods, oceans are crucial. In addition to helping to remove and breakdown garbage and pollutants, coastal ecosystems serve as storm-damage buffers. Oceans are also the main sink of carbon dioxide on Earth (Winiarska et al., 2024).

Figure 5. Preserve the blue and protect Earth (UN SDG Website)

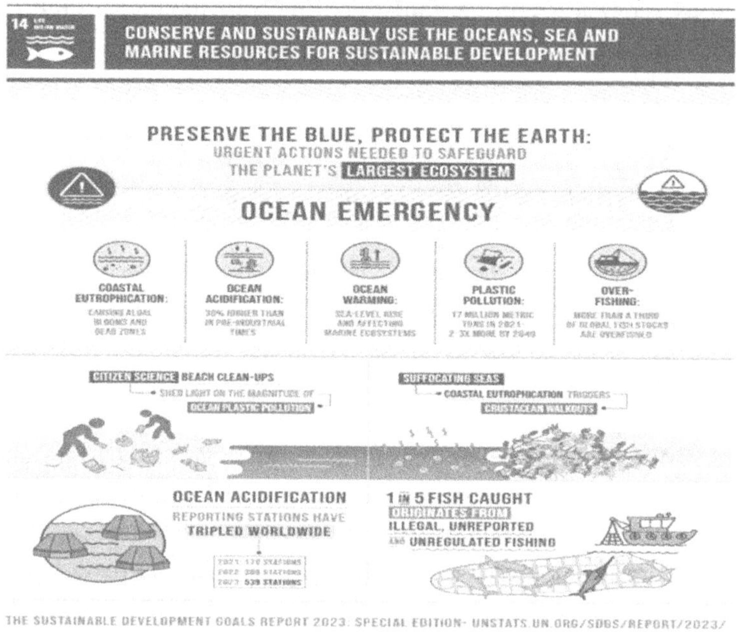

But the amount of marine pollution that is already damaging the ocean is worrying; in 2021, over 17 million metric tons were present, and by 2040, that amount is expected to treble or triple. The most harmful kind of ocean contamination is definitely plastic pollution (Singh et al., 2024). The average pH of the ocean now is 8.1, which is around 30% more acidic than it was before industrialization (Chiang et al., 2024). The food security is impacted by ocean acidification, which endangers marine life, upsets the food chain, and erodes the vital functions the ocean performs. The sustainable future depends on ensuring the responsible management of this vital global resource (Hrovat et al., 2024). This calls for bolstering financing for ocean science, stepping up conservation initiatives, and moving quickly to stop climate change and save the biggest ecosystem on Earth. The urgency with which this huge yet delicate resource has to be protected is not being met by current conservation efforts (Gilani et al., 2024).

4. BIODIVERSITY AND SPECIES-ORIENTED LEGAL INSTRUMENTS

The Convention on Biological Diversity requires consistent implementation with the law of the sea concerning the marine environment and 196 countries have ratified the Convention on Biological Diversity (CBD) which provides the international legal framework for "the preservation of biological diversity, the responsible utilization of its components and the just and equitable sharing of benefits arising from the exploitation of genetic resources." Its main objective is to encourage initiatives that support a sustainable future (Singh & Kaunert, 2024). The addressing of biodiversity conservation as a shared global concern, the Convention on Biological Diversity encompasses all aspects of biodiversity, spanning ecosystems, species, and genetic resources. It also extends its coverage to biotechnology, incorporating the Cartagena Protocol on Biosafety (Aylaz & Konar, 2024). The encompassing of a wide range of domains directly or indirectly linked to biodiversity and its role in development, the convention spans disciplines such as science, politics, education, agriculture, business, culture and more (Xu et al., 2024).

4.1 CBD COP Decision XIII/10: Provides Voluntary Practical Guidance on Mitigating Marine Debris Impacts on Biodiversity and Habitats

Any durable, produced or processed solid substance that is lost, disposed of or abandoned in the marine and coastal environment is commonly referred to as marine trash (Lakshmipraba, 2024). This includes items carried by wind, rivers, drainage or sewage systems or other means from the land into the marine environment. Marine debris comes from a variety of sources, both on land and in the ocean. Marine debris has a negative impact on marine life, damages the economy and endangers people's health and safety (Singh et al., 2024). A wealth of evidence suggests that becoming entangled in or consuming marine debris can have a detrimental effect on the physical health of marine species, possibly culminating in their death and consequent effects on populations and ecosystems (Li et al., 2024).

Plastic consumption is especially dangerous since it might be a conduit for dangerous substances to enter the food chain (Li et al., 2024). It is acknowledged that marine debris may damage, alter or deteriorate ecosystems as for example by suffocating them and might serve as a possible vector for the introduction of alien species. Adverse consequences may comprise changes in an individual's biological and ecological functioning, physical harm or death (Okeke et al., 2024). It can be difficult to determine how eating marine debris will affect a specific creature, and our knowledge of the entire effects of ingestion is still lacking. High rates of debris entanglement or ingestion may indicate that a species is susceptible to population-level impacts (Singh et al., 2024).

This might have negative consequences for tiny populations, particularly for endangered species or species that are subject to numerous stressors. An estimate of the loss of ecosystem services related to this stressor should be included in the evaluation of the ecosystem-level consequences of marine debris (Brander et al., 2024). Microplastics are persistent contaminants present in all maritime environments and are expected to grow in abundance. Plastics and hazardous chemicals may spread and accumulate more easily as a result of the trophic transmission of microplastics across benthic and pelagic food webs. There is proof that chemical additions from swallowed plastics may enter tissues, including human tissue (Li et al., 2024). The physical consequences of consuming macro- and mesoplastics, including internal abrasion, obstruction and damage are also a source of concern. These effects might open a channel for marine species to absorb toxic compounds such- additives found in plastic items (Singh, 2022). The following broad actions can be used as a means of preventing and mitigating the effects of marine debris on marine and coastal biodiversity and habitats as-

(a) In upstream regions as well as the maritime and coastal environment, emphasis should be made on preventing the discarding, dumping, loss, or abandonment of any durable, produced or processed solid item (Yi et al., 2024).
(b) When appropriate, efforts to avoid and mitigate the severe negative consequences of marine trash should make use of the platforms and collaborative tools already in place (Kim et al., 2024). With using this approach, collaborations will be strengthened, and progress made in several venues, including the Global Partnership on Marine Litter, the Global Program of Action for the Protection of the Marine Environment from Land-based Activities, and regional seas conventions and action plans (Singh, 2024).
(c) Efforts to avoid and mitigate the effects of marine debris can be supported by utilizing a wide range of existing instruments and policy responses, including as market-based mechanisms, financial incentives, and public-private partnerships (Rondon et al., 2024).

5. GOVERNANCE OF MARINE PLASTICS

Maximizing normative outcomes and equitable rights and liabilities distribution are the main goals of institutional design, and these serve as standards for assessment (Zheng et al., 2024). It evaluates societal and public issues, suggests approaches, and plans for resolving them, and makes recommendations for legislation (Singh, 2024). A system's design in this setting has its own normative aim, theoretical foundation, and creative space. An assessment of the present system should be combined with knowledge of the generative logic of the real world in order to set reasonable limits for its extension or future usage (Chen et al., 2024). The challenge of "marine jurisdiction" to the conventional notion of "freedom of the sea" is the reason behind the move toward a worldwide framework for marine governance (Christudoss et al., 2024).

State needs for resource management are linked to the expansion of the maritime economy and its associated characteristic of sovereignty. The encouraging of global human interests necessitates protecting the maritime environment, underscoring the necessity for global framework coherence and cooperation (Afrose et al., 2024). Integrated plastics policies are now created at the national and regional levels; the history of research on plastic pollution shows a shift from national to international policy frameworks (Singh, 2022). The desire for an international legally binding instrument on plastic pollution has led to a return to the paradigm of regulation of marine plastic pollution through integration into global governance structures. This transition denotes a change in governance from decentralized to centralized and in methods from soft law-based to hard law-based (Singh, 2020).

Figure 6. Impact assessment of micro and nano plastic (Original)

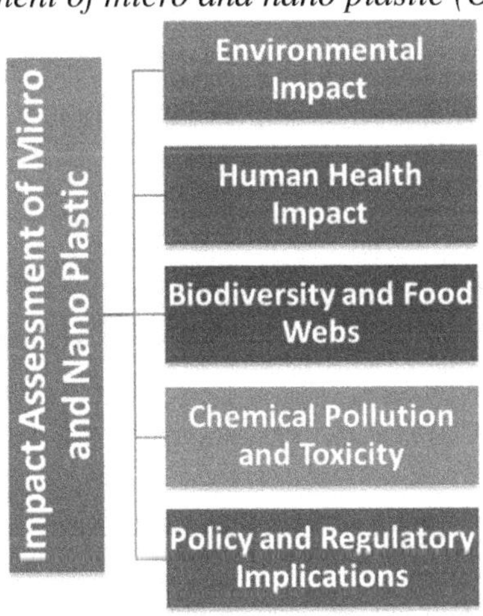

6. CASE STUDIES: COUNTRIES HAVE PROVEN SUCCESS IN EMBARK UPON PLASTIC AND MARINE POLLUTION

Oceans, seas, and marine resources are widely acknowledged to be essential to the survival of life on Earth. The fact that SDG 14 is part of the UN 2030 Agenda for Sustainable Development shows how committed the international community is to be protecting and maintaining these essential components (Zheng & Wang, 2024). "Conserving and sustainably using the oceans, seas, and marine resources" is the specific focus of this aim. In addition to encouraging the recycling of plastic debris, the UN Environment Program Clean Seas Campaign has taken proactive steps by outlawing single-use plastic items and packaging (Rehman et al., 2024).

Figure 7. Case studies (Original)

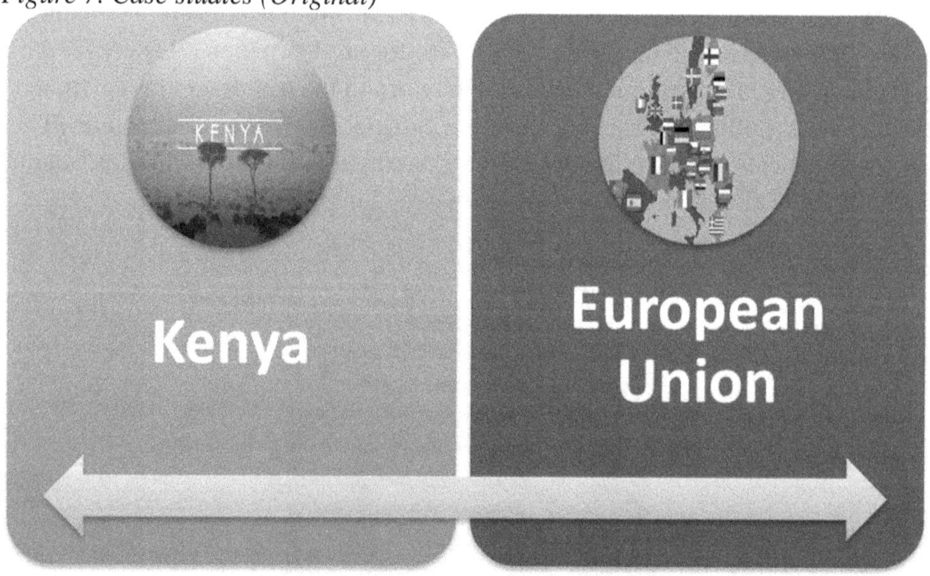

6.1 Kenya's Ban on Single-use Plastic Bags

Kenya passed a law in 2017 outlawing the use of single-use plastic bags, which are frequently given out by stores to transport goods. The negative consequences of plastic pollution served as the impetus for this rule (Tao et al., 2024). While the law was widely praised, it did not apply to bags that were used to package food or other products. The environmental effects of discarded plastic bags are substantial, which is why the United Nations and other countries have praised this innovative rule. They wind up in water bodies, cause obstructions in drainage systems, and emit toxic waste when burned in landfills (Singh, 2023). Human and fish bloodstreams have been shown to have microplastics that eventually break down from these plastic bags. Although it is still unclear how exactly microplastics affect human and animal health, research has shown some concerning possibilities (Paing et al., 2024).

6.2 European Union's Single-use Plastics Directive

Single-use plastic goods (SUPs) are disposable items that are usually used just once before being disposed of which has serious negative effects on the environment and human health [140]. SUPs are more likely to end up in the water than reusable alternatives (Yang et al., 2024). Together with fishing gear, the top 10 single-use plastic items found on beaches in the European Union account for 70% of all marine trash. With laws aimed at reducing the amount and environmental effect of particular plastic items, the European Union is dedicated to spearheading the worldwide campaign against marine litter and plastic pollution (Kutralam-Muniasamy et al., 2024).

7. CONCLUSION AND FUTURE SCOPE

The transitioning to more environmentally friendly consumption habits need all-encompassing solutions that go beyond waste management and cover every stage of a plastic product's lifespan, from design to infrastructure and domestic usage. In order to give manufacturers, consumers and legislators the proof they need to adopt technical, behavioral, and regulatory solutions to address marine plastic pollution, more funding must be allocated to research and innovation. The causes of marine plastic pollution and plastic leakage may be located, measured, and addressed using several approaches, some of which are offered by the IUCN. These tools have the potential to make a major contribution to the overall efforts in this area. The governments have a duty to protect the environment, respect citizens' rights, and address public concern over the consequences of unsustainable waste generation and insufficient waste management. These obligations drive the development of effective laws and policies for the prevention and management of marine litter. In addition, policies and laws are put in place to safeguard the high seas marine ecosystems and regions that are under the sovereign authority of other coastal nations, therefore preserving the global commonality of oceans. In order to achieve a more sustainable future and significantly lower pollution levels, society must act continuously now and continue to do so beyond 2030 in order to accomplish the Sustainable Development Goals (SDG) objectives. Prioritizing the avoidance of pollutants at their sources by policies including laws, subsidies, outreach, education, and the adoption of substitute technology is essential to reaching this aim. Reducing pollution sources is the first step toward moving toward sustainability. If pollution sources are not addressed, attempts to ameliorate the harm to the ocean and its creatures will mostly consist of remediation in the present and future.

REFERENCES

Adler, M. Y., Issoual, I., Rückert, M., Deloch, L., Meier, C., Tschernig, T., Alexiou, C., Pfister, F., Ramsperger, A. F. R. M., Laforsch, C., Gaipl, U. S., Jüngert, K., & Paulsen, F. (2024). Effect of micro-and nanoplastic particles on human macrophages. *Journal of Hazardous Materials*, 471, 134253. DOI: 10.1016/j.jhazmat.2024.134253

Afrose, S., Tran, T. K. A., O'Connor, W., Pannerselvan, L., Carbery, M., Fielder, S., Subhaschandrabose, S., & Palanisami, T. (2024). Organ-specific distribution and size-dependent toxicity of polystyrene nanoplastics in Australian bass (Macquaria novemaculeata). *Environmental Pollution*, 341, 122996. DOI: 10.1016/j.envpol.2023.122996

Ali, W., Buriro, R. S., Gandahi, J. A., Chen, Y., Aabdin, Z., Bhutto, S., Sun, J., Zhu, J., Liu, Z., & Zou, H. (2024). A critical review on male-female reproductive and developmental toxicity induced by micro-plastics and nano-plastics through different signaling pathways. *Chemico-Biological Interactions*, 394, 110976. DOI: 10.1016/j.cbi.2024.110976

Arif, Y., Mir, A. R., Zieliński, P., Hayat, S., & Bajguz, A. (2024). Microplastics and nanoplastics: Source, behavior, remediation, and multi-level environmental impact. *Journal of Environmental Management*, 356, 120618. DOI: 10.1016/j.jenvman.2024.120618

Aylaz, G., & Konar, N. M. (2024). Current publication activity on nanoplastics & microplastics: A bibliometric analysis. *Sustainable Water Resources Management*, 10(2), 95. DOI: 10.1007/s40899-024-01072-9

Bao, S., Yi, J., Xian, B., Rao, C., Xiang, D., Tang, W., & Fang, T. (2024). Global analysis of the adverse effects of micro-and nanoplastics on intestinal health and microbiota of fish. *Journal of Hazardous Materials*, 470, 134157. DOI: 10.1016/j.jhazmat.2024.134157

Brander, S. M., König, A., Almroth, B. C., & Hampton, L. T. (2024). The potential for toxicit y to fishes from micro-and nanoplastics, and their additives. *Toxicology of Fishes. CRC Press*, 362-391.

Chen, T., Jiang, H., He, Y., Shen, Y., Huang, Z., Gu, Y., Wei, Q., Zhao, J., & Chen, X. (2024). Nanoplastics and chrysene pollution: Potential new triggers for nonalcoholic fatty liver disease and hepatitis, insights from juvenile Siniperca chuatsi. *The Science of the Total Environment*, 922, 171125. DOI: 10.1016/j.scitotenv.2024.171125

Chen, X., Sun, Y., Feng, Z., Mu, C., & Zhang, T. (2024). Effects of micro-and nanoplastics on the physiology and metabolism of Portunus trituberculatus. *Clean (Weinheim)*, 52(3), 2300175. DOI: 10.1002/clen.202300175

Chiang, C. C., Yeh, H., Shiu, R. F., Chin, W. C., & Yen, T. H. (2024). Impact of microplastics and nanoplastics on liver health: Current understanding and future research directions. *World Journal of Gastroenterology*, 30(9), 1011–1017. DOI: 10.3748/wjg.v30.i9.1011

Christudoss, A. C., Chandrasekaran, N., & Mukherjee, A. (2024). Polystyrene nanoplastics alter the ecotoxicological effects of diclofenac on freshwater microalgae Scenedesmus obliquus. *Environmental Science. Processes & Impacts*, 26(1), 56–70. DOI: 10.1039/D3EM00341H

Deng, L., Cheung, S., Liu, J., Chen, J., Chen, F., Zhang, X., & Liu, H. (2024). Nanoplastics impair growth and nitrogen fixation of marine nitrogen-fixing cyanobacteria. *Environmental Pollution*, 350, 123960. DOI: 10.1016/j.envpol.2024.123960

Duncan, T. V., Khan, S. A., Patri, A. K., & Wiggins, S. (2024). Regulatory Science Perspective on the Analysis of Microplastics and Nanoplastics in Human Food. *Analytical Chemistry*, 96(11), 4343–4358. DOI: 10.1021/acs.analchem.3c05408

Fraissinet, S., De Benedetto, G. E., Malitesta, C., Holzinger, R., & Materić, D. (2024). Microplastics and nanoplastics size distribution in farmed mussel tissues. *Communications Earth & Environment*, 5(1), 128. DOI: 10.1038/s43247-024-01300-2

Gaur, V. K., Sirohi, R., Bhat, M. I., Gautam, K., Sharma, P., Srivastava, J. K., & Pandey, A. (2023). A review on the effect of micro-and nano-plastics pollution on the emergence of antimicrobial resistance. *Chemosphere*, 311, 136877. DOI: 10.1016/j.chemosphere.2022.136877

Ge, J., Jin, P., Xie, S., Beardall, J., Feng, Y., Guo, C., Ma, Z., & Gao, G. (2024). Micro-and nanoplastics interact with conventional pollutants on microalgae: Synthesis through meta-analysis. *Environmental Pollution*, 342, 123127. DOI: 10.1016/j.envpol.2023.123127

Geremia, E., Muscari Tomajoli, M. T., Murano, C., Petito, A., & Fasciolo, G. (2023). The impact of micro-and nanoplastics on aquatic organisms: Mechanisms of oxidative stress and implications for human health—a review. *Environments (Basel, Switzerland)*, 10(9), 161. DOI: 10.3390/environments10090161

Gilani, S. J., Alshawwa, S. Z., Asif, M., Fatima, K., & Khan, N. A. Microplastics and Nanoplastics in Food. In Handbook of Seafood and Seafood Products Analysis (pp. 915–926). CRC Press.

Gong, X., Shi, G., Zou, D., Wu, Z., Qin, P., Yang, Y., Hu, X., Zhou, L., & Zhou, Y. (2023). Micro-and nano-plastics pollution and its potential remediation pathway by phytoremediation. *Planta*, 257(2), 35. DOI: 10.1007/s00425-023-04069-4

Hrovat, B., Uurasjärvi, E., Viitala, M., Del Pino, A. F., Mänttäri, M., Papamatthaiakis, N., Haapala, A., Peiponen, K.-E., Roussey, M., & Koistinen, A. (2024). Preparation of synthetic micro-and nano plastics for method validation studies. *The Science of the Total Environment*, 925, 171821. DOI: 10.1016/j.scitotenv.2024.171821

Kharraz, J. A., Jia, M., Farid, M. U., Khanzada, N. K., Hilal, N., Hasan, S. W., & An, A. K. (2024). Determination of microplastic pollution in marine ecosystems and its effective removal using an advanced nanobubble flotation technique. *Journal of Water Process Engineering*, 57, 104637. DOI: 10.1016/j.jwpe.2023.104637

Kim, M. S., Lee, Y. H., Lee, Y., Jeong, H., Wang, M., Wang, D. Z., & Lee, J. S. (2024). Multigenerational effects of elevated temperature on host-microbiota interactions in the marine water flea Diaphanosoma celebensis exposed to micro-and nanoplastics. *Journal of Hazardous Materials*, 465, 132877. DOI: 10.1016/j.jhazmat.2023.132877

Kutralam-Muniasamy, G., Shruti, V. C., Pérez-Guevara, F., & Flores, J. A. (2024). The emerging field of inductively coupled plasma mass spectrometry for (micro) nanoplastic analysis:"The 3As" analysis, advances, and applications. *Trends in Analytical Chemistry*, 174, 117673. DOI: 10.1016/j.trac.2024.117673

Lakshmipraba, J. (2024). Photo/Hydro/Thermal Degradation of Macro-plastics into Micro-and Nano-forms. In Microplastics and Pollutants: Interactions, Degradations and Mechanisms (pp. 51–74). Springer Nature Switzerland. doi:DOI: 10.1007/978-3-031-54565-8_3

Lazăr, N. N., Călmuc, M., Milea, . A., Georgescu, P. L., & Iticescu, C. (2024). Micro and nano plastics in fruits and vegetables: A review. *Heliyon*, 10(6), e28291. DOI: 10.1016/j.heliyon.2024.e28291

Li, F., Huang, D., Wang, G., Cheng, M., Chen, H., Zhou, W., Xiao, R., Li, R., Du, L., & Xu, W. (2024). Microplastics/nanoplastics in porous media: Key factors controlling their transport and retention behaviors. *The Science of the Total Environment*, 926, 171658. DOI: 10.1016/j.scitotenv.2024.171658

Li, X., Zheng, Y., Lu, L., Eom, J., Ru, S., Li, Y., & Wang, J. (2024). Trophic transfer of micro-and nanoplastics and toxicity induced by long-term exposure of nanoplastics along the rotifer (Brachionus plicatilis)-marine medaka (Oryzias melastigma) food chain. *Environmental Pollution*, 346, 123599. DOI: 10.1016/j.envpol.2024.123599

Liu, L., Du, R. Y., Jia, R. L., Wang, J. X., Chen, C. Z., Li, P., Kong, L.-M., & Li, Z. H. (2024). Micro (nano) plastics in marine medaka: Entry pathways and cardiotoxicity with triphenyltin. *Environmental Pollution*, 342, 123079. DOI: 10.1016/j.envpol.2023.123079

Liu, M., Song, X., Liu, C., Cui, X., Sun, W., Li, Z., & Wang, J. (2024). Nanoplastics increase the adverse impacts of lead on the growth, morphological structure and photosynthesis of marine microalga Platymonas helgolandica. *Marine Environmental Research*, 193, 106259. DOI: 10.1016/j.marenvres.2023.106259

Lv, S., Cui, K., Zhao, S., Li, Y., Liu, R., Hu, R., Zhi, B., Gu, L., Wang, L., Wang, Q., & Shao, Z. (2024). Continuous generation and release of microplastics and nanoplastics from polystyrene by plastic-degrading marine bacteria. *Journal of Hazardous Materials*, 465, 133339. DOI: 10.1016/j.jhazmat.2023.133339

Ma, Y. B., Xie, Z. Y., Hamid, N., Tang, Q. P., Deng, J. Y., Luo, L., & Pei, D. S. (2023). Recent advances in micro (nano) plastics in the environment: Distribution, health risks, challenges and future prospects. *Aquatic Toxicology (Amsterdam, Netherlands)*, 261, 106597. DOI: 10.1016/j.aquatox.2023.106597

Marfella, R., Prattichizzo, F., Sardu, C., Fulgenzi, G., Graciotti, L., Spadoni, T., D'Onofrio, N., Scisciola, L., La Grotta, R., Frigé, C., Pellegrini, V., Municinò, M., Siniscalchi, M., Spinetti, F., Vigliotti, G., Vecchione, C., Carrizzo, A., Accarino, G., Squillante, A., & Paolisso, G. (2024). Microplastics and nanoplastics in atheromas and cardiovascular events. *The New England Journal of Medicine*, 390(10), 900–910. DOI: 10.1056/NEJMoa2309822

Meegoda, J. N., & Hettiarachchi, M. C. (2023). A path to a reduction in micro and nanoplastics pollution. *International Journal of Environmental Research and Public Health*, 20(8), 5555. DOI: 10.3390/ijerph20085555

Mojiri, A., Zhou, J. L., & Lou, Z. (2024). Micro (nano) plastics in Aquatic Environments: State of the Art and Beyond. *Water (Basel)*, 16(6), 902. DOI: 10.3390/w16060902

Mustapha, S., Tijani, J. O., Elabor, R., Salau, R. B., Egbosiuba, T. C., Amigun, A. T., Shuaib, D. T., Sumaila, A., Fiola, T., Abubakar, Y. K., Abubakar, H. L., Ossamulu, I. F., Abdulkareem, A. S., Ndamitso, M. M., Sagadevan, S., & Mohammed, A. K. (2024). Technological approaches for removal of microplastics and nanoplastics in the environment. *Journal of Environmental Chemical Engineering*, 12(2), 112084. DOI: 10.1016/j.jece.2024.112084

Naz, S., Chatha, A. M. M., Khan, N. A., Ullah, Q., Zaman, F., Qadeer, A., Khan, I. M., Danabas, D., Kiran, A., Skalickova, S., Bernatova, S., Khan, M. Z., & Horky, P. (2024). Unraveling the ecotoxicological effects of micro and nano-plastics on aquatic organisms and human health. *Frontiers in Environmental Science*, 12, 1390510. DOI: 10.3389/fenvs.2024.1390510

Ojha, M. D., & Skariyachan, S. (2024). The Fate of Micro/Nano Plastic Pollutants in the Natural Environment. In *Remediation of Plastic and Microplastic Waste* (pp. 215–239). CRC Press.

Okeke, E. S., Ezeorba, T. P. C., Chen, Y., Mao, G., Feng, W., & Wu, X. (2024). Association of tetrabromobisphenol A (TBBPA) with micro/nano-plastics: A review of recent findings on ecotoxicological and health impacts. *The Science of the Total Environment*, 927, 172308. DOI: 10.1016/j.scitotenv.2024.172308

Paing, Y. M. M., Eom, Y., Song, G. B., Kim, B., Choi, M. G., Hong, S., & Lee, S. H. (2024). Neurotoxic effects of polystyrene nanoplastics on memory and microglial activation: Insights from in vivo and in vitro studies. *The Science of the Total Environment*, 924, 171681. DOI: 10.1016/j.scitotenv.2024.171681

Pittura, L., Gorbi, S., León, V. M., Bellas, J., González, J. A. C., Albentosa, M., & Regoli, F. (2023). Microplastics and nanoplastics in the marine environment. *Contaminants of Emerging Concern in the Marine Environment*, 311-348.

Rajendran, D., Kamalakkannan, M., & Chandrasekaran, N. (2024). Surface functionalization, particle size and pharmaceutical co-contaminant dependent impact of nanoplastics on marine crustacean–Artemia salina. *Environmental Science. Processes & Impacts*, 26(7), 1130–1146. DOI: 10.1039/D4EM00010B

Rehman, A., Huang, F., Zhang, Z., Habumugisha, T., Yan, C., Shaheen, U., & Zhang, X. (2024). Nanoplastic contamination: Impact on zebrafish liver metabolism and implications for aquatic environmental health. *Environment International*, 187, 108713. DOI: 10.1016/j.envint.2024.108713

Rondon, R., Cosseau, C., Bergami, E., Cárdenas, C. A., Pérez-Toledo, C., Alvarez, D., Aldridge, J., Font, A., Garrido, I., Santa Cruz, F., Perrois, G., Balbi, T., Corsi, I., & González-Aravena, M. (2024). Exposure to nanoplastics and nanomaterials either single and combined affects the gill-associated microbiome of the Antarctic soft-shelled clam Laternula elliptica. *Marine Environmental Research*, 198, 106539. DOI: 10.1016/j.marenvres.2024.106539

Sandil, S., & Zaray, G. (2024). Impact of Microplastics and Nanoplastics in the Aquatic Environment. *Management of Micro and Nano-plastics in Soil and Biosolids: Fate, Occurrence, Monitoring, and Remedies*, 25-68.

Sharma, A., & Singh, B. (2022). Measuring Impact of E-commerce on Small Scale Business: A Systematic Review. *Journal of Corporate Governance and International Business Law*, 5(1).

Shukla, S., Pei, Y., Li, W. G., & Pei, D. S. (2024). Toxicological Research on Nano and Microplastics in Environmental Pollution: Current Advances and Future Directions. *Aquatic Toxicology (Amsterdam, Netherlands)*, 270, 106894. DOI: 10.1016/j.aquatox.2024.106894

Singh, B. (2020). Global science and jurisprudential approach concerning healthcare and illness. *Indian Journal of Health and Medical Law*, 3(1), 7–13.

Singh, B. (2022). Understanding Legal Frameworks Concerning Transgender Healthcare in the Age of Dynamism. *Electronic Journal of Social and Strategic Studies*, 3(1), 56–65. DOI: 10.47362/EJSSS.2022.3104

Singh, B. (2022). Relevance of Agriculture-Nutrition Linkage for Human Healthcare: A Conceptual Legal Framework of Implication and Pathways. *Justice and Law Bulletin*, 1(1), 44–49.

Singh, B. (2022). COVID-19 Pandemic and Public Healthcare: Endless Downward Spiral or Solution via Rapid Legal and Health Services Implementation with Patient Monitoring Program. *Justice and Law Bulletin*, 1(1), 1–7.

Singh, B. (2023). Unleashing Alternative Dispute Resolution (ADR) in Resolving Complex Legal-Technical Issues Arising in Cyberspace Lensing E-Commerce and Intellectual Property: Proliferation of E-Commerce Digital Economy. *Revista Brasileira de Alternative Dispute Resolution-Brazilian Journal of Alternative Dispute Resolution-RBADR*, 5(10), 81–105. DOI: 10.52028/rbadr.v5i10.ART04.Ind

Singh, B. (2023). Tele-Health Monitoring Lensing Deep Neural Learning Structure: Ambient Patient Wellness via Wearable Devices for Real-Time Alerts and Interventions. *Indian Journal of Health and Medical Law*, 6(2), 12–16.

Singh, B. (2023). Blockchain Technology in Renovating Healthcare: Legal and Future Perspectives. In *Revolutionizing Healthcare Through Artificial Intelligence and Internet of Things Applications* (pp. 177-186). IGI Global.

Singh, B. (2023). Federated Learning for Envision Future Trajectory Smart Transport System for Climate Preservation and Smart Green Planet: Insights into Global Governance and SDG-9 (Industry, Innovation and Infrastructure). *National Journal of Environmental Law*, 6(2), 6–17.

Singh, B. (2023). Eyes on the Sky for Real Estate Progression and Green Buildings: Addressing Sustainable Development Satellite SDG 9 (Industry, Innovation, and Infrastructure) and Policy Framework. *National Journal of Real Estate Law*, 6(2), 1–7.

Singh, B. (2024). Evolutionary Global Neuroscience for Cognition and Brain Health: Strengthening Innovation in Brain Science. In *Biomedical Research Developments for Improved Healthcare* (pp. 246-272). IGI Global.

Singh, B. (2024). Featuring Consumer Choices of Consumable Products for Health Benefits: Evolving Issues from Tort and Product Liabilities. *Journal of Law of Torts and Consumer Protection Law*, 7(1), 53–56.

Singh, B. (2024). Social Cognition of Incarcerated Women and Children: Addressing Exposure to Infectious Diseases and Legal Outcomes. In Reddy, K. (Ed.), *Principles and Clinical Interventions in Social Cognition* (pp. 236–251). IGI Global., DOI: 10.4018/979-8-3693-1265-0.ch014

Singh, B. (2024). Featuring Consumer Choices of Consumable Products for Health Benefits: Evolving Issues from Tort and Product Liabilities. *Journal of Law of Torts and Consumer Protection Law*, 7(1).

Singh, B. (2024). Green Infrastructure in Real Estate Landscapes: Pillars of Sustainable Development and Vision for Tomorrow. *National Journal of Real Estate Law*, 7(1), 4–8.

Singh, B. (2024). Cherish Growth, Advancement and Tax Structure: Addressing Social and Economic Prospects. *Journal of Taxation and Regulatory Framework*, 7(1), 7–10.

Singh, B. (2024). Legal Dynamics Lensing Metaverse Crafted for Videogame Industry and E-Sports: Phenomenological Exploration Catalyst Complexity and Future. *Journal of Intellectual Property Rights Law*, 7(1), 8–14.

Singh, B. (2024). Transformative Wave of IoMT, EHRs, RPM Technologies to Revolutionize Public Health. *Indian Journal of Health and Medical Law*, 7(2), 22–26.

Singh, B. (2024). Lensing Legal Dynamics for Examining Responsibility and Deliberation of Generative AI-Tethered Technological Privacy Concerns: Infringements and Use of Personal Data by Nefarious Actors. In Ara, A., & Ara, A. (Eds.), *Exploring the Ethical Implications of Generative AI* (pp. 146–167). IGI Global., DOI: 10.4018/979-8-3693-1565-1.ch009

Singh, B. (2024). Biosensors in Intelligent Healthcare and Integration of Internet of Medical Things (IoMT) for Treatment and Diagnosis. *Indian Journal of Health and Medical Law*, 7(1), 1–7.

Singh, B. (2024). Evolutionary Global Neuroscience for Cognition and Brain Health: Strengthening Innovation in Brain Science. In Prabhakar, P. (Ed.), *Biomedical Research Developments for Improved Healthcare* (pp. 246–272). IGI Global., DOI: 10.4018/979-8-3693-1922-2.ch012

Singh, B., Jain, V., Kaunert, C., & Vig, K. (2024). Shaping Highly Intelligent Internet of Things (IoT) and Wireless Sensors for Smart Cities. In *Secure and Intelligent IoT-Enabled Smart Cities* (pp. 117-140). IGI Global.

Singh, B., & Kaunert, C. (2024). Revealing Green Finance Mobilization: Harnessing FinTech and Blockchain Innovations to Surmount Barriers and Foster New Investment Avenues. In *Harnessing Blockchain-Digital Twin Fusion for Sustainable Investments* (pp. 265-286). IGI Global.

Singh, B., & Kaunert, C. (2024). Harnessing Sustainable Agriculture Through Climate-Smart Technologies: Artificial Intelligence for Climate Preservation and Futuristic Trends. In *Exploring Ethical Dimensions of Environmental Sustainability and Use of AI* (pp. 214-239). IGI Global.

Singh, B., & Kaunert, C. (2024). Integration of Cutting-Edge Technologies such as Internet of Things (IoT) and 5G in Health Monitoring Systems: A Comprehensive Legal Analysis and Futuristic Outcomes. *GLS Law Journal*, 6(1), 13–20. DOI: 10.69974/glslawjournal.v6i1.123

Singh, B., & Kaunert, C. (2024). Future of Digital Marketing: Hyper-Personalized Customer Dynamic Experience with AI-Based Predictive Models. In Revolutionizing the AI-Digital Landscape (pp. 189–203). Productivity Press. doi:DOI: 10.4324/9781032688305-14

Singh, B., & Kaunert, C. (2024). Salvaging Responsible Consumption and Production of Food in the Hospitality Industry: Harnessing Machine Learning and Deep Learning for Zero Food Waste. In Singh, A., Tyagi, P., & Garg, A. (Eds.), *Sustainable Disposal Methods of Food Wastes in Hospitality Operations* (pp. 176–192). IGI Global., DOI: 10.4018/979-8-3693-2181-2.ch012

Singh, B., Kaunert, C., & Vig, K. (2024). Reinventing Influence of Artificial Intelligence (AI) on Digital Consumer Lensing Transforming Consumer Recommendation Model: Exploring Stimulus Artificial Intelligence on Consumer Shopping Decisions. In Musiolik, T., Rodriguez, R., & Kannan, H. (Eds.), *AI Impacts in Digital Consumer Behavior* (pp. 141–169). IGI Global., DOI: 10.4018/979-8-3693-1918-5.ch006

Singh, B., Vig, K., & Kaunert, C. (2024). Modernizing Healthcare: Application of Augmented Reality and Virtual Reality in Clinical Practice and Medical Education. In Modern Technology in Healthcare and Medical Education: Blockchain, IoT, AR, and VR (pp. 1-21). IGI Global.

Sun, Y., Zhao, X., Sui, Q., Sun, X., Zhu, L., Booth, A. M., Chen, B., Qu, K., & Xia, B. (2024). Polystyrene nanoplastics affected the nutritional quality of Chlamys farreri through disturbing the function of gills and physiological metabolism: Comparison with microplastics. *The Science of the Total Environment*, 910, 168457. DOI: 10.1016/j.scitotenv.2023.168457

Tao, S., Li, T., Li, M., Yang, S., Shen, M., & Liu, H. (2024). Research advances on the toxicity of biodegradable plastics derived micro/nanoplastics in the environment: A review. *The Science of the Total Environment*, 916, 170299. DOI: 10.1016/j.scitotenv.2024.170299

Valiyaveettil Salimkumar, A., Kurisingal Cleetus, M. C., Ehigie, J. O., Onogbosele, C. O., Nisha, P., Kumar, B. S., ... & Rejish Kumar, V. J. (2024). Adsorption Behavior and Interaction of Micro-Nanoplastics in Soils and Aquatic Environment. *Management of Micro and Nano-plastics in Soil and Biosolids: Fate, Occurrence, Monitoring, and Remedies*, 283-311.

Varshney, A. (2024). Cleaning Up the Smallest Pollutants: The Potential of Microbial Degradation in Tackling Micro-and Nano-Plastic Pollution. In Management of Micro and Nano-plastics in Soil and Biosolids: Fate, Occurrence, Monitoring, and Remedies (pp. 367–389). Springer Nature Switzerland. doi:DOI: 10.1007/978-3-031-51967-3_15

Wang, H., Wang, Q., Lv, M., Li, J., Zhao, X., Song, Z., Wang, X., You, J., Wang, Y., & Chen, L. (2023). Marine micro (nano) plastics determination and its environmental toxicity evaluation. *Trends in Analytical Chemistry*, 168, 117332. DOI: 10.1016/j.trac.2023.117332

Winiarska, E., Jutel, M., & Zemelka-Wiacek, M. (2024). The potential impact of nano-and microplastics on human health: Understanding human health risks. *Environmental Research*, 251, 118535. DOI: 10.1016/j.envres.2024.118535

Xu, Y., Ou, Q., van der Hoek, J. P., Liu, G., & Lompe, K. M. (2024). Photo-oxidation of Micro-and Nanoplastics: Physical, Chemical, and Biological Effects in Environments. *Environmental Science & Technology*, 58(2), 991–1009. DOI: 10.1021/acs.est.3c07035

Yang, H., Ju, J., Wang, Y., Zhu, Z., Lu, W., & Zhang, Y. (2024). Micro-and nanoplastics induce kidney damage and suppression of innate immune function in zebrafish (Danio rerio) larvae. *The Science of the Total Environment*, 931, 172952. DOI: 10.1016/j.scitotenv.2024.172952

Yi, J., Ma, Y., Ruan, J., You, S., Ma, J., Yu, H., Zhao, J., Zhang, K., Yang, Q., Jin, L., Zeng, G., & Sun, D. (2024). The invisible Threat: Assessing the reproductive and transgenerational impacts of micro-and nanoplastics on fish. *Environment International*, 183, 108432. DOI: 10.1016/j.envint.2024.108432

Yu, J., Tian, J. Y., Jiang, Y., Wang, X. D., Song, X. R., Liu, L. F., & Yang, G. P. (2024). Effects of micro-and nano-plastics on growth, antioxidant system, DMS, and DMSP production in Emiliania huxleyi. *Environmental Pollution*, 351, 124084. DOI: 10.1016/j.envpol.2024.124084

Zheng, S., Tang, B. Z., & Wang, W. X. (2024). Microplastics and nanoplastics induced differential respiratory damages in tilapia fish Oreochromis niloticus. *Journal of Hazardous Materials*, 465, 133181. DOI: 10.1016/j.jhazmat.2023.133181

Zheng, S., & Wang, W. X. (2024). Single-Cell RNA Sequencing Profiling Cellular Heterogeneity and Specific Responses of Fish Gills to Microplastics and Nanoplastics. *Environmental Science & Technology*, 58(13), 5974–5986. DOI: 10.1021/acs.est.3c10338

Chapter 11
Mounting Legal-Driven Solutions for Plastic Pollution Focusing on Environment and Coastal Management:
Eradicating Marine Pollution in Alignment With SDG 14 (Life Below Water)

Bhupinder Singh
https://orcid.org/0009-0006-4779-2553
Sharda University, India

Sahil Lal
https://orcid.org/0000-0001-9827-3717
Sharda University, India

Manmeet Kaur Arora
https://orcid.org/0009-0002-5071-117X
Sharda University, India

Christian Kaunert
https://orcid.org/0000-0002-4493-2235
Dublin City University, Ireland

DOI: 10.4018/979-8-3693-9163-1.ch011

ABSTRACT

The growing danger of plastic pollution and its catastrophic effects on marine ecosystems highlight the pressing need for creative approaches to water governance as well as a thorough legal framework. Marine pollution comprises a mixture of chemicals and debris, primarily originating from terrestrial sources and finding its way into the ocean through runoff or wind dispersion. Also, the typical forms of marine debris encompass a range of plastic items such as shopping bags, beverage bottles, cigarette butts, bottle caps, food wrappers and fishing gear. In line with the objectives of Sustainable Development Goal 14 (SDG 14) Life Below Water, this chapter explores methods that can be scaled up to mitigate and eventually eradicate marine pollution by integrates policy assessments, legal framework and water governance tactics via an interdisciplinary approach to suggest practical for addressing this worldwide water pollution issue.

1. INTRODUCTION

Plastics make up a significant amount of marine trash; plastic debris, which can range in size from tiny particles to substantial chunks, keeps building up in marine environments and is a serious hazard to our natural environments (Singh & Kaunert, 2024) (Singh et al., 2024) (Arora et al., 2024). Marine life is severely harmed by the repercussions which include food chain disruption and marine species entanglement and ingestion. This chapter explores the perspective on the barriers and drivers affecting the successful implementation of national measures in line with international commitments as well as an overview of the responsibilities outlined in international and regional legal and policy frameworks pertaining to marine pollution (Yen et al., 2023). This contamination leads to adverse effects on the environment, the well-being of various organisms and the global economic structures. Marine debris refers to all human-made objects, predominantly plastic, that find their way into the ocean. This chapter's focus extends beyond dealing with marine pollution; it emphasizes the global Sustainable Development Goals 14 (SDGs) and the complete and integrated role of government in ensuring a healthy environment and sustainable livelihoods (Brusius et al., 2022).

The persistence of plastic waste poses a significant environmental challenge, as these items can take several centuries to break down. The extensive analysis of international accords, treaties and Indian Legal Provisions, this paper aims to evaluate the effectiveness of current legal frameworks in tackling the ubiquitous problem of marine plastic debris. This chapter explores the complex interactions between state-of-the-art governance models and legal tools, with an emphasis on protecting marine life and reducing plastic pollution. Legislative and policy frameworks that effectively prevent and manage marine litter are shaped by governments' obligations to protect the environment, respect citizens' rights and address public growing awareness of the negative effects of unsustainable waste generation and inadequate waste management (Devlaeminck et al., 2024). The significant factors such as- littering, storm winds and inadequate waste management collectively contribute to the buildup of this refuse. The frameworks of law and policy set forth responsibilities to preserve the world's oceans, avoid transboundary damage to high seas marine habitats and defend regions within the national sovereignty of neighboring coastal nations (Neal et al., 2023).

Figure 1. Landscape of introduction split sections (Original)

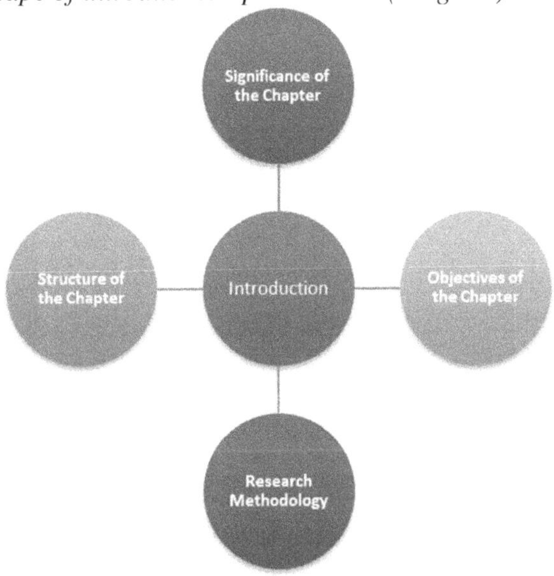

1.1 Significance of the Chapter

The health of seas and ecosystems is greatly impacted by the pressing environmental challenges of plastic waste and marine pollution. The world has become a holding tank for an alarming array of plastic waste due to the ongoing manufacture and disposal of plastic products (Rossi, S 2022). These pollutants which may range in size from microplastics to bigger objects like bottles and bags, endanger marine life, upset the delicate balance of our planet's aquatic habitats, disrupt the food chain and pose health concerns to humans (Christodoulou et al., 2024) (Ofori-Mensah et al., 2023). It safeguards the health of seas and contributes to a sustainable future for the environment and humanity by tackling the problems of plastic pollution and marine pollution (Sanabria et al., 2023). The UN Environment Assembly, which has recognized this issue in five resolutions, emphasizes how important it is. In particular, Resolution 5/14 called on the UNEP Executive Director to set up an Intergovernmental Negotiating Committee (INC) to draft a globally enforceable agreement on plastic pollution that includes maritime areas (Agnelli et al., 2022). It is intended for this instrument to take a holistic approach that addresses the manufacture, design and disposal of plastic during its whole life cycle (Albaladejo-García et al., 2023).

1.2 Objectives of the Chapter

The principal aim is to provide an extensive evaluation of the detrimental effects of plastic pollution on natural ecosystems, drawing links between the management of plastic pollution and the objectives of sustainable development (Singh B et al., 2024). The paper discusses policy efforts that use transdisciplinary methods including sustainability, life cycle assessment and the circular economy. It also highlights gaps in existing policy suggestions and expertise (Singh & Kaunert, 2024).

Study of Current Legal Provisions: Examine local, national and international laws related to plastic pollution and managing coastlines. Identify good and bad points and gaps blocking effective solutions. Understand the laws to propose realistic, large-scale solutions (Singh, B., 2024).

Research on Effective Cases Studies: Look into real-life examples where legal actions have successfully reduced marine pollution (Singh, B., 2024). Find out what made these actions successful, including regulatory rules, corrective measures and involving stakeholders. From this research, pull out teachings and best practices that can be used in different situations (Singh & Kaunert, 2024).

Model Legal Frameworks: In accordance with SDG 14, propose and create model legal frameworks that are especially suited to handle plastic pollution and coastal management. These frameworks ought to be created with the diversity of marine ecosystems and the differing levels of pollution in various areas in mind, in order to address the shortcomings and gaps that have been found. The aim is to provide scalable, realistic legal solutions that may be tailored to suit certain countries (Singh B.,2024).

Stakeholder Engagement and Capacity Building: Consider how different stakeholders, such as local communities, businesses, governments, and non-governmental organizations, may help put legal solutions into action. Provide plans for involving stakeholders and increasing their ability in order to guarantee that the suggested legislative frameworks are successfully adopted and put into practice. The purpose of this goal is to encourage teamwork and shared accountability in eradicating marine pollution (Ferraro et al., 2020).

Mechanisms for Observation and Assessment: Establish thorough monitoring and evaluation procedures to gauge the long-term effects of the legislative frameworks that have been put in place. Set up key performance indicators (KPIs) to track the development of better coastal management and a decrease in plastic pollution. This goal is essential because it offers a methodical way to monitor and modify tactics in response to actual results, so guaranteeing the scalability and long-term sustainability of legal-driven solutions (Palmer-Abbs et al., 2024).

Figure 2. Objectives of this chapter (Original)

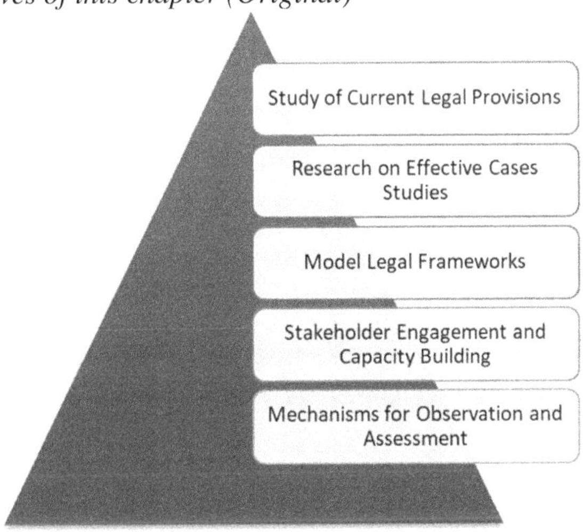

1.3 Research Methodology

The methodologies adopted for the paper is exploratory and doctrinal in nature. The literature review such as- books, articles, book chapter, commentaries', dictionaries, encyclopedia, journal papers, monographs, newspaper article, websites are used. In order to assess the environmental effects and resource consumption of plastic products over their entire life cycle, life cycle assessment and circularity are essential. The review also highlights the need for innovations in the reduction, reuse, recycling, and recovery of plastics, as well as the investigation of environmentally friendly alternatives. It is emphasized and recommended that laws be put into place that will empower and educate communities and individuals to reduce plastic pollution and embrace alternatives. In order to lessen the negative impacts of plastic pollution on ecosystems, it is imperative that we prioritize and work together to develop ways for managing plastic waste, as well as to implement regulatory changes and raise public awareness.

1.4 Structure of the Chapter

This paper comprehensively explores the diverse arena of Scaling Legal Driven Solutions for Plastic Pollution Lensing Environment and Coastal Management: Reducing and Eliminating Marine Pollution in Alignment with SDG 14 (Life Below Water). Section 2 elaborates the Marine Plastic Pollution: Causes and Impact Assessment. Section 3 expresses the Global Legal and Policy Frameworks on Marine Plastics. Section 4 lays down the Biodiversity and Species-Oriented Legal Instruments. Section 5 discusses the Extended Producer Responsibility (EPR) Regulations. Section 6 deals with Case Studies. Finally, Section 7 Conclude the Paper with Future Scope.

Figure 3. Structure/ flow of the chapter (Original)

2. MARINE PLASTIC POLLUTION: CAUSES AND IMPACT ASSESSMENT

The fishing sector, boating, and aquaculture are the main causes of plastic pollution in the ocean. Plastic degrades into tiny particles known as microplastics (smaller than 5 mm) or nanoplastics (smaller than 100 nm) when exposed to sun UV radiation, wind, currents, and other environmental variables. Marine life may readily consume them due to their small size (Sutherland et al., 2021). Many nations lack the sanitary landfills, incinerator plants, recycling capacity, infrastructure for the circular economy and appropriate waste management and disposal systems needed to tackle plastic pollution. "Plastic leakage" into rivers and seas is the consequence of this deficit. In areas where waste management systems are inadequate to limit plastic waste, the international traffic in plastic trash both legal and illicit can worsen the effects on ecosystems. The maritime ecosystem is severely threatened by plastic pollution, which puts human health, marine wildlife, food safety and quality and ocean health at risk. It also causes climate change and has a negative effect on coastal tourism (Giskes et al., 2022).

Impact on Marine Ecosystems: Several marine animals get entangled, suffocated, and swallowed by plastic waste, which is one of the most obvious effects on marine ecosystems. Sea creatures, such as fish, turtles, whales, and seabirds, frequently confuse plastic debris for food which causes them to starve to death as their stomachs fill with the material. In addition, these creatures sustain wounds including cuts, infections, decreased swimming capacity and internal damage. Floating plastics are a hazard to marine biodiversity and the entire food web because they facilitate the spread of invasive marine organisms (Chikodzi et al., 2021).

Impact on Food and Human Health: Microplastics have been found in beer, salt, tap water and samples of the world's oceans including the Arctic (Busch et al., 2021). These findings raise questions. Numerous chemicals utilized in the manufacturing of plastic are recognized to be carcinogens and endocrine system disruptors, resulting in immunological, neurological, reproductive, and developmental issues in both people and wildlife. Recent discoveries even suggest that human placentas contain microplastics, although further investigation is required to determine the full scope of this problem (Bello et al., 2022). Plastic that is exposed to saltwater for an extended period of time accumulates toxic chemicals on its surface. These pollutants enter the digestive tracts of marine creatures that consume plastic trash, gradually building up in the food chain. Research is still being done since eating seafood can expose humans to toxins that are transferred from marine animals to humans (Tyagi et al., 2024)

Impact on Tourism: Plastic trash detracts from the aesthetic attractiveness of popular tourist locations, which lowers tourism-related revenue. Significant financial expenses are also incurred for site upkeep and cleaning. The economics, wildlife and physical and mental health of a nation can all suffer from the buildup of plastic waste on its beaches (Hermawan et al., 2021)

Effects on Climate Change: The production of plastic contributes to climate change. The incineration of plastic waste releases carbon dioxide, while landfills emit methane, thereby increasing greenhouse gas emissions (Serôdio et al., 2022).

Figure 4. Impact of marine plastic pollution (Original)

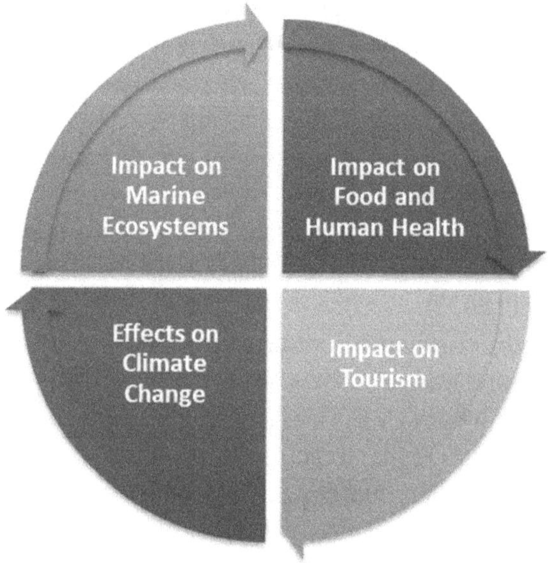

3. GLOBAL LEGAL AND POLICY FRAMEWORKS ON PLASTICS

The documents at international level specify the obligations and directives for the joint actions of the member countries. Legal and policy-based government solutions offer a chance to tackle a range of issues that contribute to marine litter at every stage of its life cycle, from waste management to product design (Akenji., et al 2020). The engagement of the industry, consumer behavior, and public knowledge are crucial elements in the mitigation of marine litter. These actions are insufficient to adequately address the problem of marine litter since they are voluntary, fragmented, and insufficient. These frameworks provide states or areas with the responsibility of preventing marine pollution. There might be regional or national variations in the social environment of national implementation measures (Williams., et al 2021).

3.1 The 1972 London Convention and 1996 Protocol: Prohibit Dumping Waste Streams Containing Plastic into Marine Environment

"London Convention" also known as the "Convention on the Prevention of Marine Pollution by Dumping of Wastes and Other Matter 1972" is one of the first international accords devoted to protecting the marine environment from human activity. Its main goals are to make it easier to regulate all sources of marine pollution and to put workable solutions in place to stop people from throwing trash and other materials into the ocean (Onyena, A. P,2021). As of right now, 87 States have joined this Convention as Parties. In 2022, the London Convention will mark its 50th anniversary. This Convention's adoption marks a major turning point that corresponds with the Stockholm United Nations Environment Conference in 1972. When taken as a whole, these occasions symbolized the first steps toward giving environmental issues top priority and taking accountability for the harm that people have caused to the ocean (Singh, B. 2024).

The London Convention and Protocol's main objective is to make it easier to regulate all possible causes of maritime pollution. According to paragraphs I and II of the Convention and article 2 of the Protocol, Contracting Parties are required to have policies in place that successfully prevent contamination of the maritime environment caused by sea dumping (Singh & Kaunert, 2024). The London Convention supervises the disposal of garbage into the sea with the goal of controlling all sources of marine pollution. It uses a "black-and-grey-list" approach to classify wastes according to the risk they cause to the environment (Singh, B. 2024).

The disposal of materials included on the blacklist is absolutely forbidden, but materials on the greylist need a special permission from a recognized national body and are subject to tight regulations and requirements. Once a general permission has been obtained, any further items or substances may be disposed of. The Convention and the Protocol have similar goals, but the Protocol takes a more stringent stance (Arisman and Yun A., 2023) Fatimah. "Waste Management in Indonesia: Strategies and Implementation of Sustainable Development Goals (SDGs) and Circular Economy." In *Circular Economy Adoption: Catalysing Decarbonisation Through Policy Instruments*, pp. 131 157. Singapore: Springer Nature Singapore, 2023. It uses a "reverse list" policy which restricts exporting wastes for marine dumping or incineration, forbids any dumping unless specifically authorized and includes a "precautionary approach" as a universal requirement. The Protocol includes provisions for technical support and extended compliance processes. Under certain restrictions, a transitional phase enables new Contracting Parties to progressively achieve Protocol compliance over a five-year period (Robin, 2023).

3.2 MARPOL (International Convention for the Prevention of Pollution from Ships): Annex V Prohibits Discharge of all Garbage (including plastics) into the Sea

Trash left over by ships is a serious hazard to marine life, maybe even more dangerous than oil or chemical spills. Given that plastic may float for a long time, it is quite dangerous. In addition to unintentionally consuming plastic, fish and marine mammals can become entangled in plastic ropes, nets, bags and seemingly innocuous items like the plastic rings used to lock down beverage cans (Graham,2023). Although onshore activities such as beachgoers discarding their trash and fisherman getting rid of undesirable items are responsible for a significant share of the debris seen along the coast, certain areas experience a significant quantity of the junk coming from passing ships. It is more convenient for ships to toss rubbish overboard in certain regions rather than properly dispose of it in ports (Chatziparaskeva et al., 2022).

With growing environmental consciousness, the long-held belief that the oceans can absorb whatever put into them has progressively changed. While certain things can be ruined by the water, it could take months or even years for this to happen. Education campaigns are necessary to discourage people from utilizing the oceans as landfills since some people continue to have the antiquated notion that the sea is capable of holding any kind of rubbish. In this endeavor, bolstering the enforcement of laws such as MARPOL Annex V is essential (Mutuku et al., 2024) The goal of MARPOL Annex V is to minimize and completely stop ships discharging trash into the ocean. Although the Annex is optional, more than 150 nations supported it when it came into effect on December 31, 1988. All waste kinds are typically forbidden by the Annex from being disposed of in the ocean, with the following exceptions: food waste, cargo residues, cleaning supplies, additives, and animal corpses (Ashai, 2021). The Marine Environment Protection Committee (MEPC) has created guidelines to help governments, ships and port operators adopt MARPOL Annex V. Port receiving facilities are essential for maintaining MARPOL Annex V compliance. Governments are required to set up sufficient receiving areas at ports and terminals so that waste may be disposed of properly without causing ships excessive delays. By use of regional agreements, Small Island Developing States (SIDS) may satisfy these obligations while adhering to International Maritime Organization (IMO) regulations (Singh et al., 2024).

3.3 Basel Convention on the Control of Transboundary Movements of Hazardous Wastes and their Disposal: Regulates the International Movement of Plastic Waste

The global agreement comprising 190 nations, the Basel Convention on the Control of Transboundary Movements of Hazardous Wastes, and their Disposal (1989) tackles the management, disposal, and cross-border transit of hazardous wastes created worldwide (Singh & Kaunert, 2024). Among the Basel, Rotterdam and Stockholm Conventions which together cover waste management and the life cycle of worldwide chemicals, it is the oldest. Destroyed materials that might endanger human health, or the environment are classified as hazardous wastes (Singh et al., 2024). Wastes that include electronics, polymers, radioactive materials, hazardous chemicals, heavy metals, infectious medical waste, and flammable, explosive or caustic materials fall under this category. It is estimated that 400 million tons of hazardous waste are created yearly or almost 13 tons per second, however exact figures may differ. Even while developed countries produce the majority of this trash, the amount produced by many emerging nations has increased quickly. For recycling or disposal, some hazardous waste is moved over international boundaries, although most of it stays in the country of origin.

Hazardous waste disposal rules began to take shape in industrialized nations in the 1970s and 1980s. As a result, several businesses looked for more affordable disposal locations, which caused a notable increase in hazardous waste exports from the North to the South (de Sousa., et al., 2021). To absorb this garbage lured developing nations and the Caribbean, but these regions sometimes lacked the administrative and technological capabilities for proper disposal. A large portion of this commerce entailed the illegal introduction of wastes into nations through bribery or mislabeling, even if some of it was legitimate [. Certain industries, like the ship-breaking and electronic waste industries in South Asia, have broken import laws or ignored and obstructed domestic environmental and human health regulations. These industries recycle products that contain hazardous materials (Chukwuemeka et al., 2024). Protecting the environment and human health from the harmful effects of hazardous wastes is the main goal of the Basel Convention. The scope of the convention includes two categories of wastes designated as “other wastes”: household garbage and incinerator ash, as well as a wide variety of wastes categorized as “hazardous wastes” according to their origin and/or content.

3.3.1 Goals and Specifications

The following major goals are at the center of the convention's provisions as- minimizing the production of hazardous waste and encouraging ecologically responsible management, wherever the disposal site may be; limiting the transboundary movement of hazardous wastes, unless it is thought to be in accordance with the principles of ecologically sound management; creating a legal framework for situations in which cross-border travel is permitted (Ghosh et al.,2022)

General regulations compel states to follow the fundamentals of ecologically sound waste management in order to accomplish the first goal (Article 4) (Wakwella et al.,2023). In order to achieve the second goal, there are prohibitions in place: it is not permitted to export hazardous wastes to Antarctica, a non-party state to the Basel Convention, or a party that has forbade the import of hazardous waste (Article 4). Parties may enter into agreements with non-parties or third parties for the management of hazardous waste, so long as those agreements are "no less environmentally sound" than the Basel Convention (Article 11). When permitted, transboundary movements must respect the convention's regulatory framework, serve as an ecologically sound solution, and uphold the values of nondiscrimination and environmentally sound management (Peel et al.,2021). The prior informed consent notion served as the foundation for the regulatory system when it was first introduced. It requires that the authorities of the exporting state notify the authorities of potential importing and transit states in advance of an export, giving them comprehensive details about the movement that is planned (Barrowclough & Birkbeck, 2022). Only after receiving formal approval from all states involved will the movement be allowed to continue (Articles 6 and 7). According to Articles 10 and 13, the convention encourages collaboration between parties, including the sharing of information pertinent to its implementation and technical support, especially for developing nations. In order to facilitate this collaboration, the Secretariat acts as a clearinghouse (Article 16) (Hilmi et al.,), Nathalie, Nicholas Theux-Lowen, and Matías Bastián Crisóstomo. "Ocean-Related Impacts of Climate Change on Economy." In *Life Below Water*, pp. 779-790. Cham: Springer International Publishing, 2022.

4. EXTENDED PRODUCER RESPONSIBILITY (EPR) REGULATIONS

This is a comprehensive legal instrument which can address plastic pollution throughout the entire life cycle using circular economy principles. The European Union's Extended Producer Responsibility (EPR) legislation apply to a number of member states and include expired pharmaceuticals and medical waste in their EPR frameworks. Eight member states combined these waste categories in 2014, producing 240,000 tons of garbage in total. Outside of official EPR recognition, there exist take-back schemes such as those for sharps and medications, wherein these products are returned to pharmacies and medical facilities (Wang et al., 2022). These programs are essential in guaranteeing that garbage related to pharmaceuticals and sharp objects is managed properly, following national guidelines, and staying out of domestic waste streams. To deter reuse, several nations require pharmaceutical waste to be burned or denaturated.

The damage that marine plastic pollution done to the ecosystem as a whole, emphasizes the importance to address as governance problem. A significant step toward global regulation was made by the Fifth United Nations Environment Assembly (UNEA) when it approved the resolution "End Plastic Pollution: Towards an International Legally Binding Instrument." This resolution highlights the cycle economy of plastics and the role of companies in promoting social responsibility, shifting the focus to a global approach to addressing marine plastic pollution (Jouffray et al., 2021). There is currently a greater focus on the Extended Producer Responsibility (EPR) system, which integrates these aspects. The harm that marine plastic waste poses to marine animals and human health makes it a major environmental concern. Marine ecosystems may be harmed by a number of things, including trash discharges, surface pollution, deep-sea mining, and crude oil spills. Plastic trash pollution is one of these issues that is very urgent. According to a UNESCO assessment, 80 percent of marine pollution is made up of plastic trash and by 2050, the weight of marine plastic garbage might equal the weight of all the fish in the ocean if it is not managed (Herrera-Franco et al., 2024). UN Sustainable Development Goal 14 which emphasizes the conservation and sustainable use of oceans and marine resources, is hampered by marine litter, which is defined as persistent, produced or processed solid wastes dumped in the marine and coastal environment.

The fact that primary and secondary microplastics are not biodegradable and enter the marine food chain to negatively affect ecosystems exacerbates the severity of the problem (Sedek, 2021). The problem's intricacy is further highlighted by the ambiguity around estimates of the amount of plastic that enters the ocean from land. Marine plastic pollution has a negative impact on the marine environment, ecosystems, human health, safety, and sustainable development due to its transboundary and extensive nature. In order to fight and eradicate plastic pollution, the Fifth United Nations Environment Assembly decided in March 2022 to create an international legally enforceable instrument on plastics. This tool takes into account plastics' whole life cycle, including material design, production, disposal, and relationships to other industries. By recognizing the relationship between land and sea and tackling the shortcomings of end-of-pipe governance, the idea of whole-process control is in line with the interconnectedness of governance in the context of marine plastic pollution. The tool also emphasizes how economic growth and environmental protection are inextricably linked in maritime governance (Mubtaker et al.,2024).

The responsibility of manufacturers in the process-oriented management of the plastics life cycle is the main emphasis of this instrument's governance of marine plastic pollution. A critical component of environmental protection, the Extended Producer Responsibility (EPR) system requires reexamination as the paradigm evolves from decentralization to specialization and from methods based on soft laws to those based on hard laws. The project is to improve the efficacy of the EPR system in controlling marine plastic pollution and address its normative posture in an international plastics agreement. In the end, this will help to foster the growth of an ocean-based economy and a better society (Palmer-Abbs et al., 2024).

4.3 Stressing Plastics' Circular Economy

International law has governed the management of marine plastics ever since the London Dumping Convention was established in 1972. Parties are required under the United Nations Convention on the Law of the Sea (UNCLOS) to "prevent, reduce, and control pollution of the marine environment" which includes dumping, ships, land-based sources, and operations on the seabed (Mishra et al., 2024). International accords that govern the discharge of regulations include the Basel Convention and the International Convention for the Prevention of Pollution from Ships. Certain plastic goods are prohibited from being produced or used under the Stockholm Convention. The two main strategies that influence international agreements on marine plastic pollution are source governance and end governance. Source governance focuses on environmental preservation through strict production and usage regulations. In order to lessen restrictions on the sector, end governance concentrates on collecting and recycling plastic waste (Ediboğlu S et al., 2024). The establishment of a plastics linear economy serves as the cornerstone for governance goals which prioritize a restorative and regenerative plastic economy built on recycling, reuse, and sustainable alternatives. It is anticipated that the worldwide agreement on plastic pollution control would heavily depend on the shift to a circular economy (Nasir et al., 2024).

Many parties are involved in the governance of marine plastic pollution, but the main decision-makers are sovereign governments and intergovernmental organizations. The public, businesses, scientists, and non-governmental organizations that make up the global civil society are all essential. Non-governmental organizations use volunteerism, education, scientific research, and awareness-raising to advance the circular economy and the sustainable development of the oceans (Chan et al.,2024). Corporate organizations bear the responsibility for design, manufacture, packaging, distribution, usage, maintenance, disposal, recycling, and reuse in the life cycle of marine plastics, as they are major participants in this process. It is the social obligation of producers to lead campaigns to reduce pollution, enhance manufacturing methods, promote recycling, and make pertinent data public.

4.4 Description of the EPR

Three themes are influencing the regulation of marine plastic pollution: the move toward a global framework, the emphasis on the circular economy, and the emphasis on corporate accountability (Mousavi et al., 2024). Plastics companies are a major third-party governance factor, and these developments point to a direction in governance and particular methodologies. Both the scientific theory of the plastics life cycle and the circular economy demand that end-of-source and end-of-pipe governance be compatible and concurrent (Sharifi et al., 2024). Systems are now designed with businesses overseeing the entire process at its center. Reevaluating the Extended Producer Responsibility (EPR) system is necessary in order to create a unified standard for plastics governance, since it functions internationally as an environmental policy.

5. CASE STUDIES IN EFFECTIVE LEGAL ACTION: SEVERAL COUNTRIES HAVE DEMONSTRATED SUCCESS IN TACKLING PLASTIC POLLUTION THROUGH ROBUST LEGAL FRAMEWORKS.

Oceans, national seas, and marine resources are all at risk of depletion due to a number of environmental issues that continue to exist despite these efforts, including plastic waste. Establishing a strong and enforceable regulatory framework that guarantees the preservation and sustainable use of the maritime environment and its ecosystems is essential to properly addressing these challenges. The oppressive effects of plastic pollution on seas and beaches highlight how urgent it is to review and improve the regulatory structure (Singh et al., 2024).

Figure 5. Case studies (Original)

5.1 Kenya's Ban on Single-use Plastic Bags

This regulation forbids the use, production, and importation of any plastic bags used for residential or commercial packaging. Plastic bags for construction, food packaging, medical waste and rubbish bin liners are among the exceptions. Of particular severity worldwide, violators who produce or use these kinds of plastic bags may be fined up to $40,000 (equal to 4.4 million Kenyan shillings) or imprisoned for a maximum of four years. The extensive dumping of plastics and insufficient waste disposal facilities, which led to catastrophic flooding, served as the catalyst for this prohibition. Government authorities were first made aware of this problem through social media awareness initiatives (Singh, 2022). Notwithstanding objections from the Kenyan Association of Manufacturers, this action was successful in reducing the usage of single-use plastic bags by 80%, in contrast to earlier failed attempts to outlaw the bags (NEMA, 2019). The success may be primarily ascribed to the enthusiastic backing of the general population and proactive involvement with relevant parties .

5.2 European Union's Single-Use Plastics Directive

A number of regulations are put into effect under the EU's Single-Use Plastics Directive, each of which is geared at a particular product. These actions are reasonable and planned to provide the best outcomes while taking the availability of more environmentally friendly options into account (Singh, 2024).

Several important requirements are covered under the Single-Use Plastics Directive, including:

Product Prohibitions: The directive forbids the use of certain single-use plastic products, including balloon sticks, cotton bud sticks, oxo-degradable plastics, plates, straws, stirrers, cutlery (forks, knives, spoons and chopsticks) and expanded polystyrene (EPS) food containers and cups . Beginning in 2021, the European Union implemented prohibitions on some commodities, including those for which non-plastic substitutes are readily accessible (Article 5).

Article 6 sets forth design objectives for manufacturers, mandating that they add 25% recycled material to PET beverage bottles by 2025 and 30% recycled content to all plastic beverage bottles by 2030 (Singh, 2024). As part of product design guidelines, lids and tops of containers must be "tethered" to the container throughout the duration of the container's life by 2024 in order to decrease environmental pollution and promote recycling (dos Muchangos et al., 2014).

Target for the Separate Collection of Plastic Beverage Bottles: Article 9 sets goals for the recycling of plastic beverage bottles that are collected separately. By 2025, member states must collect 77% of single-use plastic bottles with caps and lids; by 2029, 90% of these bottles must be collected .

Article 8 requires member states to create their own Extended Producer Responsibility (EPR) programs by 2021. The expenses of waste management are the responsibility of manufacturers of single-use plastic items such as- food containers, packages, wrappers, beverage containers and cups, tobacco products with filters, wet wipes, balloons and lightweight plastic carrying bags. This covers costs for litter removal, transportation, treatment, and awareness-raising initiatives (Wassénius et al., 2024)

Measures to Raise knowledge: The directive contains a number of actions to raise consumer knowledge about proper waste disposal, as well as labeling requirements to enhance communication about the usage of certain single-use goods (Article 4) (Article 7).

In the wake of increased concern over plastic pollution and its effects on sustainability worldwide, the European Commission developed a circular economy strategy with the goal of improving resource management that is both rigorous and efficient. But in 2015, the proposal ran into resistance and failed, mostly because of worries about how the private sector would be affected. Non-governmental groups like Greenpeace and the business community which included large companies like Nestle, Unilever and P&G reacted negatively to this.

5.2 Denmark: Agreement on Recycling of Transport Packaging

A wide range of industrial sectors are involved in transport packaging, including importers, producers, converters, packers, fillers, users, retailers and recyclers (Kumar et al.,2024). These organizations are mostly connected by virtue of their usage or sales of transport packaging. They frequently come from a variety of industrial sectors (Levy et al., 2023)

In Denmark, the industry is thought to have a high level of environmental consciousness. The public's understanding since the early 1980s, well-publicized cases of businesses breaking environmental laws, and modifications to local administrative and regulatory environmental policies are all credited with this increased awareness (Yap et al., 2024). Industrial awareness is also greatly aided by the Ministry of Environment and the Environmental Protection Agency of Denmark's strong presence. Everybody concerned acknowledges a moderate to elevated level of awareness of the ecological consequences of transportation packaging (Kalengyo et al., 2024). This is impacted by previous rules that applied to certain companies and sectors, as well as the unintended consequences of agreements, particularly those that SMEs received through trade groups. Additionally, it indicates the industry's readiness to exercise environmental consciousness, which results in a low degree of worry and constructive acts (Leddin, 2024).

6. CONCLUSION AND FUTURE SCOPE

In order to combat marine plastic pollution, it is imperative that the current international legal frameworks be actively upheld and reinforced. The 1972 London Convention, the 1996 London Protocol, and the 1978 Protocol to the International Convention for the Prevention of Pollution from Ships (MARPOL) are important examples of these. The 1972 Convention on the Prevention of Marine Pollution by Dumping Wastes and Other Matter is also known as the London Convention. National legislative frameworks centered on Extended Producer Responsibility should also be investigated by regional and national governments as they are proving to be creative and economical solutions. Circular economy-promoting policies have to be taken into account as well. Governments, academic institutions and industry collaborations are required to rethink product design, reevaluate usage and disposal, and reduce microplastic waste from products such as tires, synthetic fabrics, and pellets.

Scaling Up Legal Solutions

Plastic pollution is widespread in both land and water ecosystems, posing significant concerns for all living organisms. The environmental exposure of plastic waste exacerbates issues and its escalating production and accumulation are unprecedented, driven by indiscriminate usage, inadequate recycling, and deposition in landfills. The leakage of plastic waste into terrestrial and aquatic ecosystems is happening at an unprecedented pace, presenting a complex challenge for researchers, policymakers, citizens and other stakeholders.

Laws must be reinforced and expanded in order to reduce plastic pollution in a meaningful and long-lasting way. The following are important tactics as-

International regulatory harmonization: More uniformity between domestic and international legislation would guarantee fair play and close legal gaps.

Increasing enforcement: Strict enforcement procedures are essential for discouraging infractions and guaranteeing adherence to rules.

Encouraging openness and accountability: Accountability and public participation are encouraged by public access to information on plastic manufacturing, waste disposal, and enforcement actions.

Integration with strategies for coastal management and the environment: For a comprehensive strategy, legal frameworks on plastic pollution should be included with larger environmental and coastal management strategies.

The effectiveness of different legal instruments in addressing plastic pollution from various sources (e.g., industry, households).

The role of economic incentives and disincentives in promoting alternative materials and waste management practices.

The development of innovative technologies for plastic waste processing and recycling.

Strategies for fostering public awareness and behavior change to reduce plastic consumption and waste generation.

The control of marine litter is hampered in many developing nations worldwide by many issues. Initiatives for regional and international capacity building as well as financing possibilities supporting short-, medium-, and long-term solutions should be used to address these issues. Legislative and regulatory solutions must be created with these difficulties in mind and include steps to lessen or eliminate their effects.

Collecting pollutants prior to their entry into the ocean is important, especially for those that are currently difficult to decrease at the source. Putting in place wastewater treatment facilities and placing gross pollutant traps at point-sources like stormwater and wastewater drains are workable approaches. It's also critical to take action at the sink, focusing on locations where the greatest amount of pollution may be recovered. Examples include targeted clean-ups at beaches and coastal waterways with considerable accumulations of plastic pollution, as well as quick clean-up reactions to major pollution disasters like oil spills or flooding. These priority activities provide useful examples of mitigating marine pollution, even though they are not perfect solutions. But for them to be successful, everyone in the system—from pollution sources to sinks—must fully support them, including the government, business community, and people. Both individuals and governments should push for enforceable regulations that promote sustainable goods and activities, as well as the development of efficient enforcement strategies that comply with the laws. Actions in marine and coastal regions should be informed by both science and Indigenous knowledge, regardless of the zones under consideration.

REFERENCES

Agnelli, Alberto, and Piera Tortora. (2022) "The role of development co-operation in tackling plastic pollution: Key trends, instruments, and opportunities to scale up action.".

Akenji, Lewis, Magnus Bengtsson, Yasuhiko Hotta, Mizuki Kato, and Matthew Hengesbaugh. (2020): "Policy responses to plastic pollution in Asia: Summary of a regional gap analysis." *Plastic waste and recycling*: 531-567.

Alamoush, A. S., Ballini, F., & Ölçer, A. I. (2021). Revisiting port sustainability as a foundation for the implementation of the United Nations Sustainable Development Goals (UN SDGs). *Journal of Shipping and Trade*, 6(1), 1–40. DOI: 10.1186/s41072-021-00101-6

Albaladejo-García, J. A., & Zabala, J. A. (2023). Does the Blue Flag program contribute to the achievement of the Sustainable Development Goals? In *Blue Flag Beaches* (pp. 69–84). Routledge., DOI: 10.4324/9781003323570-6

Arisman, and Yun Arifatul Fatimah. (2023). "Waste Management in Indonesia: Strategies and Implementation of Sustainable Development Goals (SDGs) and Circular Economy.". In Circular Economy Adoption: Catalysing Decarbonisation Through Policy Instruments (pp. 131–157). Springer Nature Singapore. doi:DOI: 10.1007/978-981-99-4803-1_5

Arora, M. K., Lal, S., & Singh, B. (2024). A Way Forward to Realm of Environmental Law in India: Unveiling Judicial Activism and Enforcement Challenges. *National Journal of Environmental Law*, 7(1), 85–89.

Ashai, Perfect. "Analysis of marine plastic litter policies in Ghana." (2021).

Barrowclough, D., & Birkbeck, C. D. (2020) *Transforming the global plastics economy: the political economy and governance of plastics production and pollution.* No. 142. GEG Working Paper.

Bello, Fatai Abiola. (2022) "Critical analysis of policies on single-use plastics disposal from ships as a source of pollution to the marine environment.".

Borongan, G., & NaRanong, A. (2022). Factors in enhancing environmental governance for marine plastic litter abatement in Manila, the Philippines: A combined structural equation modeling and DPSIR framework. *Marine Pollution Bulletin*, 181, 113920. DOI: 10.1016/j.marpolbul.2022.113920

Braga, H. O., & Musiello-Fernandes, J. (2022). Types of Fisheries and Their Impact on Sustainable Development Goals. In *Life Below Water* (pp. 1075–1084). Springer International Publishing., DOI: 10.1007/978-3-319-98536-7_20

Brusius, B. K., Buss de Souza, R., & Barbieri, E. (2022). Stranding of marine animals: Effects of environmental variables. In *Life below water* (pp. 945–954). Springer International Publishing., DOI: 10.1007/978-3-319-98536-7_102

Busch, P.-O., Schulte, M. L., & Simon, N. (2021). *Strengthen the global science and knowledge base to reduce marine plastic pollution*. Nordic Council of Ministers., DOI: 10.6027/temanord2021-519

Chan, N. (2024). Voluntary commitments for ocean sustainability: Comparing the UN Ocean Conference and Our Ocean Conference platforms. *Marine Policy*, 160, 105956. DOI: 10.1016/j.marpol.2023.105956

Chatziparaskeva, G., Papamichael, I., & Zorpas, A. A. (2022). Microplastics in the coastal environment of Mediterranean and the impact on sustainability level. *Sustainable Chemistry and Pharmacy*, 29, 100768. DOI: 10.1016/j.scp.2022.100768

Chikodzi, David, Kaitano Dube, and Nonkululeko Ngcobo. (2021) "Rethinking harbours, beaches and urban estuaries waste management under climate-induced floods in South Africa." *The increasing risk of floods and tornadoes in Southern Africa* 127-140.

Christodoulou, A., & Fernández, J. E. (2021). Maritime Governance and International Maritime Organization instruments focused on sustainability in the light of United Nations' sustainable development goals. In Sustainability in the Maritime Domain: Towards Ocean Governance and Beyond (pp. 415–461). Springer International Publishing. doi:DOI: 10.1007/978-3-030-69325-1_20

Chukwuemeka, I. S., Li, G., Mo, Y., & Jacques, K. J. (2024). *Impacts of microplastics and urbanization on soil health: an urgent concern for sustainable development*. Green Analytical Chemistry.

Cowan, E., & Tiller, R. (2021). What shall we do with a sea of plastics? A systematic literature review on how to pave the road toward a global comprehensive plastic governance agreement. *Frontiers in Marine Science*, 8, 798534. DOI: 10.3389/fmars.2021.798534

de Marffy-Mantuano, A. (2021) "Legal Tools in Combating Marine Pollution and Mitigating the Effects of Acidification." In *Fulfilling the Sustainable Development Goals*, pp. 341-356. Routledge, doi:DOI: 10.4324/9781003144274-27 de Sousa, F. D. B. (2021). The role of plastic concerning the sustainable development goals: The literature point of view. *Cleaner and Responsible Consumption*, 3, 100020. DOI: 10.1016/j.clrc.2021.100020

Deshpande, P. C., & Haskins, C. (2021). Application of systems engineering and sustainable development goals towards sustainable management of fishing gear resources in Norway. *Sustainability (Basel)*, 13(9), 4914. DOI: 10.3390/su13094914

Devlaeminck, D. J. The Role of International Water Law in the Management of Marine Plastic Pollution: The Case of China and Its Transboundary Rivers. In *Plastic Pollution* (pp. 190–206). CRC Press.

dos Muchangos, L. S., & Velempini, K. M. "Plastic Waste Management in Southern African–an Overview of Mozambique and Botswana." Available at *SSRN* 4715926. doi:DOI: 10.2139/ssrn.4715926

Duarte, C. M., Agusti, S., Barbier, E., Britten, G. L., Castilla, J. C., Gattuso, J.-P., Fulweiler, R. W., Hughes, T. P., Knowlton, N., Lovelock, C. E., Lotze, H. K., Predragovic, M., Poloczanska, E., Roberts, C., & Worm, B. (2020). Rebuilding marine life. *Nature*, 580(7801), 39–51. DOI: 10.1038/s41586-020-2146-7

Ediboğlu Sakowsky, E., & Gündoğdu, S. (2024). An Imported Problem? Plastic Waste Effects on Türkiye's Environment with Specific Emphasis on Relevant Multilateral Environmental Agreements. In Plastic Waste Trade: A New Colonialist Means of Pollution Transfer (pp. 217–246). Springer Nature Switzerland. doi:DOI: 10.1007/978-3-031-51358-9_11

Farinha, I. Carlos AM Duarte, and Mafalda Gil de Carvalho. "S&S'21 Marine Plastic Pollution, Design & Circular Economy." In *UNIDCOM/IADE International Conference Senses & Sensibility*, pp. 308-320. Cham: Springer Nature Switzerland, 2021.

Ferraro, G., & Failler, P. (2020). Governing plastic pollution in the oceans: Institutional challenges and areas for action. *Environmental Science & Policy*, 112, 453–460. DOI: 10.1016/j.envsci.2020.06.015

Ghosh, S. K., Chakraborty, S. S., Rana, T., Karstensen, K. H., & Saha, P. K. (2022). Legislative Support to Curb Marine Littering through Wastes Co processing by Promoting Circular Economy. In *Effective Waste Management and Circular Economy* (pp. 27–41). CRC Press., DOI: 10.1201/9781003231608-4

Giskes, I., Baziuk, J., Pragnell-Raasch, H., & Perez Roda, A. (2022). *Report on good practices to prevent and reduce marine plastic litter from fishing activities*. Food & Agriculture Org.

Graham, Roxanne Elizabeth Donna. (2023) "Combatting the marine litter crisis in the Windward Islands: examining source-to-sea pathways and fostering multi-scale solutions.".

Hermawan, S., & Astuti, W. (2021). Analysing several ASEAN countries' policy for combating marine plastic litter. *Environmental Law Review*, 23(1), 9–22. DOI: 10.1177/1461452921991731

Herrera-Franco, G., Merchán-Sanmartín, B., Caicedo-Potosí, J., Bitar, J. B., Berrezueta, E., & Carrión-Mero, P. (2023). A systematic review of coastal zone integrated waste management for sustainability strategies. *Environmental Research*, 117968.

Hilmi, N., Theux-Lowen, N., & Crisóstomo, M. B. (2022). Ocean-Related Impacts of Climate Change on Economy. In Life Below Water (pp. 779–790). Springer International Publishing. doi:DOI: 10.1007/978-3-319-98536-7_158

Jouffray, J.-B., Blasiak, R., Nyström, M., Österblom, H., Tokunaga, K., Wabnitz, C., & Norström, A. (2021). "Blue Acceleration: An ocean of risks and opportunities." *Ocean Risk and Resilience Action Alliance (ORRAA).Report*, 42, 1–42.

Kalengyo, R. B., & Nasr, M. (2024). Water resource management to meet SDG13 on climate change. In *Water, the Environment and the Sustainable Development Goals* (pp. 157–178). Elsevier., DOI: 10.1016/B978-0-443-15354-9.00014-1

Kumar, M., Bhujbal, S. K., Kohli, K., Prajapati, R., Sharma, B. K., Sawarkar, A. D., Abhishek, K., Bolan, S., Ghosh, P., Kirkham, M. B., Padhye, L. P., Pandey, A., Vithanage, M., & Bolan, N. (2024). A review on value-addition to plastic waste towards achieving a circular economy. *The Science of the Total Environment*, 921, 171106. DOI: 10.1016/j.scitotenv.2024.171106

Leddin, D. (2024). The impact of climate change, pollution and biodiversity loss on digestive health and disease. *Gastro Hep Advances*, 3(4), 519–534. DOI: 10.1016/j.gastha.2024.01.018

Levy, S. R., Gopang, M., Ramírez, L., & Allan, B. I. (2024). Bernardo, Martin D. Ruck, and Anni Sternisko. "A human rights-based approach to climate injustices at the local, national, and international levels: Program and policy recommendations.". *Social Issues and Policy Review*, 18(1), 3–30. DOI: 10.1111/sipr.12103

Marine. (2022). "The Role of Industry in Addressing Marine Chemical Pollution."

Mishra, M., Sudarsan, D., Celso, A. G. S., Marques da Silva, R., Beja, S. K., Paul, S., Bhanja, P., & Sethy, M. (2024). Current patterns and trends of microplastic pollution in the marine environment: A bibliometric analysis. *Environmental Science and Pollution Research International*, 31(15), 1–20. DOI: 10.1007/s11356-024-32511-x

Mousavi, M., Kowsari, E., Gheibi, M., Cheshmeh, Z. A., Teymoorian, T., & Ramakrishna, S. (2024). Assessing Bioplastics' Economic, Commercial, Political, and Energy Potential with Circular Economy Modeling: A Sustainable Solution to Plastic Waste Management. *Materials Circular Economy*, 6(1), 1–36. DOI: 10.1007/s42824-023-00098-2

Mubtaker, Y. (2020). *SDGs for Everyone*. Yama Mubtaker.

Mutuku, J., Tocock, M., Yanotti, M., Tinch, D., & MacDonald, D. H. (2024). Public perceptions of the value of reducing marine plastics in Australian waters. *Ecological Economics*, 217, 108065. DOI: 10.1016/j.ecolecon.2023.108065

Nasir, M. S., Tahir, I., Ali, A., Ayub, I., Nasir, A., Abbas, N., Sajjad, U., & Hamid, K. (2024). Innovative technologies for removal of micro plastic: A review of recent advances. *Heliyon*, 10(4), e25883. DOI: 10.1016/j.heliyon.2024.e25883

Neal, A., Johnson, M., & Havrda, M. (2023). Multi-level climate action through circular supply chain management of ocean plastic. In *Handbook of Multi-Level Climate Actions* (pp. 215–241). Edward Elgar Publishing., DOI: 10.4337/9781802202458.00022

Ofori-Mensah, E. M. (2023) "The local perspective on Ocean Plastic Pollution: Identifying Areas to Channel Ocean Literacy at Great Ningo in Ghana." Master's thesis, NTNU

Onyena, A. P., Aniche, D. C., Ogbolu, B. O., Rakib, M. R. J., Uddin, J., & Walker, T. R. (2021). Governance strategies for mitigating microplastic pollution in the marine environment: A review. *Microplastics*, (1), 15–46. DOI: 10.3390/microplastics1010003

Palmer-Abbs, M., Deshpande, P., & Karl, C. W. (2024). Enabling transition thinking on complex issues (wicked problems): A framework for future circular economic transitions of plastic management in the Norwegian fisheries and aquaculture sectors. *Journal of Cleaner Production*, 449, 141420. DOI: 10.1016/j.jclepro.2024.141420

Peel, J., Godden, L., Palmer, A., & Markey-Towler, M. R. (2021). Regional Solutions: Export Measures for Plastic Recyclables to Reduce Marine Plastic Pollution in the Pacific. *Marine Plastic Pollution and the Rule of Law*, 11(1), 1.

Raubenheimer, Karen, and Niko Urho. (2023) "Science-policy interface for plastic pollution.".

Robin, R. S., Karthik, R., Nithin, A., & Purvaja, R. (2023). Removal of marine litter and its impact along the coast of India. *Records of the Zoological Survey of India*, 67–86.

Rognerud, I., Steindal, E. H., Tartiu, V. E., & Kumar, N. (2022). *International sustainability criteria for plastic products in a global agreement on plastic pollution*. Nordic Council of Ministers., DOI: 10.6027/temanord2022-509

Rossi, S. (2022). Marine Solid Pollution—From Macroplastics to Nanoplastics. In SDG 14: Life Below Water: A Machine-Generated Overview of Recent Literature (pp. 63–110). Springer International Publishing.

Sanabria, Adriana Gomez, and Florian Lindl. (2023) "Adoption of global circular waste management systems is crucial to cut leakage of waste into aquatic environments.".

Schäli, J. (2022). The Protection of the Marine Environment from Land-based Sources of Plastic Pollution in International Law. In *The Mitigation of Marine Plastic Pollution in International Law* (pp. 107–377). Brill Nijhoff., DOI: 10.1163/9789004508613_004

Sedek, S. S. S. (2021). Sustainable development of plastic pollution awareness campaigns on social media. *International Design Journal*, 11(2), 339–364. DOI: 10.21608/idj.2021.153616

Serôdio, J., & Paterson, D. M. (2022). Role of microphytobenthos in the functioning of estuarine and coastal ecosystems. In *Life Below Water* (pp. 894–906). Springer International Publishing., DOI: 10.1007/978-3-319-98536-7_11

Sharifi, A., Allam, Z., Bibri, S. E., & Khavarian-Garmsir, A. R. (2024). Smart cities and sustainable development goals (SDGs): A systematic literature review of co-benefits and trade-offs. *Cities (London, England)*, 146, 104659. DOI: 10.1016/j.cities.2023.104659

Sharma, H. B., Vanapalli, K. R., Samal, B., Cheela, V. R. S., Dubey, B. K., & Bhattacharya, J. (2021). VR Sankar Cheela, Brajesh K. Dubey, and Jayanta Bhattacharya. "Circular economy approach in solid waste management system to achieve UN-SDGs: Solutions for post-COVID recovery.". *The Science of the Total Environment*, 800, 149605. DOI: 10.1016/j.scitotenv.2021.149605

Singh, B. (2022). Relevance of Agriculture-Nutrition Linkage for Human Healthcare: A Conceptual Legal Framework of Implication and Pathways. *Justice and Law Bulletin*, 1(1), 44–49.

Singh, B. (2024). Biosensors in Intelligent Healthcare and Integration of Internet of Medical Things (IoMT) for Treatment and Diagnosis. *Indian Journal of Health and Medical Law*, 7(1), 1–7.

Singh, B. (2024). Evolutionary Global Neuroscience for Cognition and Brain Health: Strengthening Innovation in Brain Science. In *Biomedical Research Developments for Improved Healthcare* (pp. 246-272). IGI Global.

Singh, B. (2024). Evolutionary Global Neuroscience for Cognition and Brain Health: Strengthening Innovation in Brain Science. In Prabhakar, P. (Ed.), *Biomedical Research Developments for Improved Healthcare* (pp. 246–272). IGI Global., DOI: 10.4018/979-8-3693-1922-2.ch012

Singh, B. (2024). Featuring Consumer Choices of Consumable Products for Health Benefits: Evolving Issues from Tort and Product Liabilities. *Journal of Law of Torts and Consumer Protection Law*, 7(1), 53–56.

Singh, B. (2024). Lensing Legal Dynamics for Examining Responsibility and Deliberation of Generative AI-Tethered Technological Privacy Concerns: Infringements and Use of Personal Data by Nefarious Actors. In Ara, A., & Ara, A. (Eds.), *Exploring the Ethical Implications of Generative AI* (pp. 146–167). IGI Global., DOI: 10.4018/979-8-3693-1565-1.ch009

Singh, B. (2024). Social Cognition of Incarcerated Women and Children: Addressing Exposure to Infectious Diseases and Legal Outcomes. In Reddy, K. (Ed.), *Principles and Clinical Interventions in Social Cognition* (pp. 236–251). IGI Global., DOI: 10.4018/979-8-3693-1265-0.ch014

Singh, B., Arora, M. K., & Lal, S. (2024). Mapping Nature, Environment with SDG-9; Harnessing Machine Learning, Fostering Planet Earth Protection. *National Journal of Environmental Law*, 7(1).

Singh, B., Jain, V., Kaunert, C., & Vig, K. (2024). Shaping Highly Intelligent Internet of Things (IoT) and Wireless Sensors for Smart Cities. In *Secure and Intelligent IoT-Enabled Smart Cities* (pp. 117-140). IGI Global.

Singh, B., & Kaunert, C. (2024). Salvaging Responsible Consumption and Production of Food in the Hospitality Industry: Harnessing Machine Learning and Deep Learning for Zero Food Waste. In Singh, A., Tyagi, P., & Garg, A. (Eds.), *Sustainable Disposal Methods of Food Wastes in Hospitality Operations* (pp. 176–192). IGI Global., DOI: 10.4018/979-8-3693-2181-2.ch012

Singh, B., & Kaunert, C. (2024). Future of Digital Marketing: Hyper-Personalized Customer Dynamic Experience with AI-Based Predictive Models. *Revolutionizing the AI-Digital Landscape: A Guide to Sustainable Emerging Technologies for Marketing Professionals*, 189.

Singh, B., & Kaunert, C. (2024). Harnessing Sustainable Agriculture Through Climate-Smart Technologies: Artificial Intelligence for Climate Preservation and Futuristic Trends. In *Exploring Ethical Dimensions of Environmental Sustainability and Use of AI* (pp. 214-239). IGI Global.

Singh, B., & Kaunert, C. (2024). Revealing Green Finance Mobilization: Harnessing FinTech and Blockchain Innovations to Surmount Barriers and Foster New Investment Avenues. In *Harnessing Blockchain-Digital Twin Fusion for Sustainable Investments* (pp. 265-286). IGI Global.

Singh, B., Kaunert, C., & Vig, K. (2024). Reinventing Influence of Artificial Intelligence (AI) on Digital Consumer Lensing Transforming Consumer Recommendation Model: Exploring Stimulus Artificial Intelligence on Consumer Shopping Decisions. In Musiolik, T., Rodriguez, R., & Kannan, H. (Eds.), *AI Impacts in Digital Consumer Behavior* (pp. 141–169). IGI Global., DOI: 10.4018/979-8-3693-1918-5.ch006

Singh, B., Vig, K., & Kaunert, C. (2024). Modernizing Healthcare: Application of Augmented Reality and Virtual Reality in Clinical Practice and Medical Education. In Modern Technology in Healthcare and Medical Education: Blockchain, IoT, AR, and VR (pp. 1-21). IGI Global.

Sulistiawati, L. Y. (2022). Challenges in marine plastic pollution regulation in Indonesia. *Asia Pacific Journal of Environmental Law*, 25(1), 100–118.

Sutherland, D. L., McCauley, J., Labeeuw, L., Ray, P., Kuzhiumparambil, U., Hall, C., Doblin, M., Nguyen, L. N., & Ralph, P. J. (2021). How microalgal biotechnology can assist with the UN Sustainable Development Goals for natural resource management. *Current Research in Environmental Sustainability*, 3, 100050. DOI: 10.1016/j.crsust.2021.100050

Tan, D. (2022). Assessing Indonesia's Environmental Laws Pertaining to the Abatement of Marine Plastic Pollution: A Euphemism? *Jurnal Media Hukum*, 29(1), 1–33. DOI: 10.18196/jmh.v29i1.13414

Tyagi, I., Kumar, V., & Tyagi, K. (2024). Water pollution—sources and health implications of the environmental contaminants on the aquatic ecosystem and humans: approach toward sustainable development goals. In *Water, the Environment and the Sustainable Development Goals* (pp. 35–66). Elsevier., DOI: 10.1016/B978-0-443-15354-9.00008-6

Wakwella, A., Wenger, A., Jenkins, A., Lamb, J., Kuempel, C. D., Claar, D., Corbin, C., Falinski, K., Rivera, A., Grantham, H. S., & Jupiter, S. D. (2023). Integrated watershed management solutions for healthy coastal ecosystems and people. *Cambridge Prisms: Coastal Futures*, 1, e27. DOI: 10.1017/cft.2023.15

Wang, G., Li, X., Humphries, M., Chinni, V., Uthaipan, K., & Dai, M. (2022). Coastal pollution. In *Blue Economy: An Ocean Science Perspective* (pp. 251–286). Springer Nature Singapore., DOI: 10.1007/978-981-19-5065-0_9

Wassénius, E., Crona, B., & Quahe, S. (2024). Essential environmental impact variables: A means for transparent corporate sustainability reporting aligned with planetary boundaries. *One Earth*, 7(2), 211–225. DOI: 10.1016/j.oneear.2024.01.014

Williams, A. T., & Rangel-Buitrago, N. (2022). The past, present, and future of plastic pollution. *Marine Pollution Bulletin*, 176, 113429. DOI: 10.1016/j.marpolbul.2022.113429

Yap, C. K., Tze, Y. A. H., Nulit, R., Syazwan, W. M., Okamura, H., Horie, Y., & Ong, M. C. (2024). Copper in Commercial Marine Fish: From Biomonitoring to the ESG (Environment, Social, and Governance). *Method. Pollutants*, 4(1), 117–135. DOI: 10.3390/pollutants4010008

Yen, N. T. H., & Dung, N. P. (2023). Marine pollution management in response to plastic waste under international law and national law–challenges and implications for developing countries: A case study of Vietnam. *Asia Pacific Journal of Environmental Law*, 26(2), 162–193. DOI: 10.4337/apjel.2023.02.02

Chapter 12
Optimizing Public Awareness Campaigns and Behavior Change Strategies for Reducing Micro- and Nano-Plastic Pollution:
The Role of Social Media and Influencers in Fostering Sustainable Practices

Mohit Yadav
https://orcid.org/0000-0002-9341-2527
O.P. Jindal Global University, India

Ashutosh Pandey
https://orcid.org/0000-0002-8255-8459
FORE School of Management, New Delhi, India

Xuan-Hoa Nghiem
https://orcid.org/0000-0003-2292-0257
International School, Vietnam National University, Hanoi, Vietnam

Rashika Singh
Gautam Buddha University, India

DOI: 10.4018/979-8-3693-9163-1.ch012

ABSTRACT

Micro- and nano-plastic pollution poses a significant environmental and health challenge. This chapter explores the potential of public awareness campaigns and behavior change strategies in mitigating this issue. It delves into the critical role of social media and influencers in fostering sustainable practices. The chapter examines effective communication channels, messaging strategies, and audience segmentation to optimize campaign impact. It also analyzes the influence of influencers on shaping public perceptions and driving behavioral shifts. By understanding the dynamics of social media and influencer marketing, policymakers, environmental organizations, and businesses can develop targeted interventions to reduce plastic pollution and promote sustainable lifestyles. This research contributes to the growing body of knowledge on effective communication strategies for environmental challenges.

INTRODUCTION

The ubiquity of plastic has now begun to connote that a massive, insidious, and increasing ecological threat in it lies in micro- and nano-plastic pollution. These tiny fragments, smaller than five millimeters in size, pollute oceans, water bodies, and even the air that we breathe. Microplastics come from disintegrated larger plastic waste, fibers from synthetic clothing, and microbeads from personal care products. Many microplastics are intentionally designed or manufactured as such for any number of industrial applications and are further divided into a still smaller category of nano plastics. Be their origins the microplastic, good or bad retailer/consumer practices, or intentional pollution by industry or otherwise, these small-size pollutants pose a large-scale threat to the ecology of this world and potentially also to human health. Some great and damaging possibilities await in the future if they will infiltrate the food cycle. In addition to this, health impacts that could microplastics have on human beings could not be ruled out as various studies are ongoing on the same.

The multi-pronged approach is needed for such a complex issue. Necessary is the implementation of public awareness and motivational campaigns to adopt healthy and sustainable practices which reduce plastic use and encourage plastic disposal (Mahmud et al., 2022). Traditional avenues include public service announcements and educational programming, which are effective but might be just limited in terms of scope and interactivity. Here, the power of social media and influencers steps in. Social media platforms, for their part, offer instant global networking among billions of users. With an established online following, social media influencers are able to leverage the benefits of scale and credibility in influencing a high volume of public opinion and behavior. Engaging and effective campaigns can be developed by tapping into the full potential of social media and influencers to sensitize people on how to participate in sustainable practices in order to be part of the solution in combating micro- and nano-plastic pollution.

Micro- and Nano-Plastics: A Growing Threat

Micro- and nanoplastic particles are insidious environmental pollutants that are now threatening the health of our planet and possibly ourselves. These are small plastic fragments, below the size of a sesame seed (microplastics), even entering the size range of human hair (nanoplastics), and they find their way into our environment through a series of pathways. The larger plastic debris that gets lost, usually due to irresponsibility and vagaries, degrades and breaks into microplastics over a period of time through the process of weathering and decomposition (Watkins et al., 2019). Another source of microplastic pollution is from synthetic fibers released by our clothing during washing. Microbeads, used in daily personal care products such as exfoliating scrubs and face washes, have been identified as a major concern. Nanoplastics are even smaller than microplastics. They are usually purposely designed for industrial use in cosmetics and coatings, and some for medical purposes.

Micro- and nano- repetition of plastics wreaks havoc on the ecosystems. Indeed, the marine habitat is one of the more vulnerable, as the pollutants accumulate there: in oceans, seas, and waterways. Marine organisms, from plankton through to ocean giants such as whales, can mistake microplastics for food, with ingestion being a common outcome; other documented effects include blockages in digestive tracts, internal injury, and decreased nutrient absorption. Another worry is that by consuming contaminated seafood, there is a likelihood that microplastics will get into the food chain and eventually impact human health. While the finding of this study is still emerging, research has shown that microplastics are actually present in human organs. These findings have forced scientists to determine whether it is possible that health-related consequences arise from this fact.

Traditional Public Awareness Campaigns: Strengths and Limitations:

Traditional public awareness campaigns are now a significant tool in the process of awareness, related to, among other things, environmental pollutants like plastic. They are usually disseminated by means of public service announcements on television and radio, educational programs in schools and communities, and informational brochures or posters, which raise general awareness about the problem. Their strengths are their ability to dispose clear messages to the highest proportion of the audience, at the same time they cover the highest proportion (Eriksen et al., 2018). For instance, PSAs can really grab one's attention and give their recommendations in a very short period. It can help a child become a steward of the environment through the education it gives, particularly at a tender age.

However, the shortcomings of the traditional approaches are quite apparent. The message imparted through them can at times be too general in nature and not find special relevance or appeal to any specific demographics or behavior. Moreover, PSAs and brochures typically employ only one-way communication – that which is far less conducive to engagement and behavior change than a two-way approach. In addition, media which are already full in one high-competition outlet may mean that the compound effect of these campaigns will, in reality, be thin as diversification creates many alternative choices for the audience's attention. Raising awareness has become possible by using the conventional methods, but it is time to implement some additional complementary interactive measures that can really bring about behavior change.

Social media has changed the way people communicate and relay information, making for a very dynamic space far beyond what traditional media's one-way flow allowed (Walther et al., 2020). With platforms like Facebook, Instagram, YouTube, and TikTok, it is almost impossible to find someone who does not have an active account across the world (Nohara et al., 2018). It actually brings into being within this digital, connected world a powerful instrument not only of information-sharing but also of real-time persuasion of public perception and behaviors. This new class of tastemakers is found right in the throes of this dynamic online environment in the form of social media influencers.

More than just large personalities, these influencers are individuals with large followings who have cultivated dedicated online communities off specific niches or areas of expertise. The power to influence they wield comes from all the trust and rapport they develop with their audience. People see them as more approachable and genuine, unlike conventional ways of advertising. Authenticity is the key over here. Many influencers leverage the credibility they have built to sell products, brands, or even ideas to their following. Imagine this: A cosmetic influencer who cares about being environmentally friendly shares lists of green cosmetic manufacturers that she really loves (Selvaraj et al., 2022). Their real excitement, therefore, rubs off on their followers, who are likely to respond to the communication better and potentially shift their purchase behavior toward greener choices. The influence goes far beyond product endorsements. For instance, travel bloggers can display places that are green and promote responsible tourism, hence influencing their followers to travel with a smaller footprint.

The interactivity of social media further amplifies the influence of these digital tastemakers. Influencers are not one-way message channels; they ignite conversations (Rashed et al., 2023). Quality content, imaginative videos, or useful posts will set their based-on fire with discussions and actions. Consider the following powerful and moving posts on Instagram by a marine biologist about the effects of microplastics on sea creatures. This post shared amongst numerous followers and further discussed could then go on to create the awareness in them which would inspire them into finally adopting practices that reduce plastic use. Social media allows influencers to have a ripple effect where the message they forward can first reach a large number of people and thereby influence consumer behavior towards a sustainable future.

Harnessing Public Awareness Campaigns via Social Media

Social media offers a very dynamic, interactive platform to create an effective public awareness campaign. It is more effective than the traditional means in reaching a larger audience and sparking an engaging effect. The following, on how to make use of its potential for optimizing campaigns dedicated to reducing micro- and nano-plastic pollution, will be elaborated upon.

1. **Content is King and Queen**
    - **Storytelling:** Go beyond mundane statistics and numbers. Write powerful stories that will resonate deep with the audience. Show how great the ocean is and how much the wildlife that is affected by plastic pollution is going through.
    - **Visually Attractive**: The human being processes information more easily and quickly through images than through text. It is preferably using excellent images, infographics, and short videos with strong educational messages that result in creating attraction and making information easy to digest.
    - **Leverage User-Generated Content (UGC)**: Engage the public in your campaign by conducting competitions or challenges that incentivize the users to generate content on responsible use of plastics or, conversely, on the manifestation of the problem (Le, T. (2022). This creates a shared belief that the collective and society have ownership in the cause.
2. **Getting the Right Audience:**
    - **Platform Specificity:** Tailor your message and content for each platform. What resonates on Instagram might not go very well on LinkedIn. Understand each platform's demographics and content preferences to drive maximum reach and engagement.

- **Micro-targeting:** Make use of the powerful targeting tools that are on offer from social media to deliver your message to the most receptive exact demographics. This allows for crisscrossing, where focused messaging is directed to their concerns and behaviors.

3. **The Power of Calls to Action (CTAs):**
   - Let knowledge not be restricted to mere decoration. Have a directive call to action in the message that will prompt some behavior, if not petitioning, adoptions of reusable others or attending a beach cleaning event.
   - Make the point of action simple: You can provide resources, websites, or links to funds that support the cause directly through your social media posts.
4. **Building Trust and Authenticity:**
   - **Partner with credible voices:** The brands/Government can thus join hands with green organizations, scientists, or even everyday people who are passionate about saving the planet. Their experiential knowledge and real-life experiences will instill trust and give more profundity to your campaign.
   - **Be transparent:** Clearly tell what this campaign is for and how donations or participation will be utilized. It obviously builds trust and can include proper participation.
5. **Embrace the Power of Social Listening:**
   - **Hear what your audience has to say:** Monitor conversations using social listening tools, keeping track of emerging trends, and shifting your campaign messaging when necessary.
   - **Respond to comments and questions:** By showing your audience that you're taking the time to read and consider their comments, you're showing them that their voices matter and creating a sense of community around the cause.

The Role of Influencers in Fostering Sustainable Practices: A Force for Good, But with Nuances

Social media influencers nowadays have emerged as a force molding consumer behavior and attitude. Responsible influencers are hereby expected to play a critical role in promoting sustainable practices that will help reduce plastic use in the war against micro- and nano-plastic pollution. Here is how.

- **Product Endorsements and Reviews:** For Influencers, it is an opportunity to cash in on their credibilities, promoting eco-friendly alternatives to plastic products. Reviews featuring reusable water bottles, clothing brands using recycled materials for sustainable ware, or personal care products with no plastic can open the audience's eyes to enlighten their choices.

- **Inspirational Lifestyle Content:** Influencers are in the business of lifestyle representation and not just product promotion (Ertz et al., 2023). A sustainable fashion blogger can display through their lifestyle how one can clothe oneself with timeless pieces and advocate for quality over quantity by lowering textile wastage and plastic from synthetic fibers. Similarly, travel bloggers can highlight destinations committed to responsible tourism practices or showcase eco-friendly travel essentials, like reusable bags or water purifiers.
- **Educational Content with a Personal Touch:** Any influencer could create informative content that brings to light the issue of micro- and nano-plastic pollution. For instance, they could work with scientists to explain the science behind the issue or share experiences of personally witnessing plastic pollution. The idea is to change the way such information is traditionally passed on by infusing it with enactments that make it more interesting and relatable to the target group than merely feeding them the plain facts and figures on the ground.
- **Advocacy and Awareness Campaigns:** Environmental organizations can articulate their voices through the volume gained from influencers in advocacy and awareness campaigns. Influencers can do live Q and A's with experts, share information on NGO developed learning tools, or launch online challenges that spread the word about the issue and inspire action.

However, making influencers work for sustainability has some particular twists and turns of its own:

- **Authenticity Issues and Greenwashing:** A company needs to be mindful that it is supporting an influencer who genuinely cares for their cause, and it should not stir up participation for a product only in pursuit of a quick endorsement deal. Collaboration with credible and authentic voices fosters trust with the audience and delivers a more impactful message.
- **The Potential to Encourage Unsustainable Consumption:** As much as influencers can promote the use of sustainable alternatives, the path to be followed is very narrow. If the content goes over the advertising edge, it emulates and puts across the message of promoting overconsumption, irrespective of whether the products are sustainable or not.

Strategies for Effective Influencer Partnerships

- **Long-Term Collaboration:** Developing authentic relationships with influencers who are ecologically conscious thereby resulting in campaigns that are more effective and authentic.
- **Transparency and Clear Goals:** Clearly set out the goals of the campaign and the ways in which these collaborations are beneficial to all. Transparency establishes trust with the audience.
- **Creative Freedom with Brand Alignment:** Allow the influencer the right to say anything according to their style but aligned with the overall campaign goals.

By harnessing the power of social media influencers in responsible ways, we can leverage their reach and influence to inspire a shift towards sustainable practices and empower individuals to become part of the solution in tackling micro- and nano-plastic pollution (Jambeck et al., 2023).

Case Studies: Success Stories in Social Media and Influencer Campaigns

Effective social media campaigns and influencer collaborations effectively communicating the issue of plastic pollution are poised to inspire a wave of similar efforts. Here are two radically different examples that illustrate how this might happen:

1. **The #RefuseTheStraw Movement:**
   - **Platform:** Primarily organic movement gaining traction on Instagram Twitter, and Facebook.
   - **Focus:** Reducing the use of single-use plastic straws
   - **Influencers:** A mix of celebrities, environmental organizations, and everyday people passionate about sustainability.

Then in 2018, the movement went viral with such strong visuals as marine animals trapped in plastic straws. Here, celebrities like Mandy Moore and Adrian Grenier also posted pictures on their social media while refusing plastic straws in a restaurant under #RefuseTheStraw. Other environmental organizations, such as Greenpeace, got more involved in this movement by popularizing the educational part, sharing advice and alternatives that would go well with your new straw-less life. The movement was organic in nature, integrated with the reach of celebrities as well as established NGOs, giving rise to large-scale public awareness, thereby resulting in more and more restaurants providing reusable straw alternatives.

Key Takeaways:

- **Leveraging Organic Reach:** social media gives the 'common' man the power to contribute to a cause. UGC and organic sharing can be massive amplifiers of What's Up.
- **Influencer Effect:** Celebrities and established voices can help bring initial attention and add credibility to the cause.

2. **"LifeStraw Give Back" Campaign**

- **Platform:** Instagram with support from influencers
- **Objective**: Create awareness about plastic pollution in developing countries and solution to the same
- **Influencers:** Travel bloggers and other social media people emphasizing a sustainable way of life.

The portable water filter manufacturer LifeStraw made a campaign with travel influencers. The influencers were showing their travels around developing countries and how the availability of clean drinking water is problematic due to plastic waterway pollution. They started using LifeStraw products as a means of a sustainable solution; part of the sales would go to projects in these regions that work on clean water.

Key takeaways:

- **Storytelling and Personal Connection**: The influencers created compellingly relatable stories through their travel experiences.
- **Cause-Related Marketing:** With the partner influencer, the product and the message will easily blend since the values are already aligned.

These are but two instances. Plastic pollution campaigns are abundant and diverse in structure and form, just like any other successful campaign (Rice & Moxley, 2021). Spice your chapter up by featuring case studies from other areas or groups dissimilar to your initial target audience.

Challenges and Considerations: Understanding Nuances of Social Media and Influencers

Marketing through social media and influencers has so much potential to get the word out on sustainable practices and ridding the world of micro- and nano-plastic pollution. But there are different challenges and considerations that come with it. The following takes a closer look at a few key areas of consideration:

- **Misinformation and greenwashing:** social media has a fast-paced environment. One way is through unverified claims or uncorroborated facts regarding plastic pollution or sustainable alternatives, which can mislead the public. Similarly, "greenwashing"—where companies misrepresent themselves as having the interests of the environment at heart when they actually don't—will work against trust and possibilities of genuine will (Kurniawan, et al., 2023). Steps to address such challenges will include working with credible sources of information, such as scientists, to at least ensure that campaigns are grounded in fact and, at best, have transparency and clear articulation of goals at their core.

- **Influencer Authenticity and Accountability:** The impact of influencer marketing is dependent on the level of trust. The authenticity of influencers, especially those who come off as supporting sustainable products only for endorsement deals and do not really have a personal commitment to the cause, is of concern. Moreover, the very culture of desire that social media propagate could end up negating any sustainable behavior that this campaign tries to create. It is for this reason that there is a need to develop long-term relationships with those influencers who think and act green. Additionally, encouraging transparency in partnerships and allowing influencers to express their authentic self in all sponsored content ensures that partners remain authentic.
- **Measures and Impact:** Success in social media campaigns is hard to gauge. When looking at measures, like likes, shares, and growth in followers, it might present with some indication of the reach, although it needs more insight into what the campaign actually does (Ali, 2024). Website traffic generated by a campaign or changes in consumer behavior — for example, more meaningful success metrics can be seen in the increased usage of the reusable option— can be traced by ongoing surveys documenting a more significant public sentiment.
- **Equity and Accessibility Concerns**: Not everyone has access to social media, and not everyone has afforded the currently promoted alternatives that would make luxury products accessible. A powerful campaign has to take such differences into account. Consider including some low-cost, accessible solutions along with premium alternatives. Translating content into multiple languages and partnering with influencers from a great diversity of backgrounds will help in getting the message across to a more representative population.
- **The social media landscape** is changing every day. New features are driving into the town, and some established platforms are changing their algorithms. Updating and revising campaigns accordingly will always be important in maintaining efficacy. This might involve investment in social listening plat-

forms to keep abreast of trends and audience tastes, or new content forms for audiences like live streaming and interactive polling.

By acknowledging these challenges and putting in place appropriate strategies, we can make sure that social media and influencer partnerships are powerful tools for raising real public awareness to drive positive behavior change in the fight against micro- and nano-plastic pollution.

CONCLUSION AND RECOMMENDATIONS

The high level of micro- and nano-plastic pollution is a very dangerous situation for the entire health profile of our planet, and probably for us. Such a complex problem can only be addressed effectively by a multi-faceted intervention (Liu, 2024). Public awareness campaigns would serve the best interest by engaging all people in change toward sustainable behavior patterns that avoid plastic use and related discarding. While conventional techniques did what they could do, this is the time to leverage the massive potential of social media and influence through the social network to spread the message and take people in a meaningful way, deeper than ever before.

Social media provides an interactive space where engaging storytelling, strong visuals, and a strong call to action can combine effectively (Heidbreder et al., 2021). Teaming up with voices of authority, from recognized environmental groups or individual scientists, for example, can add credibility to the message. In choosing the influencer for this campaign, this combination of values and reach can be put in motion to create the provocation that drives change toward sustainability. Case studies prove that successful social media campaigns and collaboration with influencers have the potential to really make a difference through impactful storytelling and organic user-generated participation, which will further amplify their message and ignite human behavior.

Effectively, though, success in this digital environment is determined by first overcoming the obstacles. Misinformation, concerns over the authenticity of the influencer herself, and the very landscape of social media are changing at such dynamic speed that social and digital space needs due deliberation and flexibility in planning. This goes from building in transparency and working together with credible information sources to prioritizing measurement through metrics that truly demonstrate behavior change. Some are also the need to ensure messaging is accessible and equitable and keeping up with the ever-evolving state of digital.

Building on those, some **recommendations** include the following:

- Environmental organizations should partner with social media platforms and the influencers themselves to bring organizational strength, reach, and content creation expertise together to maximize the impact.
- Business cases for social media literacy and critical thinking skills education. Empowerment in finding out fact from fiction in an online environment is prime to combatting misinformation on plastic pollution and its sustainable alternatives.
- Provide a basis for developing clear and consistent labeling guidelines for sustainable products that will build consumer trust and enable informed decision-making.
- Rest further research on the efficiency of social media campaigns and influencer pairs on promoting and changing people's behavior toward sustainability. This would advance current campaigns and optimize the strategy for working best with real long-term environmental gains.

If we harness the potential of social media and influencers with a dose of reality for the existing issues, we can make far greater strides in stirring a global movement toward responsible plastic use and enable people to become proactive agents in shaping a cleaner, healthier future for our earth, with less plastic pollution involved. It is through joint efforts that such growing threats of micro- and nano-plastic pollution are to be controlled.

REFERENCES

Ali, S. S. (2024). *Nano influencers: using ordinary people for extraordinary results* (Doctoral dissertation).

Eriksen, M., Thiel, M., Prindiville, M., & Kiessling, T. (2018). Microplastic: what are the solutions. *Freshwater microplastics, 273.*

Ertz, M., Addar, W., Ouerghemmi, C., & Takaffoli, M. (2023). Overview of factors influencing consumer engagement with plastic recycling. *Wiley Interdisciplinary Reviews. Energy and Environment*, 12(6), e493. DOI: 10.1002/wene.493

Heidbreder, L. M., Lange, M., & Reese, G. (2021). # Plasticfreejuly–analyzing a worldwide campaign to reduce single-use plastic consumption with twitter. *Environmental Communication*, 15(7), 937–953. DOI: 10.1080/17524032.2021.1920447

Henderson, L. (2018). Telling stories about (micro) plastic pollution: Media images, public perceptions and social change.

Jambeck, J., Moss, E., Dubey, B., Arifin, Z., Godfrey, L., Hardesty, B. D., & Woodall, L. (2023). Leveraging multi-target strategies to address plastic pollution in the context of an already stressed ocean. In *The Blue Compendium: From Knowledge to Action for a Sustainable Ocean Economy* (pp. 141–184). Springer International Publishing., DOI: 10.1007/978-3-031-16277-0_5

Kurniawan, T. A., Haider, A., Mohyuddin, A., Fatima, R., Salman, M., Shaheen, A., Ahmad, H. M., Al-Hazmi, H. E., Othman, M. H. D., Aziz, F., Anouzla, A., & Ali, I. (2023). Tackling microplastics pollution in global environment through integration of applied technology, policy instruments, and legislation. [PubMed]. *Journal of Environmental Management*, 346, 118971. DOI: 10.1016/j.jenvman.2023.118971

Le, T. (2022). *Plastic Pollution in the Canadian Great Lakes: Drivers, Barriers and Policy Recommendations* (Master's thesis, University of Waterloo).

Liu, Y. (2024). *Microplastics: Research, Problems, and Solutions* (Doctoral dissertation, Worcester Polytechnic Institute).

Mahmud, A., Wasif, M. M., Roy, H., Mehnaz, F., Ahmed, T., Pervez, M. N., Naddco, V., & Islam, M. S. (2022). Aquatic microplastic pollution control strategies: Sustainable degradation techniques, resource recovery, and recommendations for Bangladesh. *Water (Basel)*, 14(23), 3968. DOI: 10.3390/w14233968

Nohara, N. M. L., Ariza-Tarazona, M. C., Triboni, E. R., Nohara, E. L., Villarreal-Chiu, J. F., & Cedillo-González, E. I. (2024). Are you drowned in microplastic pollution? A brief insight on the current knowledge for early career researchers developing novel remediation strategies. [PubMed]. *The Science of the Total Environment*, 918, 170382. DOI: 10.1016/j.scitotenv.2024.170382

Rashed, A. H., Yesilay, G., Hazeem, L., Rashdan, S., AlMealla, R., Kilinc, Z., Ali, F., Abdulrasool, F., & Kamel, A. H. (2023). Micro-and nano-plastics contaminants in the environment: Sources, fate, toxicity, detection, remediation, and sustainable perspectives. *Water (Basel)*, 15(20), 3535. DOI: 10.3390/w15203535

Rice, R. E., & Moxley, C. M. (2021). Plastic communication campaigns and interventions: Foci, theoretical frameworks, variables, and methods. In *The Handbook of International Trends in Environmental Communication* (pp. 132–154). Routledge., DOI: 10.4324/9780367275204-11

Selvaraj, S., Prasadh, S., Fuloria, S., Subramaniyan, V., Sekar, M., Ahmed, A. M., Bouallegue, B., Hari Kumar, D., Sharma, V. K., Maziz, M. N. H., Sathasivam, K. V., Meenakshi, D. U., & Fuloria, N. K. (2022). COVID-19 Biomedical Plastics Wastes—Challenges and Strategies for Curbing the Environmental Disaster. *Sustainability (Basel)*, 14(11), 6466. DOI: 10.3390/su14116466

Walther, B. A., Kusui, T., Yen, N., Hu, C. S., & Lee, H. (2020). Plastic pollution in East Asia: Macroplastics and microplastics in the aquatic environment and mitigation efforts by various actors. *Plastics in the aquatic environment-Part I: Current status and challenges*, 353-403.

Watkins, E., Schweitzer, J. P., Leinala, E., & Börkey, P. (2019). Policy approaches to incentivise sustainable plastic design.

Compilation of References

Abdel-Fatah, M. A. (2018). Nanofiltration systems and applications in wastewater treatment: Review article. *Ain Shams Engineering Journal*, 9(4), 3077–3092. DOI: 10.1016/j.asej.2018.08.001

Abdi, J., & Abedini, H. (2020). MOF-based polymeric nanocomposite beads as an efficient adsorbent for wastewater treatment in batch and continuous systems: Modelling and experiment. *Chemical Engineering Journal*, 400, 125862. DOI: 10.1016/j.cej.2020.125862

Abideen, Z., Hanif, M., Munir, N., & Nielsen, B. L. (2022). Impact of Nanomaterials on the Regulation of Gene Expression and Metabolomics of Plants under Salt Stress. *Plants*, 11(5), 691. DOI: 10.3390/plants11050691

Abomohra, A., & Hanelt, D. (2022). Recent Advances in Micro-/Nanoplastic (MNPs) Removal by Microalgae and Possible Integrated Routes of Energy Recovery. *Microorganisms*, 10(12), 12. Advance online publication. DOI: 10.3390/microorganisms10122400

Adeyemi, J. O., & Fawole, O. A. (2023). Metal-Based Nanoparticles in Food Packaging and Coating Technologies: A Review. *Biomolecules*, 13(7), 1092. DOI: 10.3390/biom13071092

Adler, M. Y., Issoual, I., Rückert, M., Deloch, L., Meier, C., Tschernig, T., Alexiou, C., Pfister, F., Ramsperger, A. F. R. M., Laforsch, C., Gaipl, U. S., Jüngert, K., & Paulsen, F. (2024). Effect of micro-and nanoplastic particles on human macrophages. *Journal of Hazardous Materials*, 471, 134253. DOI: 10.1016/j.jhazmat.2024.134253

Aeschlimann, M., Li, G., Kanji, Z. A., & Mitrano, D. M. (2022). Potential impacts of atmospheric microplastics and nanoplastics on cloud formation processes. *Nature Geoscience*, 15(12), 967–975. DOI: 10.1038/s41561-022-01051-9

Afrin, S., Rahman, M., Hossain, M. N., Uddin, M. K., & Malafaia, G. (2022). Are there plastic particles in my sugar? A pioneering study on the characterization of microplastics in commercial sugars and risk assessment. *The Science of the Total Environment*, 837, 155849. DOI: 10.1016/j.scitotenv.2022.155849

Afrose, S., Tran, T. K. A., O'Connor, W., Pannerselvan, L., Carbery, M., Fielder, S., Subhaschandrabose, S., & Palanisami, T. (2024). Organ-specific distribution and size-dependent toxicity of polystyrene nanoplastics in Australian bass (Macquaria novemaculeata). *Environmental Pollution*, 341, 122996. DOI: 10.1016/j.envpol.2023.122996

Agnelli, Alberto, and Piera Tortora. (2022) "The role of development co-operation in tackling plastic pollution: Key trends, instruments, and opportunities to scale up action.".

Ajith, M. P., Priyadarshini, E., & Rajamani, P. (2020). Effective and selective removal of heavy metals from industrial effluents using sustainable Si–CD conjugate based column chromatography. *Bioresource Technology*, 314, 314. DOI: 10.1016/j.biortech.2020.123786

Akenji, Lewis, Magnus Bengtsson, Yasuhiko Hotta, Mizuki Kato, and Matthew Hengesbaugh. (2020): "Policy responses to plastic pollution in Asia: Summary of a regional gap analysis." *Plastic waste and recycling*: 531-567.

Alamoush, A. S., Ballini, F., & Ölçer, A. I. (2021). Revisiting port sustainability as a foundation for the implementation of the United Nations Sustainable Development Goals (UN SDGs). *Journal of Shipping and Trade*, 6(1), 1–40. DOI: 10.1186/s41072-021-00101-6

Alam, S., Bhardwaj, L. K., Mallick, R., & Rai, S. (2023). Estimation of heavy metals and fluoride ion in vegetables grown nearby the stretch of River Yamuna, Delhi (NCR), India. *Indian Journal of Environmental Protection*, 43(1), 64–73.

Albaladejo-García, J. A., & Zabala, J. A. (2023). Does the Blue Flag program contribute to the achievement of the Sustainable Development Goals? In *Blue Flag Beaches* (pp. 69–84). Routledge., DOI: 10.4324/9781003323570-6

Ali, S. S. (2024). *Nano influencers: using ordinary people for extraordinary results* (Doctoral dissertation).

Ali, M. E., Hoque, M. E., Hossain, S. K. S., & Biswas, M. C. (2020). Nanoadsorbents for wastewater treatment: Next generation biotechnological solution. *International Journal of Environmental Science and Technology*, 17(9), 4095–4132. DOI: 10.1007/s13762-020-02755-4

Ali, S. S., Elsamahy, T., Koutra, E., Kornaros, M., El-Sheekh, M., Abdelkarim, E. A., Zhu, D., & Sun, J. (2021). Degradation of conventional plastic wastes in the environment: A review on the current status of knowledge and future perspectives of disposal. *The Science of the Total Environment*, 771, 144719. DOI: 10.1016/j.scitotenv.2020.144719

Ali, W., Buriro, R. S., Gandahi, J. A., Chen, Y., Aabdin, Z., Bhutto, S., Sun, J., Zhu, J., Liu, Z., & Zou, H. (2024). A critical review on male-female reproductive and developmental toxicity induced by micro-plastics and nano-plastics through different signaling pathways. *Chemico-Biological Interactions*, 394, 110976. DOI: 10.1016/j.cbi.2024.110976

Allen, D. (2022). Microplastics and Nanoplastics in the Marine-Atmosphere Environment. *Nature Reviews Earth & Environment,* 3(6), 393. https://doi.org/doi:10.1038/s43017-022-00292-x

Allen, S. A. (2019). Atmospheric microplastics in remote and high-altitude ecosystems. *Nature*, 569(7756), 641–645.

Allen, S., Allen, D., Karbalaei, S., Maselli, V., & Waker, T. R. (2022). Micro(Nano) Plastics Sources, Fate, and Effects: What We Know after Ten Years of Research. *Journal of Hazardous Materials Advances*, 6, 1. DOI: 10.1016/j.hazadv.2022.100057

Allouzi, M. M. A., Tang, D. Y. Y., Chew, K. W., Rinklebe, J., Bolan, N., Allouzi, S. M. A., & Show, P. L. (2021). Micro (nano) plastic pollution: The ecological influence on soil-plant system and human health. *The Science of the Total Environment*, 788, 147815. DOI: 10.1016/j.scitotenv.2021.147815

Alprol, A. E., Gaballah, M. S., & Hassaan, M. A. (2021). Micro and Nanoplastics analysis: Focus on their classification, sources, and impacts in marine environment. *Regional Studies in Marine Science*, 42, 101625. DOI: 10.1016/j.rsma.2021.101625

Al-Thawadi, S. (2020). Microplastics and nanoplastics in aquatic environments: Challenges and threats to aquatic organisms. *Arabian Journal for Science and Engineering*, 45(6), 4419–4440. DOI: 10.1007/s13369-020-04402-z

Amato-Lourenço, L. F., Galvao, L. D. S., Weger, L. A. D., Heimstra, P. S., Vijver, M. G., & Mauad, T. (2020). An Emerging Class of Air Pollutants: Potential Effects of Microplastics to Respiratory Human Health? *The Science of the Total Environment*, 749, 141676. Advance online publication. DOI: 10.1016/j.scitotenv.2020.141676

Amin, M. T., & Alabama Manzoor, U. (2014). A review of removal of pollutants from water/wastewater using different types of nanomaterials. *Advances in Materials Science and Engineering*, 825910. Advance online publication. DOI: 10.1155/2014/825910

Amobonye, A., Bhagwat, P., Raveendran, S., Singh, S., & Pillai, S. (2021). Environmental Impacts of Microplastics and Nanoplastics: A Current Overview. *Frontiers in Microbiology*, 12, 768297. Advance online publication. DOI: 10.3389/fmicb.2021.768297

Anand, U., Dey, S., Bontempi, E., Ducoli, S., Vethaak, A. D., Dey, A., & Federici, S. (2023). Biotechnological method to remove microplastic: A review. *Environmental Chemistry Letters*, 21(3), 1787–1810. DOI: 10.1007/s10311-022-01552-4

Arif, Y., Mir, A. R., Zieliński, P., Hayat, S., & Bajguz, A. (2024). Microplastics and nanoplastics: Source, behavior, remediation, and multi-level environmental impact. *Journal of Environmental Management*, 356, 120618. DOI: 10.1016/j.jenvman.2024.120618

Arisman, and Yun Arifatul Fatimah. (2023). "Waste Management in Indonesia: Strategies and Implementation of Sustainable Development Goals (SDGs) and Circular Economy.". In Circular Economy Adoption: Catalysing Decarbonisation Through Policy Instruments (pp. 131–157). Springer Nature Singapore. doi:DOI: 10.1007/978-981-99-4803-1_5

Arora, M. K., Lal, S., & Singh, B. (2024). A Way Forward to Realm of Environmental Law in India: Unveiling Judicial Activism and Enforcement Challenges. *National Journal of Environmental Law*, 7(1), 85–89.

Arshadi, M., Soleymanzadeh, M., & Salvacion, J., & SalimiVahid, F. (2014). Nanoscale Zero-Valent Iron (NZVI) supported on singular waste for Pb (II) removal from aqueous solution: Kinetics, thermodynamic and mechanism. *Journal of Colloid and Interface Science*, 426, 241–251. DOI: 10.1016/j.jcis.2014.04.014

Ashai, Perfect. "Analysis of marine plastic litter policies in Ghana." (2021).

Ashfaq, A., Khursheed, N., Fatima, S., Anjum, Z., & Younis, K. (2022). Application of nanotechnology in food packaging: Pros and Cons. *Journal of Agriculture and Food Research*, 7, 100270. DOI: 10.1016/j.jafr.2022.100270

Asselt, V. A., & Te Giffel, M. C. (2005). Pathogen resistance, and adaptation to disinfectants and sanitisers. In *Understanding Pathogen Behaviour* (pp. 484–506). Elsevier., DOI: 10.1533/9781845690229.3.484

Astner, A. F., Hayes, D. G., O'Neill, H. M., Evans, B. R., Pingali, S. V., Urban, V. S., & Young, T. M. (2022). Forming micro-and nano-plastics from agricultural plastic films for employment in fundamental research studies. [Journal of Visualized Experiments] [PubMed]. *Journal of Visualized Experiments*, (185), e64112.

Auta, H. S., Emenike, C. U., & Fauziah, S. H. (2017). Screening of Bacillus strains isolated from mangrove ecosystems in Peninsular Malaysia for microplastic degradation. *Environmental Pollution*, 231, 1552–1559. DOI: 10.1016/j.envpol.2017.09.043

Aylaz, G., & Konar, N. M. (2024). Current publication activity on nanoplastics & microplastics: A bibliometric analysis. *Sustainable Water Resources Management*, 10(2), 95. DOI: 10.1007/s40899-024-01072-9

Azeem, I., Adeel, M., Ahmad, M. A., Shakoor, N., Jiangcuo, G. D., Azeem, K., Ishfaq, M., Shakoor, A., Ayaz, M., Xu, M., & Rui, Y. (2021). Uptake and Accumulation of Nano/Microplastics in Plants: A Critical Review. *Nanomaterials (Basel, Switzerland)*, 11(11), 2935. DOI: 10.3390/nano11112935

Banerjee, T., & Srivastava, R. K. (2012). Plastics waste management and resource recovery in India. *International Journal of Environment and Waste Management*, 10(1), 90–111. DOI: 10.1504/IJEWM.2012.048153

Bao, S., Yi, J., Xian, B., Rao, C., Xiang, D., Tang, W., & Fang, T. (2024). Global analysis of the adverse effects of micro-and nanoplastics on intestinal health and microbiota of fish. *Journal of Hazardous Materials*, 470, 134157. DOI: 10.1016/j.jhazmat.2024.134157

Barage, S., Lakkakula, J., Sharma, A., Roy, A., Alghamdi, S., Almehmadi, M., Hossain, M. J., Allahyani, M., & Abdulaziz, O. (2022). Nanomaterial in Food Packaging: A Comprehensive Review. *Journal of Nanomaterials*, 2022(1), 1–12. DOI: 10.1155/2022/6053922

Barnes, J. K. (2020). Accumulation and fragmentation of plastic debris in marine environments. *Environmental Pollution*, 269(110915), 1–17.

Barría, C., Brandts, I., Tort, L., Oliveira, M., & Teles, M. (2020). Effect of nanoplastics on fish health and performance: A review. [PubMed]. *Marine Pollution Bulletin*, 151, 110791. DOI: 10.1016/j.marpolbul.2019.110791

Barros, J., Kumar, S., & Seena, S. (2023). Does functionalised nanoplastics modulate the cellular and physiological responses of aquatic fungi to metals? [PubMed]. *Environmental Pollution*, 337, 122549. DOI: 10.1016/j.envpol.2023.122549

Barrowclough, D., & Birkbeck, C. D. (2020) *Transforming the global plastics economy: the political economy and governance of plastics production and pollution.* No. 142. GEG Working Paper.

Baruah, A., Chaudhary, V., Malik, R., & Tomer, V. K. (2019). Nanotechnology Based Solutions for Wastewater Treatment. In Nanotechnology in Water and Wastewater Treatment: Theory and Applications (pp. 337–368). Elsevier. doi:DOI: 10.1016/B978-0-12-813902-8.00017-4

Basavegowda, N., Mandal, T. K., & Baek, K. H. (2019). Bimetallic and Trimetallic Nanoparticles for Active Food Packaging Applications: A Review. *Food and Bioprocess Technology*, 13(1), 30–44. DOI: 10.1007/s11947-019-02370-3

Bello, Fatai Abiola. (2022) "Critical analysis of policies on single-use plastics disposal from ships as a source of pollution to the marine environment.".

Benfer, S., Popp, U., Richter, H., Siewert, C., & Tomandl, G. (2001). Development and characterization of ceramic nanofiltration membranes. *Separation and Purification Technology*, 22(1-2), 231–237. DOI: 10.1016/S1383-5866(00)00133-7

Berber, M.R. (2020). Current Advances of Polymer Composites for Water Treatment and Desalination. *J. Chem., 2020 (Article ID 7608423)*, 19 pages. doi:DOI: 10.1155/2020/7608423

Bermúdez, J. R., & Swarzenski, P. W. (2021). A microplastic size classification scheme aligned with universal plankton survey methods. *MethodsX*, 8, 101516. DOI: 10.1016/j.mex.2021.101516

Bhardwaj, L. K. (2022). Evaluation of bis (2-ethylhexyl) phthalate (DEHP) in the PET bottled mineral water of different brands and impact of heat by GC–MS/MS. *Chemistry Africa*, 5(4), 929–942. DOI: 10.1007/s42250-022-00385-6

Bhardwaj, L. K. (2023). Occurrence of microplastics (MPs) in Antarctica and its impact on the health of organisms. *Maritime Technology and Research*, 6(2), 265418. DOI: 10.33175/mtr.2024.265418

Bhardwaj, L. K., & Jindal, T. (2020). Persistent organic pollutants in lakes of Grovnes Peninsula at Larsemann Hill area, East Antarctica. *Earth Systems and Environment*, 4(2), 349–358. DOI: 10.1007/s41748-020-00154-w

Bhardwaj, L. K., Kumar, D., & Kumar, A. (2023b). Phytoremediation Potential of Ocimum Sanctum: A Sustainable Approach for Remediation of Heavy Metals. In *Phytoremediation Potential of Medicinal and Aromatic Plants* (pp. 46-57). CRC Press. doi:DOI: 10.20944/preprints202308.0593.v1

Bhardwaj, L. K., Rath, P., & Choudhury, M. (2023a). A comprehensive review on the classification, uses, sources of nanoparticles (NPs) and their toxicity on health. *Aerosol Science and Engineering*, 7(1), 69–86. DOI: 10.1007/s41810-022-00163-4

Bhardwaj, L. K., Rath, P., Yadav, P., & Gupta, U. (2024). Microplastic contamination, an emerging threat to the freshwater environment: A systematic review. *Environmental Systems Research*, 13(1), 1–17. DOI: 10.1186/s40068-024-00338-7

Bhardwaj, L. K., & Sharma, A. (2021a). Estimation of physico-chemical, trace metals, microbiological and phthalate in PET bottled water. *Chemistry Africa*, 4(4), 981–991. DOI: 10.1007/s42250-021-00267-3

Bhardwaj, L. K., & Sharma, A. (2021b). Microplastics (MPs) in Drinking Water: Uses, Sources & Transport. https://doi.org/doi:10.20944/preprints202104.0498.v1

Bhardwaj, L., Sharma, S., Ranjan, A., & Jindal, T. (2019). Persistent organic pollutants in lakes of Broknes peninsula at Larsemann Hills area, East Antarctica. [PubMed]. *Ecotoxicology (London, England)*, 28(5), 589–596. DOI: 10.1007/s10646-019-02045-x

Blackburn, K., & Green, D. (2022). The potential effects of microplastics on human health: What is known and what is unknown. *Ambio*, 51(3), 518–530. DOI: 10.1007/s13280-021-01589-9

Bodzek, M., Konieczny, K., & Kwiecińska-Mydlak, A. (2020). Nanotechnology in water and wastewater treatment. Graphene–the nanomaterial for the next generation of semipermeable membranes. *Critical Reviews in Environmental Science and Technology*, 50(15), 1515–1579. DOI: 10.1080/10643389.2019.1664258

Borongan, G., & NaRanong, A. (2022). Factors in enhancing environmental governance for marine plastic litter abatement in Manila, the Philippines: A combined structural equation modeling and DPSIR framework. *Marine Pollution Bulletin*, 181, 113920. DOI: 10.1016/j.marpolbul.2022.113920

Borrelle, S. B., Ringma, J., Law, K. L., Monnahan, C. C., Lebreton, L., McGivern, A., & Rochman, C. M. (2020). Predicted growth in plastic waste exceeds efforts to mitigate plastic pollution. *Science*, 369(6510), 1515–1518. DOI: 10.1126/science.aba3656

Bostan, N., Ilyas, N., Akhtar, N., Mehmood, S., Saman, R. U., Sayyed, R. Z., & Pandiaraj, S. (2023). Toxicity assessment of microplastic (MPs); a threat to the ecosystem. *Environmental Research*, 116523, 116523. Advance online publication. DOI: 10.1016/j.envres.2023.116523

Boucher, J., & Friot, D. (2017). *Primary microplastics in the oceans: A global evaluation of sources*. IUCN International Union for Conservation of Nature., DOI: 10.2305/IUCN.CH.2017.01

Bourdages, M. P. T., Provencher, J. F., Baak, J. E., Mallory, M. L., & Vermaire, J. C. (2021). Breeding seabirds as vectors of microplastics from sea to land: Evidence from colonies in Arctic Canada. *The Science of the Total Environment*, 764, 142808. DOI: 10.1016/j.scitotenv.2020.142808

Braga, H. O., & Musiello-Fernandes, J. (2022). Types of Fisheries and Their Impact on Sustainable Development Goals. In *Life Below Water* (pp. 1075–1084). Springer International Publishing., DOI: 10.1007/978-3-319-98536-7_20

Brahney, J. H. (2020). Microplastics in the marine environment. *Environmental Science & Technology*, 54(19), 11490–11514.

Brander, S. M., König, A., Almroth, B. C., & Hampton, L. T. (2024). The potential for toxicit y to fishes from micro-and nanoplastics, and their additives. *Toxicology of Fishes. CRC Press*, 362-391.

Bruggen, B., Hoek, E. M., & Tarabara, V. V. (2013). *Nanofiltration. Encyclopedia of Membrane Science and Technology*. John Wiley & Sons.

Brusius, B. K., Buss de Souza, R., & Barbieri, E. (2022). Stranding of marine animals: Effects of environmental variables. In *Life below water* (pp. 945–954). Springer International Publishing., DOI: 10.1007/978-3-319-98536-7_102

Burke, D. J., Zhu, S., Pablico-Lansigan, M. P., Hewins, C. R., & Samia, A. C. S. (2014). Titanium oxide nanoparticle effects on composition of soil microbial communities and plant performance. *Biology and Fertility of Soils*, 50(7), 1169–1173. DOI: 10.1007/s00374-014-0938-3

Busch, P.-O., Schulte, M. L., & Simon, N. (2021). *Strengthen the global science and knowledge base to reduce marine plastic pollution*. Nordic Council of Ministers., DOI: 10.6027/temanord2021-519

Cai, Z., Li, M., Zhu, Z., Wang, X., Huang, Y., Li, T., Gong, H., & Yan, M. (2023). Biological Degradation of Plastics and Microplastics: A Recent Perspective on Associated Mechanisms and Influencing Factors. *Microorganisms*, 11(7), 1661. DOI: 10.3390/microorganisms11071661

Campanale, C., Galafassi, S., Savino, I., Massarelli, C., Ancona, V., Volta, P., & Uricchio, V. F. (2022). Microplastics pollution in the terrestrial environments: Poorly known diffuse sources and implications for plants. *The Science of the Total Environment*, 805, 150431. DOI: 10.1016/j.scitotenv.2021.150431

Campanale, C., Massarelli, C., Savino, I., Locaputo, V., & Uricchio, V. F. (2020). A Detailed Review Study on Potential Effects of Microplastics and Additives of Concern on Human Health. *International Journal of Environmental Research and Public Health*, 17(4), 1212. DOI: 10.3390/ijerph17041212

Caracci, E., Vega-Herrera, A., Dachs, J., Berrojalbiz, N., Buonanno, G., Abad, E., Llorca, M., Moreno, T., & Farré, M. (2023). Micro (nano) plastics in the atmosphere of the Atlantic Ocean. [PubMed]. *Journal of Hazardous Materials*, 450, 131036. DOI: 10.1016/j.jhazmat.2023.131036

Cárdenas-Alcaide, M. F., Godínez-Alemán, J. A., González-González, R. B., Iqbal, H. M., & Parra-Saldívar, R. (2022). Environmental impact and mitigation of micro (nano) plastics pollution using green catalytic tools and green analytical methods. *Green Analytical Chemistry*, 3, 100031. DOI: 10.1016/j.greeac.2022.100031

Carr, S. A. (2016). Removal of microplastics by wastewater treatment plants: Does it really matter? *Water Research*, 92, 90–99.

Chadha, U., Bhardwaj, P., Selvaraj, S. K., Arasu, K., Praveena, S., Pavan, A., Khanna, M., Singh, P., Singh, S., Chakravorty, A., Badoni, B., Banavoth, M., Sonar, P., & Paramasivam, V. (2022). Current Trends and Future Perspectives of Nanomaterials in Food Packaging Application. *Journal of Nanomaterials*, 2022(1), 1–32. DOI: 10.1155/2022/2745416

Chai, M., Li, R., Li, B., Wu, H., & Yu, L. (2023). Responses of mangrove (*Kandelia obovate* L.) growth, photosynthesis, and rhizosphere soil properties to microplastic pollution. *Marine Pollution Bulletin*, 189, 114827. DOI: 10.1016/j.marpolbul.2023.114827

Chamas, A., Moon, H., Zheng, J., Qiu, Y., Tabassum, T., Jang, J. H., & Suh, S. (2020). Degradation rates of plastics in the environment. *ACS Sustainable Chemistry & Engineering*, 8(9), 3494–3511. DOI: 10.1021/acssuschemeng.9b06635

Chan, C. K. M., Park, C., Chan, K. M., Mak, D. C. W., Fang, J. K. H., & Mitrano, D. M. (2021). Microplastic fibre releases from industrial wastewater effluent: A textile wet-processing mill in China. *Environmental Chemistry*, 18(3), 93–100. DOI: 10.1071/EN20143

Chan, N. (2024). Voluntary commitments for ocean sustainability: Comparing the UN Ocean Conference and Our Ocean Conference platforms. *Marine Policy*, 160, 105956. DOI: 10.1016/j.marpol.2023.105956

Charcosset, C. (2012). Some perspectives. Membrane Processes in Biotechnology and Pharmaceutics. Elsevier.

Chatziparaskeva, G., Papamichael, I., & Zorpas, A. A. (2022). Microplastics in the coastal environment of Mediterranean and the impact on sustainability level. *Sustainable Chemistry and Pharmacy*, 29, 100768. DOI: 10.1016/j.scp.2022.100768

Chaudhary, P., Fatima, F., & Kumar, A. (2020). Relevance of Nanomaterials in Food Packaging and its Advanced Future Prospects. *Journal of Inorganic and Organometallic Polymers and Materials*, 30(12), 5180–5192. DOI: 10.1007/s10904-020-01674-8

Chaudhary, V., Thakur, N., Chaudhary, S., & Bangar, S. P. (2023). Remediation plan of nano/microplastic toxicity in food. [). Academic Press.] [PubMed]. *Advances in Food and Nutrition Research*, 103, 397–442. DOI: 10.1016/bs.afnr.2022.07.004

Chausali, N., Saxena, J., & Prasad, R. (2022). Recent trends in nanotechnology applications of bio-based packaging. *Journal of Agriculture and Food Research*, 7, 100257. DOI: 10.1016/j.jafr.2021.100257

Chen, E.-Y., Lin, K.-T., Jung, C.-C., Chang, C.-L., & Chen, C.-Y. (2022). Characteristics and influencing factors of airborne microplastics in nail salons. *The Science of the Total Environment*, 806(Pt 4), 151472. DOI: 10.1016/j.scitotenv.2021.151472

Chen, H., Wang, Y., Sun, X., Peng, Y., & Xiao, L. (2020). Mixing effect of polylactic acid microplastic and straw residue on soil property and ecological function. *Chemosphere*, 243, 125271. DOI: 10.1016/j.chemosphere.2019.125271

Chen, L. Y., Li, Y., Liu, S., Junaid, M., & Wang, J. (2022). Effects of micro(nano) plastics on higher plants and the rhizosphere environment. *The Science of the Total Environment*, 807, 1–9. DOI: 10.1016/j.scitotenv.2021.150841

Chen, Q., Gao, J., Yu, H., Su, H., Yang, Y., Cao, Y., Zhang, Q., Ren, Y., Holleret, H., Shi, H., Chen, C., & Liu, H. (2022). An emerging role of microplastics in the etiology of lung ground glass nodules. *Environmental Sciences Europe*, 34(1), 25. DOI: 10.1186/s12302-022-00605-3

Chen, T., Jiang, H., He, Y., Shen, Y., Huang, Z., Gu, Y., Wei, Q., Zhao, J., & Chen, X. (2024). Nanoplastics and chrysene pollution: Potential new triggers for nonalcoholic fatty liver disease and hepatitis, insights from juvenile Siniperca chuatsi. *The Science of the Total Environment*, 922, 171125. DOI: 10.1016/j.scitotenv.2024.171125

Chen, X., Sun, Y., Feng, Z., Mu, C., & Zhang, T. (2024). Effects of micro-and nanoplastics on the physiology and metabolism of Portunus trituberculatus. *Clean (Weinheim)*, 52(3), 2300175. DOI: 10.1002/clen.202300175

Chen, Y., Jing, S., Wang, Y., Song, Z., Xie, L., Shang, X., Fu, H., Yang, X., Wang, H., Wu, M., Chen, Y., Li, Q., Zhang, Y., Wang, W., Zhang, L., Wang, R., Fang, M., Zhang, Y., Li, W., & Chen, J. (2024). Quantification and Characterization of Fine Plastic Particles as Considerable Components in Atmospheric Fine Particles. *Environmental Science & Technology*, 58(10), 4691–4703. DOI: 10.1021/acs.est.3c06832

Cheriyamundath, S., & Vavilala, S. L. (2021). Nanotechnology-based wastewater treatment. *Water and Environment Journal : the Journal / the Chartered Institution of Water and Environmental Management*, 35(1), 123–132. DOI: 10.1111/wej.12610

Chiang, C. C., Yeh, H., Shiu, R. F., Chin, W. C., & Yen, T. H. (2024). Impact of microplastics and nanoplastics on liver health: Current understanding and future research directions. *World Journal of Gastroenterology*, 30(9), 1011–1017. DOI: 10.3748/wjg.v30.i9.1011

Chikodzi, David, Kaitano Dube, and Nonkululeko Ngcobo. (2021) "Rethinking harbours, beaches and urban estuaries waste management under climate-induced floods in South Africa." *The increasing risk of floods and tornadoes in Southern Africa* 127-140.

Christodoulou, A., & Fernández, J. E. (2021). Maritime Governance and International Maritime Organization instruments focused on sustainability in the light of United Nations' sustainable development goals. In Sustainability in the Maritime Domain: Towards Ocean Governance and Beyond (pp. 415–461). Springer International Publishing. doi:DOI: 10.1007/978-3-030-69325-1_20

Christudoss, A. C., Chandrasekaran, N., & Mukherjee, A. (2024). Polystyrene nanoplastics alter the ecotoxicological effects of diclofenac on freshwater microalgae Scenedesmus obliquus. *Environmental Science. Processes & Impacts*, 26(1), 56–70. DOI: 10.1039/D3EM00341H

Chukwuemeka, I. S., Li, G., Mo, Y., & Jacques, K. J. (2024). *Impacts of microplastics and urbanization on soil health: an urgent concern for sustainable development.* Green Analytical Chemistry.

Cohen, N., & Radian, A. (2022). Microplastic textile fibers accumulate in sand and are potential sources of micro (nano) plastic pollution. [PubMed]. *Environmental Science & Technology*, 56(24), 17635–17642. DOI: 10.1021/acs.est.2c05026

Connected, S. (2020, March 25). Will We Ever Find a Microplastics Alternative? *Science Connected Magazine*. https://magazine.scienceconnected.org/2020/03/will-we-ever-find-a-microplastics-alternative/

Contera, S., De La Serna, J. B., & Tetley, T. D. (2020). Biotechnology, nanotechnology, and medicine. *Emerging Topics in Life Sciences*, 4(6), 551–554. DOI: 10.1042/ETLS20200350

Cooley, H., Ajami, N., Ha, M. L., Srinivasan, V., Morrison, J., Donnelly, K., & Christian-Smith, J. (2014). Global water governance in the twenty-first century. The World's Water. Springer.

Cowan, E., & Tiller, R. (2021). What shall we do with a sea of plastics? A systematic literature review on how to pave the road toward a global comprehensive plastic governance agreement. *Frontiers in Marine Science*, 8, 798534. DOI: 10.3389/fmars.2021.798534

Cox, K. D., Covernton, G. A., Davies, H. L., Dower, J. F., Juanes, F., & Dudas, S. E. (2019). Human Consumption of Microplastics. *Environmental Science & Technology*, 53(12), 7071. DOI: 10.1021/acs.est.9b01517

Crawford, C. B., & Quinn, B. (2017). The interactions of microplastics and chemical pollutants. In *Microplastic Pollutants* (pp. 131–157). Elsevier., DOI: 10.1016/B978-0-12-809406-8.00006-2

da Costa, J. P. (2018). Micro-and nanoplastics in the environment: Research and policymaking. *Current Opinion in Environmental Science & Health*, 1, 12–16. DOI: 10.1016/j.coesh.2017.11.002

da Costa, J. P., Santos, P. S., Duarte, A. C., & Rocha-Santos, T. (2016). (Nano)plastics in the environment–sources, fates and effects. *The Science of the Total Environment*, 566, 15–26. DOI: 10.1016/j.scitotenv.2016.05.041

Danion, A., Disdier, J., Guillard, C., Abdelmalek, F., & Jaffrezic-Renault, N. (2004). Characterization and study of a single-TiO_2-coated optical fiber reactor. *Applied Catalysis B: Environmental*, 52(3), 213–223. DOI: 10.1016/j.apcatb.2004.04.005

de Marffy-Mantuano, A. (2021) "Legal Tools in Combating Marine Pollution and Mitigating the Effects of Acidification." In *Fulfilling the Sustainable Development Goals*, pp. 341-356. Routledge, doi:DOI: 10.4324/9781003144274-27

de Sousa, F. D. B. (2021). The role of plastic concerning the sustainable development goals: The literature point of view. *Cleaner and Responsible Consumption*, 3, 100020. DOI: 10.1016/j.clrc.2021.100020

De Sousa, M. S., Schlogl, A. E., Estanislau, F. R., Souza, V. G. L., Coimbra, J. S. D. R., & Santos, I. J. B. (2023). Nanotechnology in Packaging for Food Industry: Past, Present, and Future. *Coatings*, 13(8), 1411. DOI: 10.3390/coatings13081411

De Souza Machado, A. A., Lau, C. W., Kloas, W., Bergmann, J., Bachelier, J. B., Faltin, E., & Rillig, M. C. (2019). Microplastics can change soil properties and affect plant performance. *Environmental Science & Technology*, 53(10), 6044–6052. DOI: 10.1021/acs.est.9b01339

de Souza, A. M., Santos, A. L., Araújo, D. S., de Barros Magalhães, R. R., & Rocha, T. L. (2022). Micro (nano) plastics as a vector of pharmaceuticals in aquatic ecosystem: Historical review and future trends. *Journal of Hazardous Materials Advances*, 6, 100068. DOI: 10.1016/j.hazadv.2022.100068

Dehghanian, Z., Asgari Lajayer, B., Biglari Quchan Atigh, Z., Nayeri, S., Ahmadabadi, M., Taghipour, L., Senapathi, V., Astatkie, T., & Price, G. W. (2023). Micro (nano) plastics uptake, toxicity and detoxification in plants: Challenges and prospects. *Ecotoxicology and Environmental Safety*, 268, 115676. DOI: 10.1016/j.ecoenv.2023.115676

Delhi, P. I. B. (2022) Ban on identified Single Use Plastic Items from 1st July 2022. Ministry of Environment, forest, and climate change. https://www.unep.org/news-and-stories/story/microplastics-long-legacy-left-behind-plastic-pollution

Deme, G. G., Ewusi-Mensah, D., Olagbaju, O. A., Okeke, E. S., Okoye, C. O., Odii, E. C., & Sanganyado, E. (2022). Macro problems from microplastics: Toward a sustainable policy framework for managing microplastic waste in Africa. *The Science of the Total Environment*, 804, 150170. DOI: 10.1016/j.scitotenv.2021.150170

Deng, L., Cheung, S., Liu, J., Chen, J., Chen, F., Zhang, X., & Liu, H. (2024). Nanoplastics impair growth and nitrogen fixation of marine nitrogen-fixing cyanobacteria. *Environmental Pollution*, 350, 123960. DOI: 10.1016/j.envpol.2024.123960

Derco, J., & Vrana, B. (2018). Introductory Chapter: Biosorption. In *Biosorption*. IntechOpen: London, UK.

Deshpande, P. C., & Haskins, C. (2021). Application of systems engineering and sustainable development goals towards sustainable management of fishing gear resources in Norway. *Sustainability (Basel)*, 13(9), 4914. DOI: 10.3390/su13094914

Dessì, C., Okoffo, E. D., O'Brien, J. W., Gallen, M., Samanipour, S., Kaserzon, S., Rauert, C., Wang, X., & Thomas, K. V. (2021). Plastics contamination of store-bought rice. *Journal of Hazardous Materials*, 416, 125778. DOI: 10.1016/j.jhazmat.2021.125778

Devlaeminck, D. J. The Role of International Water Law in the Management of Marine Plastic Pollution: The Case of China and Its Transboundary Rivers. In *Plastic Pollution* (pp. 190–206). CRC Press.

Di Bella, G., Corsino, S. F., De Marines, F., Lopresti, F., La Carrubba, V., Torregrossa, M., & Viviani, G. (2022). Occurrence of Microplastics in Waste Sludge of Wastewater Treatment Plants: Comparison between Membrane Bioreactor (MBR) and Conventional Activated Sludge (CAS) Technologies. [PubMed]. *Membranes*, 12(4), 4. Advance online publication. DOI: 10.3390/membranes12040371

Diallo, M., Duncan, J., Savage, N., Street, A., & Sustich, R. (2008). Nanotechnology Applications for Clean Water: Solutions for Improving Water Quality, 0815519737.

Diaz-Basantes, M. F., Conesa, J. A., & Fullana, A. (2020). Microplastics in Honey, Beer, Milk and Refreshments in Ecuador as Emerging Contaminants. *Sustainability (Basel)*, 12(14), 14. Advance online publication. DOI: 10.3390/su12145514

Di, M., Liu, X., Wang, W., & Wang, J. (2019). Corrigendum to" Pollution in drinking water source areas: Microplastics in the Danjiangkou Reservoir, China [Environ. Toxicol. Pharmacol. 65 (January) 2019 82-89] [PubMed]. *Environmental Toxicology and Pharmacology*, 66, 133–133. DOI: 10.1016/j.etap.2019.01.002

Ding, H., Luo, X., Zhang, X., & Yang, H. (2019). Alginate-immobilized Aspergillus niger: Characterization and biosorption removal of thorium ions from radioactive wastewater. *Colloids and Surfaces. A, Physicochemical and Engineering Aspects*, 562, 186–195. DOI: 10.1016/j.colsurfa.2018.11.032

Ding, L., Mao, R., Guo, X., Yang, X., Zhang, Q., & Yang, C. (2019). Microplastics in surface waters and sediments of the Wei River, in the northwest of China. [PubMed]. *The Science of the Total Environment*, 667, 427–434. DOI: 10.1016/j.scitotenv.2019.02.332

Do Atn, H. Y. (2022). Leaching of microplastic-associated additives in aquatic environments: A critical review. *Environ Pollut.*, •••, 1–14.

Donkers, J. M., Hoppener, E. M., Grigorieve, I., Will, L., Melgret, B. N., Zan, B. V. D., Steeg, E. V. D., & Kooter, I. M. (2022). Advanced Epithelial Lung and Gut Barrier Models Demonstrate Passage of Microplastic Particles. *Microplastics and Nanoplastics*, 2(1), 6. DOI: 10.1186/s43591-021-00024-w

dos Muchangos, L. S., & Velempini, K. M. "Plastic Waste Management in Southern African–an Overview of Mozambique and Botswana." Available at *SSRN* 4715926. doi:DOI: 10.2139/ssrn.4715926

Dris, R. G. (2017). Synthetic microfiber pollution in airborne dust: Potential for human inhalation exposure. *Marine Pollution Bulletin*, 120(1-2), 49–53.

Du, C., Liang, H., Li, Z., & Gong, J. (2020). Pollution Characteristics of Microplastics in Soils in Southeastern Suburbs of Baoding City, China. *nternational Journal of Environmental Research and Public Health*, 1-14.

Duarte, C. M., Agusti, S., Barbier, E., Britten, G. L., Castilla, J. C., Gattuso, J.-P., Fulweiler, R. W., Hughes, T. P., Knowlton, N., Lovelock, C. E., Lotze, H. K., Predragovic, M., Poloczanska, E., Roberts, C., & Worm, B. (2020). Rebuilding marine life. *Nature*, 580(7801), 39–51. DOI: 10.1038/s41586-020-2146-7

Dube, E., & Okuthe, G. E. (2023). Plastics and Micro/Nano-Plastics (MNPs) in the Environment: Occurrence, Impact, and Toxicity. *International Journal of Environmental Research and Public Health*, 20(17), 17. Advance online publication. DOI: 10.3390/ijerph20176667

Duis, K., & Coors, A. (2016). Microplastics in the aquatic and terrestrial environment: Sources (with a specific focus on personal care products), fate and effects. *Environmental Sciences Europe*, 28(1), 2. DOI: 10.1186/s12302-015-0069-y

Du, J., Guo, Z., Li, R., Ali, A., Guo, D., Lahori, A. H., Wang, P., Liu, X., Wang, X., & Zhang, Z. (2020). Screening of Chinese mustard (Brassica juncea L.) cultivars for the phytoremediation of Cd and Zn based on the plant physiological mechanisms. *Environmental Pollution*, 261, 114213. DOI: 10.1016/j.envpol.2020.114213

Duncan, T. V., Khan, S. A., Patri, A. K., & Wiggins, S. (2024). Regulatory Science Perspective on the Analysis of Microplastics and Nanoplastics in Human Food. *Analytical Chemistry*, 96(11), 4343–4358. DOI: 10.1021/acs.analchem.3c05408

Ediboğlu Sakowsky, E., & Gündoğdu, S. (2024). An Imported Problem? Plastic Waste Effects on Türkiye's Environment with Specific Emphasis on Relevant Multilateral Environmental Agreements. In Plastic Waste Trade: A New Colonialist Means of Pollution Transfer (pp. 217–246). Springer Nature Switzerland. doi:DOI: 10.1007/978-3-031-51358-9_11

Elgarahy, A. M., Akhdhar, A., & Elwakeel, K. Z. (2021). Microplastics prevalence, interactions, and remediation in the aquatic environment: A critical review. *Journal of Environmental Chemical Engineering*, 9(5), 106224. DOI: 10.1016/j.jece.2021.106224

Elizabath, A. (2019). Application of Nanotechnology in Agriculture. *International Journal of Pure & Applied Bioscience*, 7(2), 131–139. DOI: 10.18782/2320-7051.6493

Elizalde-Velázquez, G. A., & Gómez-Oliván, L. M. (2021). Microplastics in aquatic environments: A review on occurrence, distribution, toxic effects, and implications for human health. [PubMed]. *The Science of the Total Environment*, 780, 146551. DOI: 10.1016/j.scitotenv.2021.146551

El-Saadony, M. T., Saad, A. M., Soliman, S. M., Salem, H. M., Desoky, E.-S. M., Babalghith, A. O., El-Tahan, A. M., Ibrahim, O. M., Ebrahim, A. A. M., Abd El-Mageed, T. A., Elrys, A. S., Elbadawi, A. A., El-Tarabily, K. A., & AbuQamar, S. F. (2022). Role of Nanoparticles in Enhancing Crop Tolerance to Abiotic Stress: A Comprehensive Review. *Frontiers in Plant Science*, 13, 946717. DOI: 10.3389/fpls.2022.946717

El-Sayed, M. E. (2020). Nanoadsorbents for water and wastewater remediation. *The Science of the Total Environment*, 739, 139903. DOI: 10.1016/j.scitotenv.2020.139903

Enyoh, C. E. (2019). Microplastic pollution in Nigeria: a call for concern. 26th meeting of the Imo State chapter of the Chemical Society of Nigeria (CSN) held at Imo State University, Nigeria. https://doi.org/DOI: 10.13140/RG.2.2.21027.9190

Enyoh, C. E., Verla, A. W., Verla, E. N., Ibe, F. C., & Amaobi, C. E. (2019). Airborne microplastics: A review study on method for analysis, occurrence, movement and risks. *Environmental Monitoring and Assessment*, 191(11), 668. DOI: 10.1007/s10661-019-7842-0

Epelle, E. I., Emmerson, A., Nekrasova, M., Macfarlane, A., Cusack, M., Burns, A., Mackay, W., & Yaseen, M. (2022b). Microbial Inactivation: Gaseous or Aqueous Ozonation? *Industrial & Engineering Chemistry Research*, 61(27), 9600–9610. DOI: 10.1021/acs.iecr.2c01551

Epelle, E. I., Macfarlane, A., Cusack, M., Burns, A., Amaeze, N., Richardson, K., Mackay, W., Rateb, M. E., & Yaseen, M. (2022a). Stabilisation of Ozone in Water for Microbial Disinfection. *Environments (Basel, Switzerland)*, 9(4), 45. DOI: 10.3390/environments9040045

Eriksen, M., Thiel, M., Prindiville, M., & Kiessling, T. (2018). Microplastic: what are the solutions. *Freshwater microplastics, 273.*

Ertz, M., Addar, W., Ouerghemmi, C., & Takaffoli, M. (2023). Overview of factors influencing consumer engagement with plastic recycling. *Wiley Interdisciplinary Reviews. Energy and Environment*, 12(6), e493. DOI: 10.1002/wene.493

Facciolà, A., Giuseppa, V., Ciarello, M. P., & Pietro A. D. (2021). Newly Emerging Airborne Pollutants. Breathing Plastic: The Health Impacts of Invisible Plastics in the Air - Center for International Environmental Law (ciel.org). doi:DOI: 10.3390/ijerph18062997

Fard, R. F., Azimi, A., & Bidhendi, G. N. (2011). Batch kinetics and isotherms for biosorption of cadmium onto biosolids. *Desalination and Water Treatment*, 28, 69–74. DOI: 10.5004/dwt.2011.2203

Farinha, I. Carlos AM Duarte, and Mafalda Gil de Carvalho. "S&S'21 Marine Plastic Pollution, Design & Circular Economy." In *UNIDCOM/IADE International Conference Senses & Sensibility*, pp. 308-320. Cham: Springer Nature Switzerland, 2021.

Feng, W., Deng, Y., Cao, Y., Liu, J., Han, Y., Liu, J., & Giesy, J. P. (2023). Biotechnology remediation and environmental behavior of microplastics in soils: A review. *Reviews of Environmental Contamination and Toxicology*, 261(1), 13. DOI: 10.1007/s44169-023-00038-w

Ferraro, G., & Failler, P. (2020). Governing plastic pollution in the oceans: Institutional challenges and areas for action. *Environmental Science & Policy*, 112, 453–460. DOI: 10.1016/j.envsci.2020.06.015

Fraissinet, S., De Benedetto, G. E., Malitesta, C., Holzinger, R., & Materić, D. (2024). Microplastics and nanoplastics size distribution in farmed mussel tissues. *Communications Earth & Environment*, 5(1), 128. DOI: 10.1038/s43247-024-01300-2

Galstyan, V., Bhandari, M., Sberveglieri, V., Sberveglieri, G., & Comini, E. (2018). Metal Oxide Nanostructures in Food Applications: Quality Control and Packaging. *Chemosensors (Basel, Switzerland)*, 6(2), 16. DOI: 10.3390/chemosensors6020016

Gao, B., Yao, H., Li, Y., & Zhu, Y. (2021). Microplastic addition alters the microbial community structure and stimulates soil carbon dioxide emissions in vegetable-growing soil. *Environmental Toxicology and Chemistry*, 40(2), 352–365. DOI: 10.1002/etc.4916

Gao, D., Junaid, M., Chen, X., Liao, H., Chen, G., & Wang, J. (2023). Interaction of micro (nano) plastics and bisphenols in the environment: A recent perspective on adsorption mechanisms, influencing factors and ecotoxic impacts. *Trends in Analytical Chemistry*, 165, 117132. DOI: 10.1016/j.trac.2023.117132

Gao, D., Liu, X., Junaid, M., Liao, H., Chen, G., Wu, Y., & Wang, J. (2022). Toxicological impacts of micro (nano) plastics in the benthic environment. [PubMed]. *The Science of the Total Environment*, 836, 155620. DOI: 10.1016/j.scitotenv.2022.155620

Gasperi, J., Wright, S. L., Dris, R., Collard, F., Mandin, C., Guerouache, M., Langlois, V., Kelly, F. J., & Tassin, B. (2018). Microplastics in Air: Are We Breathing It In? *Current Opinion in Environmental Science & Health*, 1, 1–5. DOI: 10.1016/j.coesh.2017.10.002

Gaur, V. K., Sirohi, R., Bhat, M. I., Gautam, K., Sharma, P., Srivastava, J. K., & Pandey, A. (2023). A review on the effect of micro-and nano-plastics pollution on the emergence of antimicrobial resistance. *Chemosphere*, 311, 136877. DOI: 10.1016/j.chemosphere.2022.136877

Ge, J., Jin, P., Xie, S., Beardall, J., Feng, Y., Guo, C., Ma, Z., & Gao, G. (2024). Micro-and nanoplastics interact with conventional pollutants on microalgae: Synthesis through meta-analysis. *Environmental Pollution*, 342, 123127. DOI: 10.1016/j.envpol.2023.123127

Geremia, E., Muscari Tomajoli, M. T., Murano, C., Petito, A., & Fasciolo, G. (2023). The Impact of Micro-and Nanoplastics on Aquatic Organisms: Mechanisms of Oxidative Stress and Implications for Human Health—A Review. *Environments (Basel, Switzerland)*, 10(9), 161. DOI: 10.3390/environments10090161

Geyer, R. K. (2015). Leaked: The story of lost and mismanaged plastic from plastic production to plastic pollution. *Environmental Science & Technology*, 49(22), 12925–12932.

Geyer, R., Jambeck, J. R., & Law, K. L. (2017). Production, use, and fate of all plastics ever made. [PubMed]. *Science Advances*, 3(7), e1700782. DOI: 10.1126/sciadv.1700782

Ghosh, S. K., Chakraborty, S. S., Rana, T., Karstensen, K. H., & Saha, P. K. (2022). Legislative Support to Curb Marine Littering through Wastes Co-processing by Promoting Circular Economy. In *Effective Waste Management and Circular Economy* (pp. 27–41). CRC Press., DOI: 10.1201/9781003231608-4

Gilani, S. J., Alshawwa, S. Z., Asif, M., Fatima, K., & Khan, N. A. Microplastics and Nanoplastics in Food. In Handbook of Seafood and Seafood Products Analysis (pp. 915–926). CRC Press.

Giskes, I., Baziuk, J., Pragnell-Raasch, H., & Perez Roda, A. (2022). *Report on good practices to prevent and reduce marine plastic litter from fishing activities.* Food & Agriculture Org.

Gleick, P. H. (2003). Global freshwater resources: Soft-path solutions for the 21st century. *Science*, 302(5650), 1524–1528. DOI: 10.1126/science.1089967

Gong, X., Shi, G., Zou, D., Wu, Z., Qin, P., Yang, Y., & Zhou, Y. (2023). Micro-and nano-plastics pollution and its potential remediation pathway by phytoremediation. *Planta*, 257(2), 35. DOI: 10.1007/s00425-023-04069-4

Gouin, T. M. (2019). Microplastics in commercial laundry detergents: Washing clothes as a source of microfibre pollution. *Microplastics and Nanoplastics*, 1(4), 259–267.

Graham, Roxanne Elizabeth Donna. (2023) "Combatting the marine litter crisis in the Windward Islands: examining source-to-sea pathways and fostering multi-scale solutions.".

Gugushe, A. S., Mpupaa, A., & Nomngongoabc, P. N. (2019). Ultrasound-assisted magnetic solid phase extraction of lead and thallium in complex environmental samples using magnetic multi-walled carbon nanotubes/zeolite nanocomposite. *Microchemical Journal*, 149, 103960. DOI: 10.1016/j.microc.2019.05.060

Guidetti, G., Giuri, D., Zanna, N., Calvaresi, M., Montalti, M., & Tomasini, C. (2018). Biocompatible and Light-Penetrating Hydrogels for Water Decontamination. *ACS Omega*, 3(7), 8122–8128. DOI: 10.1021/acsomega.8b01037

Gupta, V., Mohapatra, S., Mishra, H., Farooq, U., Kumar, K., Ansari, M., Aldawsari, M., Alalaiwe, A., Mirza, M., & Iqbal, Z. (2022). Nanotechnology in Cosmetics and Cosmeceuticals-A Review of Latest Advancements. *Gels (Basel, Switzerland)*, 8(3), 173. DOI: 10.3390/gels8030173

Gusain, R., Kumar, N., & Ray, S. S. (2020). Recent Advances in Carbon Nanomaterial-Based Adsorbents for Water Purification. *Coordination Chemistry Reviews*, 405, 213111. DOI: 10.1016/j.ccr.2019.213111

Habibi, N., Uddin, S., Fowler, S. W., & Behbehani, M. (2022). Microplastics in the atmosphere: A review. Journal of Environmental Exposure Assessment, 1(1). doi:DOI: 10.20517/jeea.2021.07

Habumugisha, T., Zhang, Z., Uwizewe, C., Yan, C., Ndayishimiye, J. C., Rehman, A., & Zhang, X. (2024). Toxicological review of micro- and nano-plastics in aquatic environments: Risks to ecosystems, food web dynamics and human health. *Ecotoxicology and Environmental Safety*, 278, 116426. DOI: 10.1016/j.ecoenv.2024.116426

Haldar, S., Muralidaran, Y., Míguez, D., Mulla, S. I., & Mishra, P. (2023). Ecotoxicity of nano-plastics and its implication on human metabolism: Current and future perspective. *The Science of the Total Environment*, 861, 160571. DOI: 10.1016/j.scitotenv.2022.160571

Haleem, A., Javaid, M., Singh, R. P., Rab, S., & Suman, R. (2023). Applications of nanotechnology in medical field: A brief review. *Global Health Journal (Amsterdam, Netherlands)*, 7(2), 70–77. DOI: 10.1016/j.glohj.2023.02.008

Han, Y., Xu, Z., & Gao, C. (2013). Ultrathin Graphene Nanofiltration Membrane for Water Purification. *Advanced Functional Materials*, 23(29), 3693–3700. DOI: 10.1002/adfm.201202601

Hartmann, N. B., Rist, S., Bodin, J., Jensen, L. H., Schmidt, S. N., Mayer, P., & Baun, A. (2017). Microplastics as vectors for environmental contaminants: Exploring sorption, desorption, and transfer to biota. *Integrated Environmental Assessment and Management*, 13(3), 488–493. DOI: 10.1002/ieam.1904

He, D., Zhang, Y., & Gao, W. (2021). Micro (nano) plastic contaminations from soils to plants: Human food risks. *Current Opinion in Food Science*, 41, 116–121. DOI: 10.1016/j.cofs.2021.04.001

Heidbreder, L. M., Lange, M., & Reese, G. (2021). # Plasticfreejuly–analyzing a worldwide campaign to reduce single-use plastic consumption with twitter. *Environmental Communication*, 15(7), 937–953. DOI: 10.1080/17524032.2021.1920447

He, L., Wang, L., Zhu, H., Wang, Z., Zhang, L., Yang, L., Dai, Y., Mo, H., Zhang, J., & Shen, J. (2021). A reusable Fe3O4/GO-COOH nano adsorbent for Ca^{2+} and Cu^{2+} removal from oilfield wastewater. *Chemical Engineering Research & Design*, 166, 248–258. DOI: 10.1016/j.cherd.2020.12.019

hen G, L. Y. (2022). Effects of micro(nano)plastics on higher plants and the rhizosphere environment. *Sci Total Environ.*, 1-13.

Henderson, L. (2018). Telling stories about (micro) plastic pollution: Media images, public perceptions and social change.

Hermawan, S., & Astuti, W. (2021). Analysing several ASEAN countries' policy for combating marine plastic litter. *Environmental Law Review*, 23(1), 9–22. DOI: 10.1177/1461452921991731

Herrera-Franco, G., Merchán-Sanmartín, B., Caicedo-Potosí, J., Bitar, J. B., Berrezueta, E., & Carrión-Mero, P. (2023). A systematic review of coastal zone integrated waste management for sustainability strategies. *Environmental Research*, 117968.

Hess, S. M. (2012). Pesticide container disposal practices in developing and newly industrialized countries: A review. *Waste Management (New York, N.Y.)*, 32(5), 899–912.

Hilmi, N., Theux-Lowen, N., & Crisóstomo, M. B. (2022). Ocean-Related Impacts of Climate Change on Economy. In Life Below Water (pp. 779–790). Springer International Publishing. doi:DOI: 10.1007/978-3-319-98536-7_158

Hong Yu, Y. Z. (2022). Microplastics as an Emerging. *Frontiers in Environmental Science*, •••, 1–18.

Horton, T. L. (2017). Microplastic in terrestrial environments: Evaluating the evidence for source-to-sink fluxes. *Marine Pollution Bulletin*, 114(1-2), 201–216.

Hoseinnejad, M., Jafari, S. M., & Katouzian, I. (2017). Inorganic and metal nanoparticles and their antimicrobial activity in food packaging applications. *Critical Reviews in Microbiology*, 44(2), 161–181. DOI: 10.1080/1040841X.2017.1332001

Hossain, N., Mobarak, M. H., Mimona, M. A., Islam, M. A., Hossain, A., Zohura, F. T., & Chowdhury, M. A. (2023). Advances and significances of nanoparticles in semiconductor applications – A review. *Results in Engineering*, 19, 101347. DOI: 10.1016/j.rineng.2023.101347

Hou, J., Xu, X., Yu, H., Xi, B., & Tan, W. (2021). Comparing the long-term responses of soil microbial structures and diversities to polyethylene microplastics in different aggregate fractions. *Environment International*, 149, 106398. DOI: 10.1016/j.envint.2021.106398

Hrovat, B., Uurasjärvi, E., Viitala, M., Del Pino, A. F., Mänttäri, M., Papamatthaiakis, N., Haapala, A., Peiponen, K.-E., Roussey, M., & Koistinen, A. (2024). Preparation of synthetic micro-and nano plastics for method validation studies. *The Science of the Total Environment*, 925, 171821. DOI: 10.1016/j.scitotenv.2024.171821

Huang, D., Tao, J., Cheng, M., Deng, R., Chen, S., Yin, L., & Li, R. (2021). Microplastics and nanoplastics in the environment: Macroscopic transport and effects on creatures. *Journal of Hazardous Materials*, 407, 124399. DOI: 10.1016/j.jhazmat.2020.124399

Huerta-Lwanga, H. R.-O. (2016). The use of seed coating technologies to enhance water uptake and germination of soybean seeds. *Industrial Crops and Products*, 93, 155–167.

Ibrahim, R. K., Hayyan, M., AlSaadi, M. A., Hayyan, A., & Ibrahim, S. (2016). Environmental application of nanotechnology: Air, soil, and water. *Environmental Science and Pollution Research International*, 23(14), 13754–13788. DOI: 10.1007/s11356-016-6457-z

Iqbal, B., Zhao, T., Yin, W., Zhao, X., Xie, Q., Khan, K. Y., Zhao, X., Nazar, M., Li, G., & Du, D. (2023). Impacts of soil microplastics on crops: A review. *Applied Soil Ecology*, 181, 104680. DOI: 10.1016/j.apsoil.2022.104680

Iqbal, S., Xu, J., Allen, S. D., Khan, S., Nadir, S., Arif, M. S., & Yasmeen, T. (2020). Unraveling consequences of soil micro-and nano-plastic pollution on soil-plant system: Implications for nitrogen (N) cycling and soil microbial activity. *Chemosphere*, 260, 127578. DOI: 10.1016/j.chemosphere.2020.127578

Jain, H. (2023). Groundwater vulnerability and risk mitigation: A comprehensive review of the techniques and applications. *Groundwater for Sustainable Development*, 22, 100968. DOI: 10.1016/j.gsd.2023.100968

Jain, H., Yadav, V., Rajput, V. D., Minkina, T., Agarwal, S., & Garg, M. C. (2022). An eco-sustainable green approach for biosorption of methylene blue dye from textile industry wastewater by sugarcane bagasse, peanut hull, and orange peel: A comparative study through response surface methodology, isotherms, kinetic, and thermodynamics. *Water, Air, and Soil Pollution*, 233(6), 187. DOI: 10.1007/s11270-022-05655-0

Jain, R., Gaur, A., Suravajhala, R., Chauhan, U., Pant, M., Tripathi, V., & Pant, G. (2023). Microplastic pollution: Understanding microbial degradation and strategies for pollutant reduction. *The Science of the Total Environment*, 905, 167098. DOI: 10.1016/j.scitotenv.2023.167098

Jaiswal, K. K., Dutta, S., Banerjee, I., Pohrmen, C. B., Singh, R. K., Das, H. T., Dubey, S., & Kumar, V. (2022). Impact of aquatic microplastics and nanoplastics pollution on ecological systems and sustainable remediation strategies of biodegradation and photodegradation. *The Science of the Total Environment*, 806, 151358. DOI: 10.1016/j.scitotenv.2021.151358

Jambeck, J., Moss, E., Dubey, B., Arifin, Z., Godfrey, L., Hardesty, B. D., & Woodall, L. (2023). Leveraging multi-target strategies to address plastic pollution in the context of an already stressed ocean. In *The Blue Compendium: From Knowledge to Action for a Sustainable Ocean Economy* (pp. 141–184). Springer International Publishing., DOI: 10.1007/978-3-031-16277-0_5

Jasińska, A. R. S., Różalska, S., Rusetskaya, V., Słaba, M., & Bernat, P. (2022). Microplastic-Induced Oxidative Stress in Metolachlor-Degrading Filamentous Fungus Trichoderma harzianum. *International Journal of Molecular Sciences*, 23(21), 1–12. DOI: 10.3390/ijms232112978

Jenner, L. C., Rotchell, J. M., Bennett, R. T., Cowen, M., Tentzeris, V., & Sadofsky, L. R. (2022). Detection of microplastics in human lung tissue using μFTIR spectroscopy. *Science of The Total Environment.* 831, 154907.

Jia, L., Liu, L., Zhang, Y., Fu, W., Liu, X., Wang, Q., Tanveer, M., & Huang, L. (2023). Microplastic stress in plants: Effects on plant growth and their remediations. *Frontiers in Plant Science*, 14, 1226484. DOI: 10.3389/fpls.2023.1226484

Jiang, B., Kauffman, A. E., Li, L., McFee, W., Cai, B., Weinstein, J., Lead, J. R., Chatterjee, S., Scott, G. I., & Xiao, S. (2020). Health impacts of environmental contamination of micro-and nanoplastics: A review. *Environmental Health and Preventive Medicine*, 25(1), 1–15. DOI: 10.1186/s12199-020-00870-9

Johnson, A. C., Ball, H., Cross, R., Horton, A. A., Jürgens, M. D., Read, D. S., Vollertsen, J., & Svendsen, C. (2020). Identification and quantification of microplastics in potable water and their sources within water treatment works in England and Wales. [PubMed]. *Environmental Science & Technology*, 54(19), 12326–12334. DOI: 10.1021/acs.est.0c03211

Jones, N. (2023). Microplastics Are Filling the Skies. Will They Affect the Climate? YaleEnvironment360 published at Yale school of environment. 2023. https://e360.yale.edu/features/plastic-waste-atmosphere-climate-weather

Joo, S. H., & Tansel, B. (2015). Novel technologies for reverse osmosis concentrate treatment: A review. *Journal of Environmental Management*, 150, 322–335. DOI: 10.1016/j.jenvman.2014.10.027

Joos, D. T., & De Tender, C. (2022). The importance of soil life and how it is influenced by (micro)plastic pollution. *Computational and Structural Biotechnology Journal*, 20, 1–21. DOI: 10.1016/j.csbj.2022.03.041

Jouffray, J.-B., Blasiak, R., Nyström, M., Österblom, H., Tokunaga, K., Wabnitz, C., & Norström, A. (2021). "Blue Acceleration: An ocean of risks and opportunities." *Ocean Risk and Resilience Action Alliance (ORRAA).Report*, 42, 1–42.

Junaid, M., Abbas, Z., Siddiqui, J. A., Liu, S., Tabraiz, S., Yue, Q., & Wang, J. (2023). Ecotoxicological impacts associated with the interplay between micro (nano) plastics and pesticides in aquatic and terrestrial environments. *Trends in Analytical Chemistry*, 165, 117133. DOI: 10.1016/j.trac.2023.117133

Junaid, M., & Wang, J. (2022). Interaction of micro (nano) plastics with extracellular and intracellular biomolecules in the freshwater environment. *Critical Reviews in Environmental Science and Technology*, 52(23), 4241–4265. DOI: 10.1080/10643389.2021.2002078

K.L., P., K.R., R., Joseph, C. J., M.S., I., Srinivas, R., & S., H. (2022). Fate, transport and degradation pathway of microplastics in aquatic environment—A critical review. *Regional Studies in Marine Science*, 56, 102647. DOI: 10.1016/j.rsma.2022.102647

Kalengyo, R. B., & Nasr, M. (2024). Water resource management to meet SDG13 on climate change. In *Water, the Environment and the Sustainable Development Goals* (pp. 157–178). Elsevier., DOI: 10.1016/B978-0-443-15354-9.00014-1

Kalhor, M. S., Aliniaeifard, S., Seif, M., Asayesh, E. J., Bernard, F., Hassani, B., & Li, T. (2018). Title: Enhanced salt tolerance and photosynthetic performance: Implication of γ-amino butyric acid application in salt-exposed lettuce (*Lactuca sativa* L.) plants. *Plant Physiology and Biochemistry*, 130, 157–172. DOI: 10.1016/j.plaphy.2018.07.003

Karmakar, M. (2023, July 24). *Managing microplastic pollution is important for meeting sustainable development goals in India*. Down To Earth. https://www.downtoearth.org.in/water/managing-microplastic-pollution-is-important-for-meeting-sustainable-development-goals-in-india-90789

Kavya, A. N. L., Sundarrajan, S., & Ramakrishna, S. (2020). Identification and characterization of micro-plastics in the marine environment: A mini review. [PubMed]. *Marine Pollution Bulletin*, 160, 111704. DOI: 10.1016/j.marpolbul.2020.111704

Khalid, N., Aqeel, M., Noman, A., & Rizvi, Z. F. (2023). Impact of plastic mulching as a major source of microplastics in agroecosystems. *Journal of Hazardous Materials*, 445, 130455. DOI: 10.1016/j.jhazmat.2022.130455

Khan, N. A., Khan, A. H., López-Maldonado, E. A., Alam, S. S., López, J. R. L., Herrera, P. F. M., & Singh, L. (2022). Microplastics: Occurrences, treatment methods, regulations and foreseen environmental impacts. [PubMed]. *Environmental Research*, 215, 114224. DOI: 10.1016/j.envres.2022.114224

Kharraz, J. A., Jia, M., Farid, M. U., Khanzada, N. K., Hilal, N., Hasan, S. W., & An, A. K. (2024). Determination of microplastic pollution in marine ecosystems and its effective removal using an advanced nanobubble flotation technique. *Journal of Water Process Engineering*, 57, 104637. DOI: 10.1016/j.jwpe.2023.104637

Khezerlou, A., Tavassoli, M., Sani, M. A., Mohammadi, K., Ehsani, A., & McClements, D. J. (2021). Application of Nanotechnology to Improve the Performance of Biodegradable Biopolymer-Based Packaging Materials. *Polymers*, 13(24), 4399. DOI: 10.3390/polym13244399

Kim, M. S., Lee, Y. H., Lee, Y., Jeong, H., Wang, M., Wang, D. Z., & Lee, J. S. (2024). Multigenerational effects of elevated temperature on host-microbiota interactions in the marine water flea Diaphanosoma celebensis exposed to micro- and nanoplastics. *Journal of Hazardous Materials*, 465, 132877. DOI: 10.1016/j.jhazmat.2023.132877

Kiran, B. R., Kopperi, H., & Venkata Mohan, S. (2022). Micro/nano-plastics occurrence, identification, risk analysis and mitigation: Challenges and perspectives. *Reviews in Environmental Science and Biotechnology*, 21(1), 169–203. DOI: 10.1007/s11157-021-09609-6

Koelmans, A. A., Nor, N. H. M., Hermsen, E., Kooi, M., Mintenig, S. M., & De France, J. (2019). Microplastics in freshwaters and drinking water: Critical review and assessment of data quality. [PubMed]. *Water Research*, 155, 410–422. DOI: 10.1016/j.watres.2019.02.054

Kögel, T., Bjorøy, Ø., Toto, B., Bienfait, A. M., & Sanden, M. (2020). Micro-and nanoplastic toxicity on aquatic life: Determining factors. [PubMed]. *The Science of the Total Environment*, 709, 136050. DOI: 10.1016/j.scitotenv.2019.136050

Koycheva, M. L.-S., Mihaylova, L. V., Todorova, M. N., Balcheva-Sivenova, Z. P., Alipieva, K., Ferrante, C., Orlando, G., & Georgiev, M. I. (2021). Leucosceptoside A from Devil's Claw Modulates Psoriasis-like Inflammation via Suppression of the PI3K/AKT Signaling Pathway in Keratinocytes. *Molecules (Basel, Switzerland)*, 26(22), 1–16. DOI: 10.3390/molecules26227014

Kumar, A., Choudhary, A., Kaur, H., Mehta, S., & Husen, A. (2021). Metal-based nanoparticles, sensors, and their multifaceted application in food packaging. *Journal of Nanobiotechnology*, 19(1), 256. Advance online publication. DOI: 10.1186/s12951-021-00996-0

Kumar, M., Bhujbal, S. K., Kohli, K., Prajapati, R., Sharma, B. K., Sawarkar, A. D., Abhishek, K., Bolan, S., Ghosh, P., Kirkham, M. B., Padhye, L. P., Pandey, A., Vithanage, M., & Bolan, N. (2024). A review on value-addition to plastic waste towards achieving a circular economy. *The Science of the Total Environment*, 921, 171106. DOI: 10.1016/j.scitotenv.2024.171106

Kumar, M., Chen, H., Sarsaiya, S., Qin, S., Liu, H., Awasthi, M. K., Kumar, S., Singh, L., Zhang, Z., Bolan, N. S., Pandey, A., Varjani, S., & Taherzadeh, M. J. (2021). Current research trends on micro-and nano-plastics as an emerging threat to global environment: A review. [PubMed]. *Journal of Hazardous Materials*, 409, 124967. DOI: 10.1016/j.jhazmat.2020.124967

Kumar, P., Inamura, Y., Bao, P. N., Abeynayaka, A., Dasgupta, R., & Abeynayaka, H. D. (2022). Microplastics in freshwater environment in Asia: A systematic scientific review. *Water (Basel)*, 14(11), 1737. DOI: 10.3390/w14111737

Kumar, S., Ahlawat, W., Bhanjana, G., Heydarifard, S., Nazhad, M. M., & Dilbaghi, N. (2014). Nanotechnology-Based Water Treatment Strategies. *Journal of Nanoscience and Nanotechnology*, 14(2), 1838–1858. DOI: 10.1166/jnn.2014.9050

Kundu, A., Shetti, N. P., Basu, S., Reddy, K. R., Nadagouda, M. N., & Aminabhavi, T. M. (2021). Identification and removal of micro-and nano-plastics: Efficient and cost-effective methods. [PubMed]. *Chemical Engineering Journal*, 421, 129816. DOI: 10.1016/j.cej.2021.129816

Kurniawan, T. A., Haider, A., Mohyuddin, A., Fatima, R., Salman, M., Shaheen, A., Ahmad, H. M., Al-Hazmi, H. E., Othman, M. H. D., Aziz, F., Anouzla, A., & Ali, I. (2023). Tackling microplastics pollution in global environment through integration of applied technology, policy instruments, and legislation. [PubMed]. *Journal of Environmental Management*, 346, 118971. DOI: 10.1016/j.jenvman.2023.118971

Kutralam-Muniasamy, G., Shruti, V. C., Pérez-Guevara, F., & Flores, J. A. (2024). The emerging field of inductively coupled plasma mass spectrometry for (micro) nanoplastic analysis:"The 3As" analysis, advances, and applications. *Trends in Analytical Chemistry*, 174, 117673. DOI: 10.1016/j.trac.2024.117673

Lai, H., Liu, X., & Qu, M. (2022). Nanoplastics and Human Health: Hazard Identification and Biointerface. *Nanomaterials (Basel, Switzerland)*, 12(8), 1298. DOI: 10.3390/nano12081298

Lakhiar, I. A., Yan, H., Zhang, J., Wang, G., Deng, S., Bao, R., Zhang, C., Syed, T. N., Wang, B., Zhou, R., & Wang, X. (2024). Plastic Pollution in Agriculture as a Threat to Food Security, the Ecosystem, and the Environment: An Overview. *Agronomy (Basel)*, 14(3), 548. DOI: 10.3390/agronomy14030548

Lakshmipraba, J. (2024). Photo/Hydro/Thermal Degradation of Macro-plastics into Micro-and Nano-forms. In Microplastics and Pollutants: Interactions, Degradations and Mechanisms (pp. 51–74). Springer Nature Switzerland. doi:DOI: 10.1007/978-3-031-54565-8_3

Lam, C. S., Ramanathan, S., Carbery, M., Gray, K., Vanka, K. S., Maurin, C., & Palanisami, T. (2018). A comprehensive analysis of plastics and microplastic legislation worldwide. *Water, Air, and Soil Pollution*, 229(11), 1–19. DOI: 10.1007/s11270-018-4002-z

Law, K. L., & Thompson, R. C. (2014). Microplastics in the seas. *Science*, 345(6193), 144–145. DOI: 10.1126/science.1254065

Lazăr, N. N., Călmuc, M., Milea, . A., Georgescu, P. L., & Iticescu, C. (2024). Micro and nano plastics in fruits and vegetables: A review. *Heliyon*, 10(6), e28291. DOI: 10.1016/j.heliyon.2024.e28291

Le, T. (2022). *Plastic Pollution in the Canadian Great Lakes: Drivers, Barriers and Policy Recommendations* (Master's thesis, University of Waterloo).

Lebreton, L. C. (2011). Future trends in plastic pollution and debris flow into the marine environment. *Environmental Science & Technology*, 45(6), 9052–9059.

Leddin, D. (2024). The impact of climate change, pollution and biodiversity loss on digestive health and disease. *Gastro Hep Advances*, 3(4), 519–534. DOI: 10.1016/j.gastha.2024.01.018

Levy, S. R., Gopang, M., Ramírez, L., & Allan, B. I. (2024). Bernardo, Martin D. Ruck, and Anni Sternisko. "A human rights-based approach to climate injustices at the local, national, and international levels: Program and policy recommendations.". *Social Issues and Policy Review*, 18(1), 3–30. DOI: 10.1111/sipr.12103

Li, F., Huang, D., Wang, G., Cheng, M., Chen, H., Zhou, W., Xiao, R., Li, R., Du, L., & Xu, W. (2024). Microplastics/nanoplastics in porous media: Key factors controlling their transport and retention behaviors. *The Science of the Total Environment*, 926, 171658. DOI: 10.1016/j.scitotenv.2024.171658

Li, H. H., Chen, X. W., Zhai, F. H., Li, Y. T., Zhao, H. M., Mo, C. H., & Li, H. (2024). Arbuscular Mycorrhizal Fungus Alleviates Charged Nanoplastic Stress in Host Plants via Enhanced Defense-Related Gene Expressions and Hyphal Capture. *Environmental Science & Technology*, 58(14), 6258–6273. DOI: 10.1021/acs.est.3c07850

Li, J. S. (2020). Sorption behaviors of microplastics in soil and water: A critical review. *The Science of the Total Environment*, 715, 1–17.

Li, J. Y., Yu, S., Yu, Y., & Xu, M. (2022). Effects of Microplastics on Higher Plants: A Review. *Bulletin of Environmental Contamination and Toxicology*, 109(2), 1–18. DOI: 10.1007/s00128-022-03566-8

Li, L., Zhou, Q., Yin, N., Tu, C., & Luo, Y. (2019). Uptake and accumulation of microplastics in an edible plant. *Kexue Tongbao*, 64(9), 928–934. DOI: 10.1360/N972018-00845

Lima, J. Z., Cassaro, R., Ogura, A. P., & Vianna, M. M. G. R. (2023). A systematic review of the effects of microplastics and nanoplastics on the soil-plant system. *Sustainable Production and Consumption*, 38, 266–282. DOI: 10.1016/j.spc.2023.04.010

Lin, D., Yang, G., Dou, P., Qian, S., Zhao, L., Yang, Y., & Fanin, N. (2020). Microplastics negatively affect soil fauna but stimulate microbial activity: insights from a field-based microplastic addition experiment. *Proceedings of the Royal Society B, 287*(1934), 20201268. https://doi.org/doi:10.1098/rspb.2020.1268

Lionetto, F., Esposito Corcione, C., Rizzo, A., & Maffezzoli, A. (2021). Production and characterization of polyethylene terephthalate nanoparticles. [PubMed]. *Polymers*, 13(21), 3745. DOI: 10.3390/polym13213745

Li, Q., Feng, Z., Zhang, T., Ma, C., & Shi, H. (2020). Microplastics in the commercial seaweed nori. *Journal of Hazardous Materials*, 388, 122060. DOI: 10.1016/j.jhazmat.2020.122060

Li, Q., Mahendra, S., Lyon, D. Y., Brunet, L., Liga, M. V., Li, D., & Alvarez, P. J. (2008). Antimicrobial nanomaterials for water disinfection and microbial control: Potential applications and implications. *Water Research*, 42(18), 4591–4602. DOI: 10.1016/j.watres.2008.08.015

Liu, Y. (2024). *Microplastics: Research, Problems, and Solutions* (Doctoral dissertation, Worcester Polytechnic Institute).

Liu, K., Wang, X., Fang, T., Xu, P., Zhu, L., & Li, D. (2019). Source and potential risk assessment of suspended atmospheric microplastics in Shanghai. *The Science of the Total Environment*, 675, 462–471. DOI: 10.1016/j.scitotenv.2019.04.110

Liu, L. Y. (2022). Uptake of microplastics by rice (Oryza sativa L.) from soil with or without fertilizer amendment. *Chemosphere*, 310(135809), 1–8.

Liu, L., Du, R. Y., Jia, R. L., Wang, J. X., Chen, C. Z., Li, P., Kong, L.-M., & Li, Z. H. (2024). Micro (nano) plastics in marine medaka: Entry pathways and cardiotoxicity with triphenyltin. *Environmental Pollution*, 342, 123079. DOI: 10.1016/j.envpol.2023.123079

Liu, M., Song, X., Liu, C., Cui, X., Sun, W., Li, Z., & Wang, J. (2024). Nanoplastics increase the adverse impacts of lead on the growth, morphological structure and photosynthesis of marine microalga Platymonas helgolandica. *Marine Environmental Research*, 193, 106259. DOI: 10.1016/j.marenvres.2023.106259

Liu, S., Wang, J., Zhu, J., Wang, J., Wang, H., & Zhan, X. (2021). The joint toxicity of polyethylene microplastic and phenanthrene to wheat seedlings. *Chemosphere*, 282, 130967. DOI: 10.1016/j.chemosphere.2021.130967

Liu, X., Li, M.-C., Lu, Y., Li, Z., Liu, C., Liu, Z., Mei, C., & Wu, Q. (2024). Cellulose nanofiber-coated delignified wood as an efficient filter for microplastic removal. *Progress in Natural Science*, 34(1), 162–171. DOI: 10.1016/j.pnsc.2024.02.010

Liu, Y., Shao, H., Liu, J., Cao, R., Shang, E., Liu, S., & Li, Y. (2021). Transport and transformation of microplastics and nanoplastics in the soil environment: A critical review. *Soil Use and Management*, 37(2), 224–242. DOI: 10.1111/sum.12709

Liu, Y., Zeng, G., Zhong, H., Wang, Z., Liu, Z., Cheng, M., Liu, G., Yang, X., & Liu, S. (2017). Effect of rhamnolipid solubilization on hexadecane bioavailability: Enhancement or reduction? *Journal of Hazardous Materials*, 322, 394–401. DOI: 10.1016/j.jhazmat.2016.10.025

Liu, Z. Q., Huang, C., Li, J.-Y., Yang, J., Qu, B., Yang, S.-Q., Cui, Y.-H., Yan, Y., Sun, S., & Wu, X. (2021). Activated carbon catalytic ozonation of reverse osmosis concentrate after coagulation pretreatment from coal gasification wastewater reclamation for zero liquid discharge. *Journal of Cleaner Production*, 286, 124951. DOI: 10.1016/j.jclepro.2020.124951

Li, W., Wang, J., Xiong, Z., & Li, D. (2023). Micro/nanorobots for efficient removal and degradation of micro/nanoplastics. *Cell Reports. Physical Science*, 4(11), 101639. DOI: 10.1016/j.xcrp.2023.101639

Li, W., Wufuer, R., Duo, J., Wang, S., Luo, Y., Zhang, D., & Pan, X. (2020). Microplastics in agricultural soils: Extraction and characterization after different periods of polythene film mulching in an arid region. *The Science of the Total Environment*, 749, 141420. DOI: 10.1016/j.scitotenv.2020.141420

Li, X., Wang, R., Dai, W., Luan, Y., & Li, J. (2023). Impacts of Micro(nano)plastics on Terrestrial Plants: Germination, Growth, and Litter. *Plants*, 12(20), 3554. DOI: 10.3390/plants12203554

Li, X., Zheng, Y., Lu, L., Eom, J., Ru, S., Li, Y., & Wang, J. (2024). Trophic transfer of micro-and nanoplastics and toxicity induced by long-term exposure of nanoplastics along the rotifer (Brachionus plicatilis)-marine medaka (Oryzias melastigma) food chain. *Environmental Pollution*, 346, 123599. DOI: 10.1016/j.envpol.2024.123599

Li, Y., Zhou, B., Zheng, G., Liu, X., Li, T., Yan, C., Cheng, C., Dai, K., Liu, C., Shen, C., & Guo, Z. (2017). Continuously prepared highly conductive and stretchable SWNT/MWNT synergistically composited electrospun thermoplastic polyurethane yarns for wearable sensing. *Journal of Materials Chemistry. C, Materials for Optical and Electronic Devices*, 6(9), 2258–2269. DOI: 10.1039/C7TC04959E

Li, Z., Li, Q., Li, R., Zhou, J., & Wang, G. (2021). The distribution and impact of polystyrene nanoplastics on cucumber plants. *Environmental Science and Pollution Research International*, 28(13), 16042–16053. DOI: 10.1007/s11356-020-11702-2

Llorca, M., Álvarez-Muñoz, D., Ábalos, M., Rodríguez-Mozaz, S., Santos, L. H., León, V. M., Campillo, J. A., Martínez-Gómez, C., Abad, E., & Farré, M. (2020). Microplastics in Mediterranean coastal area: toxicity and impact for the environment and human health.

Lu, H., Wang, J., Stoller, M., Wang, T., Bao, Y., & Hao, H. (2016). An Overview of Nanomaterials for Water and Wastewater Treatment. *Advances in Materials Science and Engineering*, 2016, 1–10. DOI: 10.1155/2016/4964828

Luo, W., Su, L., Craig, N. J., Du, F., Wu, C., & Shi, H. (2019). Comparison of microplastic pollution in different water bodies from urban creeks to coastal waters. [PubMed]. *Environmental Pollution*, 246, 174–182. DOI: 10.1016/j.envpol.2018.11.081

Lv, S., Cui, K., Zhao, S., Li, Y., Liu, R., Hu, R., Zhi, B., Gu, L., Wang, L., Wang, Q., & Shao, Z. (2024). Continuous generation and release of microplastics and nanoplastics from polystyrene by plastic-degrading marine bacteria. *Journal of Hazardous Materials*, 465, 133339. DOI: 10.1016/j.jhazmat.2023.133339

Maddela, N. R., Ramakrishnan, B., Kadiyala, T., Venkateswarlu, K., & Megharaj, M. (2023). Do Microplastics and Nanoplastics Pose Risks to Biota in Agricultural Ecosystems? *Soil Systems*, 7(1), 19. DOI: 10.3390/soilsystems7010019

Madhura, L., Singh, S., Kanchi, S., Sabela, M., & Bisetty, K., & Inamuddin. (2019). Nanotechnology-based water quality management for wastewater treatment. *Environmental Chemistry Letters*, 17(1), 65–121. DOI: 10.1007/s10311-018-0778-8

Mahesh, S., Gowda, N. K., & Mahesh, S. (2023). Identification of microplastics from urban informal solid waste landfill soil; MP associations with COD and chloride. *Water Science and Technology: A Journal of the International Association on Water Pollution Research, 87*(1), 115–129. doi:DOI: 10.2166/wst.2022.412

Mahmud, A., Wasif, M. M., Roy, H., Mehnaz, F., Ahmed, T., Pervez, M. N., Naddeo, V., & Islam, M. S. (2022). Aquatic microplastic pollution control strategies: Sustainable degradation techniques, resource recovery, and recommendations for Bangladesh. *Water (Basel)*, 14(23), 3968. DOI: 10.3390/w14233968

Marfella, R., Prattichizzo, F., Sardu, C., Fulgenzi, G., Graciotti, L., Spadoni, T., D'Onofrio, N., Scisciola, L., La Grotta, R., Frigé, C., Pellegrini, V., Municinò, M., Siniscalchi, M., Spinetti, F., Vigliotti, G., Vecchione, C., Carrizzo, A., Accarino, G., Squillante, A., & Paolisso, G. (2024). Microplastics and nanoplastics in atheromas and cardiovascular events. *The New England Journal of Medicine*, 390(10), 900–910. DOI: 10.1056/NEJMoa2309822

Margan, P., & Haghighi, M. (2018). Sono-coprecipitation synthesis and physicochemical characterization of CdO-ZnO nano photo-catalyst for removal of acid orange 7 from wastewater. *Ultrasonics Sonochemistry*, 40, 323–332. DOI: 10.1016/j.ultsonch.2017.07.003

Marine. (2022). "The Role of Industry in Addressing Marine Chemical Pollution."

Massahi, T., Amin, A. A., Meshabaz, R. A., Pirsaheb, M., Terry, L., Makhdoumi, P., Kianpour, S., Zamani, F., & Hossini, H. (2024). Microplastic occurrence and its potential role as a carrier for SARS-CoV-2 in health center wastewater treatment plant and surface water. *Journal of Sea Research*, 198, 102477. DOI: 10.1016/j.seares.2024.102477

Mauter, M. S., Zucker, I., Perreault, F., Werber, J. R., Kim, J. H., & Elimelech, M. (2018). The role of nanotechnology in tackling global water challenges. *Nature Sustainability*, 1(4), 166–175. DOI: 10.1038/s41893-018-0046-8

Ma, Y. B., Xie, Z. Y., Hamid, N., Tang, Q. P., Deng, J. Y., Luo, L., & Pei, D. S. (2023). Recent advances in micro (nano) plastics in the environment: Distribution, health risks, challenges and future prospects. *Aquatic Toxicology (Amsterdam, Netherlands)*, 261, 106597. DOI: 10.1016/j.aquatox.2023.106597

Meegoda, J. N., & Hettiarachchi, M. C. (2023). A Path to a Reduction in Micro and Nanoplastics Pollution. [PubMed]. *International Journal of Environmental Research and Public Health*, 20(8), 5555. DOI: 10.3390/ijerph20085555

Meng, H., Xue, M., Xia, T., Zhao, Y. L., Tamanoi, F., Stoddart, J. F., Zink, J., & Nel, A. E. (2010). Autonomous in vitro anticancer drug release from mesoporous silica nanoparticles by pH-sensitive nano valves. *Journal of the American Chemical Society*, 132(36), 12690–12697. DOI: 10.1021/ja104501a

Mintz, E. D., Reiff, F. M., & Tauxe, R. V. (1995). Safe water treatment and storage in the home: A practical new strategy to prevent waterborne disease. *Journal of the American Medical Association*, 273(12), 948–953. DOI: 10.1001/jama.1995.03520360062040

Mishra, M., Sudarsan, D., Celso, A. G. S., Marques da Silva, R., Beja, S. K., Paul, S., Bhanja, P., & Sethy, M. (2024). Current patterns and trends of microplastic pollution in the marine environment: A bibliometric analysis. *Environmental Science and Pollution Research International*, 31(15), 1–20. DOI: 10.1007/s11356-024-32511-x

Mitrano, D. M., Wick, P., & Nowack, B. (2021). Placing nanoplastics in the context of global plastic pollution. *Nature Nanotechnology*, 16(5), 491–500. DOI: 10.1038/s41565-021-00888-2

Mitrano, D. M., & Wohlleben, W. (2020). Microplastic regulation should be more precise to incentivize both innovation and environmental safety. *Nature Communications*, 11(1), 5324. DOI: 10.1038/s41467-020-19069-1

Modi, S., Prajapati, R., Inwati, G. K., Deepa, N., Tirth, V., Yadav, V. K., Yadav, K. K., Islam, S., Gupta, P., Kim, D., & Jeon, B. (2021). Recent Trends in Fascinating Applications of Nanotechnology in Allied Health Sciences. *Crystals*, 12(1), 39. DOI: 10.3390/cryst12010039

Moeck, C., Davies, G., Krause, S., & Schneidewind, U. (2022). Microplastics and nanoplastics in agriculture—A potential source of soil and groundwater contamination? *Grundwasser*. Advance online publication. DOI: 10.1007/s00767-022-00533-2

Mofijur, M., Ahmed, S. F., Rahman, S. A., Siddiki, S. Y. A., Islam, A. S., Shahabuddin, M., & Show, P. L. (2021). Source, distribution and emerging threat of micro-and nanoplastics to marine organism and human health: Socio-economic impact and management strategies. *Environmental Research*, 195, 110857. DOI: 10.1016/j.envres.2021.110857

Mohajerani, A., & Karabatak, B. (2020). Microplastics and pollutants in biosolids have contaminated agricultural soils: An analytical study and a proposal to cease the use of biosolids in farmlands and utilise them in sustainable bricks. *Waste Management (New York, N.Y.)*, 107, 252–265. DOI: 10.1016/j.wasman.2020.04.021

Mohana, A. A., Farhad, S. M., Haque, N., & Pramanik, B. K. (2021). Understanding the fate of nano-plastics in wastewater treatment plants and their removal using membrane processes. [PubMed]. *Chemosphere*, 284, 131430. DOI: 10.1016/j.chemosphere.2021.131430

Moharrami, P., & Motamedi, E. (2020). Application of cellulose nanocrystals prepared from agricultural wastes for the synthesis of starch-based hydrogel nanocomposites: Efficient and selective nano adsorbent for removal of cationic dyes from water. *Bioresource Technology*, 313, 123661. DOI: 10.1016/j.biortech.2020.123661

Mojiri, A., Zhou, J. L., & Lou, Z. (2024). Micro (nano) plastics in Aquatic Environments: State of the Art and Beyond. *Water (Basel)*, 16(6), 902. DOI: 10.3390/w16060902

Moradi, O., Fakhri, A., Adami, S., & Adami, S. (2013). Isotherm, thermodynamic, kinetics, and adsorption mechanism studies of Ethidium bromide by single-walled carbon nanotube and carboxylate group functionalised single-walled carbon nanotube. *Journal of Colloid and Interface Science*, 395, 224–229. DOI: 10.1016/j.jcis.2012.11.013

Mostafavi, S., Mehrnia, M., & Rashidi, A. (2009). Preparation of nanofilter from carbon nanotubes for application in virus removal from water. *Desalination*, 238(1-3), 271–280. DOI: 10.1016/j.desal.2008.02.018

Mousavi, M., Kowsari, E., Gheibi, M., Cheshmeh, Z. A., Teymoorian, T., & Ramakrishna, S. (2024). Assessing Bioplastics' Economic, Commercial, Political, and Energy Potential with Circular Economy Modeling: A Sustainable Solution to Plastic Waste Management. *Materials Circular Economy*, 6(1), 1–36. DOI: 10.1007/s42824-023-00098-2

Mubtaker, Y. (2020). *SDGs for Everyone*. Yama Mubtaker.

Muhammad Sajjad, Q. H. (2022). Microplastics in the soil environment: A critical review. *Environmental Technology & Innovation*, •••, 1–24.

Mulyanti, R., & Susanto, H. (2018). Wastewater treatment by nanofiltration membranes. *IOP Conference Series. Earth and Environmental Science*, 142, 012017. DOI: 10.1088/1755-1315/142/1/012017

Mustapha, S., Tijani, J. O., Elabor, R., Salau, R. B., Egbosiuba, T. C., Amigun, A. T., Shuaib, D. T., Sumaila, A., Fiola, T., Abubakar, Y. K., Abubakar, H. L., Ossamulu, I. F., Abdulkareem, A. S., Ndamitso, M. M., Sagadevan, S., & Mohammed, A. K. (2024). Technological approaches for removal of microplastics and nanoplastics in the environment. *Journal of Environmental Chemical Engineering*, 12(2), 112084. DOI: 10.1016/j.jece.2024.112084

Mutuku, J., Tocock, M., Yanotti, M., Tinch, D., & MacDonald, D. H. (2024). Public perceptions of the value of reducing marine plastics in Australian waters. *Ecological Economics*, 217, 108065. DOI: 10.1016/j.ecolecon.2023.108065

Nagy, E. (2019). *Nanofiltration. Basic Equations of Mass Transport Through a Membrane Layer*. Elsevier.

Napper, I. E. (2016). Assessment of microplastic pollution on land and ocean. *Environmental Pollution*, 212, 150–159.

Nasir, M. S., Tahir, I., Ali, A., Ayub, I., Nasir, A., Abbas, N., Sajjad, U., & Hamid, K. (2024). Innovative technologies for removal of micro plastic: A review of recent advances. *Heliyon*, 10(4), e25883. DOI: 10.1016/j.heliyon.2024.e25883

Nasir, S., Hussein, M. Z., Zainal, Z., & Yusof, N. A. (2018). Carbon-Based Nanomaterials/Allotropes: A Glimpse of Their Synthesis, Properties and Some Applications. *Materials (Basel)*, 11(2), 295. DOI: 10.3390/ma11020295

Naz, S., Chatha, A. M. M., Khan, N. A., Ullah, Q., Zaman, F., Qadeer, A., Khan, I. M., Danabas, D., Kiran, A., Skalickova, S., Bernatova, S., Khan, M. Z., & Horky, P. (2024). Unraveling the ecotoxicological effects of micro and nano-plastics on aquatic organisms and human health. *Frontiers in Environmental Science*, 12, 1390510. DOI: 10.3389/fenvs.2024.1390510

Neal, A., Johnson, M., & Havrda, M. (2023). Multi-level climate action through circular supply chain management of ocean plastic. In *Handbook of Multi-Level Climate Actions* (pp. 215–241). Edward Elgar Publishing., DOI: 10.4337/9781802202458.00022

Ng, E. L., Lwanga, E. H., Eldridge, S. M., Johnston, P., Hu, H. W., & Geissen, V. (2018). An overview of microplastic and nanoplastic pollution in agroecosystems. *The Science of the Total Environment*, 627, 1377–1388. DOI: 10.1016/j.scitotenv.2018.01.341

Nithya, P. (2021). Alginate coated layered double hydroxide/reduced graphene oxide nanocomposites for removal of toxic As (V) from wastewater. *Phys. E: Low-Dimens.Physica E, Low-Dimensional Systems and Nanostructures*, 127, 114527. DOI: 10.1016/j.physe.2020.114527

Nohara, N. M. L., Ariza-Tarazona, M. C., Triboni, E. R., Nohara, E. L., Villarreal-Chiu, J. F., & Cedillo-González, E. I. (2024). Are you drowned in microplastic pollution? A brief insight on the current knowledge for early career researchers developing novel remediation strategies. [PubMed]. *The Science of the Total Environment*, 918, 170382. DOI: 10.1016/j.scitotenv.2024.170382

Nowack, B. (2008). Pollution prevention and treatment using nanotechnology. In H. Krug (Ed.), *Nanotechnology general, Nanotechnology, Volume 2: environmental aspects* (1–15). Wiley-VCH. doi:DOI: 10.1002/9783527628155.nanotech010

Ofori-Mensah, E. M. (2023) "The local perspective on Ocean Plastic Pollution: Identifying Areas to Channel Ocean Literacy at Great Ningo in Ghana." Master's thesis, NTNU

Ojha, M. D., & Skariyachan, S. (2024). The Fate of Micro/Nano Plastic Pollutants in the Natural Environment. In *Remediation of Plastic and Microplastic Waste* (pp. 215–239). CRC Press.

Ojinnaka, D., & Aw, M. (2020). Micro and nano plastics: A consumer perception study on the environment, food safety threat and control systems. *Biomedical Journal of Scientific & Technical Research*, 32(2). Advance online publication. DOI: 10.26717/BJSTR.2020.31.005064

Okeke, E. S., Ezeorba, T. P. C., Chen, Y., Mao, G., Feng, W., & Wu, X. (2024). Association of tetrabromobisphenol A (TBBPA) with micro/nano-plastics: A review of recent findings on ecotoxicological and health impacts. *The Science of the Total Environment*, 927, 172308. DOI: 10.1016/j.scitotenv.2024.172308

Oliveri Conti, G., Ferrante, M., Banni, M., Favara, C., Nicolosi, I., Cristaldi, A., Fiore, M., & Zuccarello, P. (2020). Micro- and nano-plastics in edible fruit and vegetables. The first diet risks assessment for the general population. *Environmental Research*, 187, 109677. DOI: 10.1016/j.envres.2020.109677

Onyena, A. P., Aniche, D. C., Ogbolu, B. O., Rakib, M. R. J., Uddin, J., & Walker, T. R. (2021). Governance strategies for mitigating microplastic pollution in the marine environment: A review. *Microplastics*, (1), 15–46. DOI: 10.3390/microplastics1010003

Osuna-Laveaga, D. R., Ojeda-Castillo, V., Flores-Payán, V., Gutiérrez-Becerra, A., & Moreno-Medrano, E. D. (2023). Micro-and nanoplastics current status: Legislation, gaps, limitations and socio-economic prospects for future. *Frontiers in Environmental Science*, 11, 1241939. DOI: 10.3389/fenvs.2023.1241939

Ouda, M., Banat, F., Hasan, S. W., & Karanikolos, G. N. (2023). Recent advances on nanotechnology-driven strategies for remediation of microplastics and nanoplastics from aqueous environments. *Journal of Water Process Engineering*, 52, 103543. DOI: 10.1016/j.jwpe.2023.103543

Ouyang, W., Zhang, Y., Wang, L., Barceló, D., Wang, Y., & Lin, C. (2020). Seasonal relevance of agricultural diffuse pollutant with microplastic in the bay. *Journal of Hazardous Materials*, 396, 122602. DOI: 10.1016/j.jhazmat.2020.122602

Paço, A., Duarte, K., da Costa, J. P., Santos, P. S., Pereira, R., Pereira, M. E., & Rocha-Santos, T. A. (2017). Biodegradation of polyethylene microplastics by the marine fungus Zalerion maritimum. *The Science of the Total Environment*, 586, 10–15. DOI: 10.1016/j.scitotenv.2017.02.017

Paing, Y. M. M., Eom, Y., Song, G. B., Kim, B., Choi, M. G., Hong, S., & Lee, S. H. (2024). Neurotoxic effects of polystyrene nanoplastics on memory and microglial activation: Insights from in vivo and in vitro studies. *The Science of the Total Environment*, 924, 171681. DOI: 10.1016/j.scitotenv.2024.171681

Pakhomova, S., Zhdanov, I., & van Bavel, B. (2020). Polymer type identification of marine plastic litter using a miniature near-infrared spectrometer (MicroNIR). *Applied Sciences (Basel, Switzerland)*, 10(23), 8707. DOI: 10.3390/app10238707

Palmer-Abbs, M., Deshpande, P., & Karl, C. W. (2024). Enabling transition thinking on complex issues (wicked problems): A framework for future circular economic transitions of plastic management in the Norwegian fisheries and aquaculture sectors. *Journal of Cleaner Production*, 449, 141420. DOI: 10.1016/j.jclepro.2024.141420

Pandey, N., Shukla, S. K., & Singh, N. B. (2017). Water purification by polymer nanocomposites: An overview. *Nanocomposites*, 3(2), 47–66. DOI: 10.1080/20550324.2017.1329983

Pandey, P. K., Sharma, S. K., & Sambi, S. S. (2015). Removal of lead (II) from wastewater on zeolite-NaX. *Journal of Environmental Chemical Engineering*, 3(4), 2604–2610. DOI: 10.1016/j.jece.2015.09.008

Parham, H., Bates, S., Xia, Y., & Zhu, Y. (2013). A highly efficient and versatile carbon nanotube/ceramic composite filter. *Carbon*, 54, 215–223. DOI: 10.1016/j.carbon.2012.11.032

Parkinson, S. J., Tungsirisurp, S., Joshi, C., Richmond, B. L., Gifford, M. L., Sikder, A., Lynch, I., O'Reilly, R. K., & Napier, R. M. (2022). Polymer nanoparticles pass the plant interface. *Nature Communications*, 13(1), 7385. DOI: 10.1038/s41467-022-35066-y

Pascall, M. A., Zabik, M. E., Zabik, M. J., & Hernandez, R. J. (2005). Uptake of Polychlorinated Biphenyls (PCBs) from an Aqueous Medium by Polyethylene, Polyvinyl Chloride, and Polystyrene Films. [PubMed]. *Journal of Agricultural and Food Chemistry*, 53(1), 164–169. DOI: 10.1021/jf048978t

Peel, J., Godden, L., Palmer, A., & Markey-Towler, M. R. (2021). Regional Solutions: Export Measures for Plastic Recyclables to Reduce Marine Plastic Pollution in the Pacific. *Marine Plastic Pollution and the Rule of Law*, 11(1), 1.

Peidaei, F., Ahari, H., Anvar, S. A., & Ataee, M. (2021). Nanotechnology in Food Packaging and Storage: A Review. *DOAJ (DOAJ: Directory of Open Access Journals)*.https://doi.org/doi:10.22059/ijvm.2021.310466.1005130

Peiponen, K. E., Räty, J., Ishaq, U., Pélisset, S., & Ali, R. (2019). Outlook on optical identification of micro-and nanoplastics in aquatic environments. [PubMed]. *Chemosphere*, 214, 424–429. DOI: 10.1016/j.chemosphere.2018.09.111

Peng, J., Feng, Y., Wang, X., Li, J., Xu, G., Phonenasay, S., Luo, Q., Han, Z., & Lu, W. (2021). Effects of nitrogen application rate on the photosynthetic pigment, leaf fluorescence characteristics, and yield of indica hybrid rice and their interrelations. *Scientific Reports*, 11(1), 7485. DOI: 10.1038/s41598-021-86858-z

Peng, L., Fu, D., Qi, H., Lan, C. Q., Yu, H., & Ge, C. (2020). Micro-and nano-plastics in marine environment: Source, distribution and threats—A review. [PubMed]. *The Science of the Total Environment*, 698, 134254. DOI: 10.1016/j.scitotenv.2019.134254

Peralta, M. E., Mártire, D. O., Moreno, M. S., Parolo, M. E., & Carlos, L. (2021). Versatile nano adsorbents based on magnetic mesostructured silica nanoparticles with tailored surface properties for organic pollutants removal. *Journal of Environmental Chemical Engineering*, 9(1), 104841. DOI: 10.1016/j.jece.2020.104841

Pérez-Reverón, R., Álvarez-Méndez, S. J., Kropp, R. M., Perdomo-González, A., Hernández-Borges, J., & Díaz-Peña, F. J. (2022). Microplastics in Agricultural Systems: Analytical Methodologies and Effects on Soil Quality and Crop Yield. *Agriculture*, 12(8), 1162. DOI: 10.3390/agriculture12081162

Pico, Y., & Barcelo, D. (2020). Pyrolysis gas chromatography-mass spectrometry in environmental analysis: Focus on organic matter and microplastics. *Trends in Analytical Chemistry*, 130, 115964. DOI: 10.1016/j.trac.2020.115964

Pignattelli, B. A. (2021). Stress Effect Induced by Microplastics Coupled with Acid Rain. *Annals of Agricultural & Crop Sciences*, •••, 1–12.

Pignattelli, S., Broccoli, A., & Renzi, M. (2020). Physiological responses of garden cress (L. sativum) to different types of microplastics. *The Science of the Total Environment*, 727, 138609. DOI: 10.1016/j.scitotenv.2020.138609

Pittura, L., Gorbi, S., León, V. M., Bellas, J., González, J. A. C., Albentosa, M., & Regoli, F. (2023). Microplastics and nanoplastics in the marine environment. *Contaminants of Emerging Concern in the Marine Environment*, 311-348.

Pourebrahimi, S., & Pirooz, M. (2023). Microplastic pollution in the marine environment: A review. *Journal of Hazardous Materials Advances*, 10, 100327. DOI: 10.1016/j.hazadv.2023.100327

Prata, J. C., da Costa, J. P., Lopes, I., Duarte, A. C., & Rocha-Santos, T. (2020). Environmental exposure to microplastics: An overview on possible human health effects. *The Science of the Total Environment*, 702, 134455. DOI: 10.1016/j.scitotenv.2019.134455

Primozic, M., Knez, E., & Leitgeb, M. (2021). (Bio)Nanotechnology in Food Science Food Packaging. *Nanomaterials (Basel, Switzerland)*, 11(2), 292. DOI: 10.3390/nano11020292

Qin, L., Zeng, Z., Zeng, G., Lai, C., Duan, A., Xiao, R., Huang, D., Fu, Y., Yi, H., Li, B., Liu, X., Liu, S., Zhang, M., & Jiang, D. (2019). Cooperative catalytic performance of bimetallic Ni-Au nanocatalyst for highly efficient hydrogenation of nitroaromatics and corresponding mechanism insight. *Applied Catalysis B: Environmental*, 259, 118035. DOI: 10.1016/j.apcatb.2019.118035

Qu, X., Alvarez, P. J., & Li, Q. (2013). Applications of nanotechnology in water and wastewater treatment. *Water Research*, 47(12), 3931–3946. DOI: 10.1016/j.watres.2012.09.058

Rahman, L., Mallach, G., Kulka, R., & Halappanavar, S. (2021). Microplastics and nanoplastics science: Collecting and characterizing airborne microplastics in fine particulate matter. *Nanotoxicology*, 15(9), 1253–1278. DOI: 10.1080/17435390.2021.2018065

Rai, M., Pant, G., Pant, K., Aloo, B., Kumar, G., & Singh, H. (2023). icroplastic Pollution in Terrestrial Ecosystems and Its Interaction with Other Soil Pollutants: A Potential Threat to Soil Ecosystem Sustainability. *Resources*, •••, 1–14.

Rai, M., Pant, G., Pant, K., Aloo, B., Kumar, G., Singh, H., & Tripathi, V. (2023). Microplastic Pollution in Terrestrial Ecosystems and Its Interaction with Other Soil Pollutants: A Potential Threat to Soil Ecosystem Sustainability. *Resources*, 12(6), 1–14. DOI: 10.3390/resources12060067

Rai, P. K., Lee, J., Brown, R. J., & Kim, K. H. (2021). Micro-and nano-plastic pollution: Behavior, microbial ecology, and remediation technologies. *Journal of Cleaner Production*, 291, 125240. DOI: 10.1016/j.jclepro.2020.125240

Rajendran, D., Kamalakkannan, M., & Chandrasekaran, N. (2024). Surface functionalization, particle size and pharmaceutical co-contaminant dependent impact of nanoplastics on marine crustacean–Artemia salina. *Environmental Science. Processes & Impacts*, 26(7), 1130–1146. DOI: 10.1039/D4EM00010B

Ramasamy, B. S. S., & Palanisamy, S. (2021). A review on occurrence, characteristics, toxicology and treatment of nanoplastic waste in the environment. *Environmental Science and Pollution Research International*, 28(32), 43258–43273. DOI: 10.1007/s11356-021-14883-6

Randhawa, J. (2023). Advanced analytical techniques for microplastics in the environment: A review. *Bulletin of the National Research Center*, 47(1), 1–14. DOI: 10.1186/s42269-023-01148-0

Râpă, M., Darie-Ni ă, R. N., Matei, E., Predescu, A. M., Berbecaru, A. C., & Predescu, C. (2023). Insights into Anthropogenic Micro-and Nanoplastic Accumulation in Drinking Water Sources and Their Potential Effects on Human Health. [PubMed]. *Polymers*, 15(11), 2425. DOI: 10.3390/polym15112425

Rashed, A. H., Yesilay, G., Hazeem, L., Rashdan, S., AlMealla, R., Kilinc, Z., Ali, F., Abdulrasool, F., & Kamel, A. H. (2023). Micro-and nano-plastics contaminants in the environment: Sources, fate, toxicity, detection, remediation, and sustainable perspectives. *Water (Basel)*, 15(20), 3535. DOI: 10.3390/w15203535

Raubenheimer, Karen, and Niko Urho. (2023) "Science-policy interface for plastic pollution.".

Rehman, A., Huang, F., Zhang, Z., Habumugisha, T., Yan, C., Shaheen, U., & Zhang, X. (2024). Nanoplastic contamination: Impact on zebrafish liver metabolism and implications for aquatic environmental health. *Environment International*, 187, 108713. DOI: 10.1016/j.envint.2024.108713

Rehm, R., Zeyer, T., Schmidt, A., & Fiener, P. (2021). Soil erosion as transport pathway of microplastic from agriculture soils to aquatic ecosystems. *The Science of the Total Environment*, 795, 148774. DOI: 10.1016/j.scitotenv.2021.148774

Rice, R. E., & Moxley, C. M. (2021). Plastic communication campaigns and interventions: Foci, theoretical frameworks, variables, and methods. In *The Handbook of International Trends in Environmental Communication* (pp. 132–154). Routledge., DOI: 10.4324/9780367275204-11

Riffat, R. (2012). *Fundamentals of Wastewater Treatment and Engineering*. CRC Press., DOI: 10.1201/b12746

Rikta, S. Y. (2019). Application of Nanoparticles for Disinfection and Microbial Control of Water and Wastewater. In *Nanotechnology in Water and Wastewater Treatment* (pp. 159–176). Elsevier., DOI: 10.1016/B978-0-12-813902-8.00009-5

Rillig, M. C. (2020). Microplastic pollution in soil: Are plants and soil biota at risk? *Environmental Science & Technology*, 54(16), 9409–9419.

Robin, R. S., Karthik, R., Nithin, A., & Purvaja, R. (2023). Removal of marine litter and its impact along the coast of India. *Records of the Zoological Survey of India*, 67–86.

Rognerud, I., Steindal, E. H., Tartiu, V. E., & Kumar, N. (2022). *International sustainability criteria for plastic products in a global agreement on plastic pollution.* Nordic Council of Ministers., DOI: 10.6027/temanord2022-509

Rondon, R., Cosseau, C., Bergami, E., Cárdenas, C. A., Pérez-Toledo, C., Alvarez, D., Aldridge, J., Font, A., Garrido, I., Santa Cruz, F., Perrois, G., Balbi, T., Corsi, I., & González-Aravena, M. (2024). Exposure to nanoplastics and nanomaterials either single and combined affects the gill-associated microbiome of the Antarctic soft-shelled clam Laternula elliptica. *Marine Environmental Research*, 198, 106539. DOI: 10.1016/j.marenvres.2024.106539

Rossi, S. (2022). Marine Solid Pollution—From Macroplastics to Nanoplastics. In SDG 14: Life Below Water: A Machine-Generated Overview of Recent Literature (pp. 63–110). Springer International Publishing.

Ross, M. S., Loutan, A., Groeneveld, T., Molenaar, D., Kroetch, K., Bujaczek, T., Kolter, S., Moon, S., Huynh, A., Khayam, R., Franczak, B. C., Camm, E., Arnold, V. I., & Ruecker, N. J. (2023). Estimated discharge of microplastics via urban stormwater during individual rain events. *Frontiers in Environmental Science*, 11, 1090267. Advance online publication. DOI: 10.3389/fenvs.2023.1090267

Roy, P., Mohanty, A. K., & Misra, M. (2022). Microplastics in ecosystems: Their implications and mitigation pathways. (Perspective). *Environmental Science. Advances*, 1(1), 9–29. DOI: 10.1039/D1VA00012H

Roy, T., Dey, T. K., & Jamal, M. (2023). Microplastic/nanoplastic toxicity in plants: An imminent concern. *Environmental Monitoring and Assessment*, 195(1), 27. DOI: 10.1007/s10661-022-10654-z

Russo, M., Oliva, M., Hussain, M. I., & Muscolo, A. (2023). The hidden impacts of micro/nanoplastics on soil, crop and human health. *Journal of Agriculture and Food Research*, 14, 100870. DOI: 10.1016/j.jafr.2023.100870

Rytelewska, S., & Dąbrowska, A. (2022). The Raman spectroscopy approach to different freshwater microplastics and quantitative characterization of polyethylene aged in the environment. *Microplastics*, 1(2), 263–281. DOI: 10.3390/microplastics1020019

Sadegh, H., Ali, G. A. M., Gupta, V. K., Makhlouf, A. S. H., Shahryari-Ghoshekandi, R., Nadagouda, M. N., Sillanpää, M., & Megiel, E. (2017). The role of nanomaterials as effective adsorbents and their applications in wastewater treatment. *Journal of Nanostructure in Chemistry*, 7(1), 1–14. DOI: 10.1007/s40097-017-0219-4

Samal, D. (2017). Use of Nanotechnology in the Food Industry: A review. *International Journal of Environment, Agriculture and Biotechnology*, 2(4), 2270–2278. DOI: 10.22161/ijeab/2.4.90

Sanabria, Adriana Gomez, and Florian Lindl. (2023) "Adoption of global circular waste management systems is crucial to cut leakage of waste into aquatic environments.".

Sandil, S., & Zaray, G. (2024). Impact of Microplastics and Nanoplastics in the Aquatic Environment. *Management of Micro and Nano-plastics in Soil and Biosolids: Fate, Occurrence, Monitoring, and Remedies*, 25-68.

Saravanan, R., Gracia, F., & Stephen, A. (2017). Basic Principles, Mechanism, and Challenges of Photocatalysis. In M.M. Khan, D. Pradhan, Y. Sohn, Y. (Eds.), *Nanocomposites for Visible Light-Induced Photocatalysis*. Book series: Springer Series on Polymer and Composite Materials. Springer.

Sarkar, B., Dissanayake, P. D., Bolan, N. S., Dar, J. Y., Kumar, M., Haque, M. N., & Ok, Y. S. (2022). Challenges and opportunities in sustainable management of microplastics and nanoplastics in the environment. *Environmental Research*, 207, 112179. DOI: 10.1016/j.envres.2021.112179

Schäli, J. (2022). The Protection of the Marine Environment from Land-based Sources of Plastic Pollution in International Law. In *The Mitigation of Marine Plastic Pollution in International Law* (pp. 107–377). Brill Nijhoff., DOI: 10.1163/9789004508613_004

Schmid, C., Cozzarini, L., & Zambello, E. (2021). Microplastic's story. *Marine Pollution Bulletin*, 162, 111820. DOI: 10.1016/j.marpolbul.2020.111820

Schwaferts, C., Niessner, R., Elsner, M., & Ivleva, N. P. (2019). Methods for the analysis of submicrometer-and nanoplastic particles in the environment. *Trends in Analytical Chemistry*, 112, 52–65. DOI: 10.1016/j.trac.2018.12.014

Seay, J., & Ternes, M. E. (2022). A review of current challenges and legal advances in the global management of plastic waste. *Clean Technologies and Environmental Policy*, 24(3), 731–738. DOI: 10.1007/s10098-022-02289-y

Sedek, S. S. S. (2021). Sustainable development of plastic pollution awareness campaigns on social media. *International Design Journal*, 11(2), 339–364. DOI: 10.21608/idj.2021.153616

Selvaraj, S., Prasadh, S., Fuloria, S., Subramaniyan, V., Sekar, M., Ahmed, A. M., Bouallegue, B., Hari Kumar, D., Sharma, V. K., Maziz, M. N. H., Sathasivam, K. V., Meenakshi, D. U., & Fuloria, N. K. (2022). COVID-19 Biomedical Plastics Wastes—Challenges and Strategies for Curbing the Environmental Disaster. *Sustainability (Basel)*, 14(11), 6466. DOI: 10.3390/su14116466

Seung, J. K., Jae, H. C., Tae, Y. K., & Sung, Y. C. (2006). Biosorption of heavy metals and cyanide complexes on biomass. In New Developments and Application in Chemical Reaction Engineering; Studies in Surface Science and Catalysis (pp. 141–144). Elsevier.

Seuront, L., Zardi, G. I., Uguen, M., Bouchet, V. M., Delaeter, C., Henry, S., Spilmont, N., & Nicastro, K. R. (2022). A whale of a plastic tale: A plea for interdisciplinary studies to tackle micro-and nanoplastic pollution in the marine realm. *The Science of the Total Environment*, 846, 157187. DOI: 10.1016/j.scitotenv.2022.157187

Shah, M. A., Pirzada, B. M., Price, G., Shibiru, A. L., & Qurashi, A. (2022). Applications of nanotechnology in smart textile industry: A critical review. *Journal of Advanced Research*, 38, 55–75. DOI: 10.1016/j.jare.2022.01.008

Shamsuyeva, M., & Endres, H. J. (2021). Plastics in the context of the circular economy and sustainable plastics recycling: comprehensive review on research development, standardisation and market. *Composites Part C: Open Access,* p 100168. doi:DOI: 10.1016/j.jcomc.2021.100168

Shariati, S., Abedi, M., Saedi, A., Yazdani-Chamzini, A., Tamosaitiene, J., Saparauskas, J., & Stupak, S. (2017). Critical factors of the application of nanotechnology in construction industry by using anp technique under fuzzy intuitionistic environment. *Journal of Civil Engineering and Management*, 23(7), 914–925. DOI: 10.3846/13923730.2017.1343202

Sharifi, A., Allam, Z., Bibri, S. E., & Khavarian-Garmsir, A. R. (2024). Smart cities and sustainable development goals (SDGs): A systematic literature review of co-benefits and trade-offs. *Cities (London, England)*, 146, 104659. DOI: 10.1016/j.cities.2023.104659

Sharma, A., & Singh, B. (2022). Measuring Impact of E-commerce on Small Scale Business: A Systematic Review. *Journal of Corporate Governance and International Business Law*, 5(1).

Sharma, H. B., Vanapalli, K. R., Samal, B., Cheela, V. R. S., Dubey, B. K., & Bhattacharya, J. (2021). VR Sankar Cheela, Brajesh K. Dubey, and Jayanta Bhattacharya. "Circular economy approach in solid waste management system to achieve UN-SDGs: Solutions for post-COVID recovery.". *The Science of the Total Environment*, 800, 149605. DOI: 10.1016/j.scitotenv.2021.149605

Sheng, Y., Liu, Y., Wang, K., Cizdziel, J. V., Wu, Y., & Zhou, Y. (2021). Ecotoxicological effects of micronized car tire wear particles and their heavy metals on the earthworm (Eisenia fetida) in soil. *The Science of the Total Environment*, 793, 148613. DOI: 10.1016/j.scitotenv.2021.148613

Shi, M., Sun, Y., Wang, Z., He, G., Quan, H., & He, H. (2019). Plastic film mulching increased the accumulation and human health risks of phthalate esters in wheat grains. *Environmental Pollution*, 250, 1–7. DOI: 10.1016/j.envpol.2019.03.064

Shruti, V. C., Pérez-Guevara, F., Elizalde-Martínez, I., & Kutralam-Muniasamy, G. (2020). First study of its kind on the microplastic contamination of soft drinks, cold tea and energy drinks—Future research and environmental considerations. *The Science of the Total Environment*, 726, 138580. DOI: 10.1016/j.scitotenv.2020.138580

Shukla, S., Pei, Y., Li, W. G., & Pei, D. S. (2024). Toxicological Research on Nano and Microplastics in Environmental Pollution: Current Advances and Future Directions. *Aquatic Toxicology (Amsterdam, Netherlands)*, 270, 106894. DOI: 10.1016/j.aquatox.2024.106894

Simple Kitchen Swaps to Reduce Your Exposure to Microplastics. (2024, February 5). https://www.healthcentral.com/chronic-health/simple-kitchen-swaps-to-reduce-your-exposure-to-microplastics

Singh, B. (2023). Blockchain Technology in Renovating Healthcare: Legal and Future Perspectives. In *Revolutionizing Healthcare Through Artificial Intelligence and Internet of Things Applications* (pp. 177-186). IGI Global.

Singh, B. (2024). Evolutionary Global Neuroscience for Cognition and Brain Health: Strengthening Innovation in Brain Science. In *Biomedical Research Developments for Improved Healthcare* (pp. 246-272). IGI Global.

Singh, B., & Kaunert, C. (2024). Future of Digital Marketing: Hyper-Personalized Customer Dynamic Experience with AI-Based Predictive Models. In Revolutionizing the AI-Digital Landscape (pp. 189–203). Productivity Press. doi:DOI: 10.4324/9781032688305-14

Singh, B., & Kaunert, C. (2024). Future of Digital Marketing: Hyper-Personalized Customer Dynamic Experience with AI-Based Predictive Models. *Revolutionizing the AI-Digital Landscape: A Guide to Sustainable Emerging Technologies for Marketing Professionals*, 189.

Singh, B., & Kaunert, C. (2024). Harnessing Sustainable Agriculture Through Climate-Smart Technologies: Artificial Intelligence for Climate Preservation and Futuristic Trends. In *Exploring Ethical Dimensions of Environmental Sustainability and Use of AI* (pp. 214-239). IGI Global.

Singh, B., & Kaunert, C. (2024). Revealing Green Finance Mobilization: Harnessing FinTech and Blockchain Innovations to Surmount Barriers and Foster New Investment Avenues. In *Harnessing Blockchain-Digital Twin Fusion for Sustainable Investments* (pp. 265-286). IGI Global.

Singh, B., Jain, V., Kaunert, C., & Vig, K. (2024). Shaping Highly Intelligent Internet of Things (IoT) and Wireless Sensors for Smart Cities. In *Secure and Intelligent IoT-Enabled Smart Cities* (pp. 117-140). IGI Global.

Singh, B., Vig, K., & Kaunert, C. (2024). Modernizing Healthcare: Application of Augmented Reality and Virtual Reality in Clinical Practice and Medical Education. In Modern Technology in Healthcare and Medical Education: Blockchain, IoT, AR, and VR (pp. 1-21). IGI Global.

Singh, B. (2020). Global science and jurisprudential approach concerning healthcare and illness. *Indian Journal of Health and Medical Law*, 3(1), 7–13.

Singh, B. (2022). COVID-19 Pandemic and Public Healthcare: Endless Downward Spiral or Solution via Rapid Legal and Health Services Implementation with Patient Monitoring Program. *Justice and Law Bulletin*, 1(1), 1–7.

Singh, B. (2022). Relevance of Agriculture-Nutrition Linkage for Human Healthcare: A Conceptual Legal Framework of Implication and Pathways. *Justice and Law Bulletin*, 1(1), 44–49.

Singh, B. (2022). Understanding Legal Frameworks Concerning Transgender Healthcare in the Age of Dynamism. *Electronic Journal of Social and Strategic Studies*, 3(1), 56–65. DOI: 10.47362/EJSSS.2022.3104

Singh, B. (2023). Eyes on the Sky for Real Estate Progression and Green Buildings: Addressing Sustainable Development Satellite SDG 9 (Industry, Innovation, and Infrastructure) and Policy Framework. *National Journal of Real Estate Law*, 6(2), 1–7.

Singh, B. (2023). Federated Learning for Envision Future Trajectory Smart Transport System for Climate Preservation and Smart Green Planet: Insights into Global Governance and SDG-9 (Industry, Innovation and Infrastructure). *National Journal of Environmental Law*, 6(2), 6–17.

Singh, B. (2023). Tele-Health Monitoring Lensing Deep Neural Learning Structure: Ambient Patient Wellness via Wearable Devices for Real-Time Alerts and Interventions. *Indian Journal of Health and Medical Law*, 6(2), 12–16.

Singh, B. (2023). Unleashing Alternative Dispute Resolution (ADR) in Resolving Complex Legal-Technical Issues Arising in Cyberspace Lensing E-Commerce and Intellectual Property: Proliferation of E-Commerce Digital Economy. *Revista Brasileira de Alternative Dispute Resolution-Brazilian Journal of Alternative Dispute Resolution-RBADR*, 5(10), 81–105. DOI: 10.52028/rbadr.v5i10.ART04.Ind

Singh, B. (2024). Biosensors in Intelligent Healthcare and Integration of Internet of Medical Things (IoMT) for Treatment and Diagnosis. *Indian Journal of Health and Medical Law*, 7(1), 1–7.

Singh, B. (2024). Cherish Growth, Advancement and Tax Structure: Addressing Social and Economic Prospects. *Journal of Taxation and Regulatory Framework*, 7(1), 7–10.

Singh, B. (2024). Evolutionary Global Neuroscience for Cognition and Brain Health: Strengthening Innovation in Brain Science. In Prabhakar, P. (Ed.), *Biomedical Research Developments for Improved Healthcare* (pp. 246–272). IGI Global., DOI: 10.4018/979-8-3693-1922-2.ch012

Singh, B. (2024). Featuring Consumer Choices of Consumable Products for Health Benefits: Evolving Issues from Tort and Product Liabilities. *Journal of Law of Torts and Consumer Protection Law*, 7(1), 53–56.

Singh, B. (2024). Green Infrastructure in Real Estate Landscapes: Pillars of Sustainable Development and Vision for Tomorrow. *National Journal of Real Estate Law*, 7(1), 4–8.

Singh, B. (2024). Legal Dynamics Lensing Metaverse Crafted for Videogame Industry and E-Sports: Phenomenological Exploration Catalyst Complexity and Future. *Journal of Intellectual Property Rights Law*, 7(1), 8–14.

Singh, B. (2024). Lensing Legal Dynamics for Examining Responsibility and Deliberation of Generative AI-Tethered Technological Privacy Concerns: Infringements and Use of Personal Data by Nefarious Actors. In Ara, A., & Ara, A. (Eds.), *Exploring the Ethical Implications of Generative AI* (pp. 146–167). IGI Global., DOI: 10.4018/979-8-3693-1565-1.ch009

Singh, B. (2024). Social Cognition of Incarcerated Women and Children: Addressing Exposure to Infectious Diseases and Legal Outcomes. In Reddy, K. (Ed.), *Principles and Clinical Interventions in Social Cognition* (pp. 236–251). IGI Global., DOI: 10.4018/979-8-3693-1265-0.ch014

Singh, B. (2024). Transformative Wave of IoMT, EHRs, RPM Technologies to Revolutionize Public Health. *Indian Journal of Health and Medical Law*, 7(2), 22–26.

Singh, B., Arora, M. K., & Lal, S. (2024). Mapping Nature, Environment with SDG-9; Harnessing Machine Learning, Fostering Planet Earth Protection. *National Journal of Environmental Law*, 7(1).

Singh, B., & Kaunert, C. (2024). Integration of Cutting-Edge Technologies such as Internet of Things (IoT) and 5G in Health Monitoring Systems: A Comprehensive Legal Analysis and Futuristic Outcomes. *GLS Law Journal*, 6(1), 13–20. DOI: 10.69974/glslawjournal.v6i1.123

Singh, B., & Kaunert, C. (2024). Salvaging Responsible Consumption and Production of Food in the Hospitality Industry: Harnessing Machine Learning and Deep Learning for Zero Food Waste. In Singh, A., Tyagi, P., & Garg, A. (Eds.), *Sustainable Disposal Methods of Food Wastes in Hospitality Operations* (pp. 176–192). IGI Global., DOI: 10.4018/979-8-3693-2181-2.ch012

Singh, B., Kaunert, C., & Vig, K. (2024). Reinventing Influence of Artificial Intelligence (AI) on Digital Consumer Lensing Transforming Consumer Recommendation Model: Exploring Stimulus Artificial Intelligence on Consumer Shopping Decisions. In Musiolik, T., Rodriguez, R., & Kannan, H. (Eds.), *AI Impacts in Digital Consumer Behavior* (pp. 141–169). IGI Global., DOI: 10.4018/979-8-3693-1918-5.ch006

Slootmaekers, B., Carteny, C. C., Belpaire, C., Saverwyns, S., Fremout, W., Blust, R., & Bervoets, L. (2019). Microplastic contamination in gudgeons (Gobio gobio) from Flemish rivers (Belgium). [PubMed]. *Environmental Pollution*, 244, 675–684. DOI: 10.1016/j.envpol.2018.09.136

Soares, J., Miguel, I., Venâncio, C., Lopes, I., & Oliveira, M. (2020). Perspectives on Micro (Nano) Plastics in the marine environment: Biological and societal considerations. *Water (Basel)*, 12(11), 3208. DOI: 10.3390/w12113208

Soares, M. D. O., Matos, E., Lucas, C., Rizzo, L., Allcock, L., & Rossi, S. (2020). Microplastics in corals: An emergent threat. *Marine Pollution Bulletin*, 161, 111810. DOI: 10.1016/j.marpolbul.2020.111810

Song, J., Wang, C., & Li, G. (2024). Defining Primary and Secondary Microplastics: A Connotation Analysis. *ACS ES&T Water*, 4(6), 2330–2332. DOI: 10.1021/acsestwater.4c00316

Sood, M., Kapoor, D., Kumar, V., Kalia, N., Bhardwaj, R., Sidhu, G. P. S., & Sharma, A. (2021). Mechanisms of Plant Defense Under Pathogen Stress: A Review. *Current Protein & Peptide Science*, 22(5), 376–395. DOI: 10.2174/1389203722666210125122827

Sulistiawati, L. Y. (2022). Challenges in marine plastic pollution regulation in Indonesia. *Asia Pacific Journal of Environmental Law*, 25(1), 100–118.

Sultana, A., Zare, M., Thomas, V., Kumar, T. S. S., & Ramakrishna, S. (2022). Nano-based drug delivery systems: Conventional drug delivery routes, recent developments and future prospects. *Medicine in Drug Discovery*, 15, 100134. DOI: 10.1016/j.medidd.2022.100134

Sun, J., Dai, X., Wang, Q., Van Loosdrecht, M. C. M., & Ni, B.-J. (2019). Microplastics in wastewater treatment plants: Detection, occurrence and removal. *Water Research*, 152, 21–37. DOI: 10.1016/j.watres.2018.12.050

Sun, X. D., Yuan, X. Z., Jia, Y., Feng, L. J., Zhu, F. P., Dong, S. S., & Xing, B. (2020). Differentially charged nanoplastics demonstrate distinct accumulation in Arabidopsis thaliana. *Nature Nanotechnology*, 15(9), 755–760. DOI: 10.1038/s41565-020-0707-4

Sun, Y., Qian, Y., Geng, S., Wang, P., Zhang, L., & Yang, Z. (2023). Joint effects of microplastics and ZnO nanoparticles on the life history parameters of rotifers and the ability of rotifers to eliminate harmful phaeocystis. [PubMed]. *Chemosphere*, 310, 136939. DOI: 10.1016/j.chemosphere.2022.136939

Sun, Y., Zhao, X., Sui, Q., Sun, X., Zhu, L., Booth, A. M., Chen, B., Qu, K., & Xia, B. (2024). Polystyrene nanoplastics affected the nutritional quality of Chlamys farreri through disturbing the function of gills and physiological metabolism: Comparison with microplastics. *The Science of the Total Environment*, 910, 168457. DOI: 10.1016/j.scitotenv.2023.168457

Surendran, U., Jayakumar, M., Raja, P., Gopinath, G., & Chellam, P. V. (2023). Microplastics in terrestrial ecosystem: Sources and migration in soil environment. *Chemosphere*, 318, 137946. DOI: 10.1016/j.chemosphere.2023.137946

Sutherland, D. L., McCauley, J., Labeeuw, L., Ray, P., Kuzhiumparambil, U., Hall, C., Doblin, M., Nguyen, L. N., & Ralph, P. J. (2021). How microalgal biotechnology can assist with the UN Sustainable Development Goals for natural resource management. *Current Research in Environmental Sustainability*, 3, 100050. DOI: 10.1016/j.crsust.2021.100050

Su, Y., Hu, X., Tang, H., Lu, K., Li, H., Liu, S., Xing, B., & Ji, R. (2022). Steam disinfection releases micro(nano)plastics from silicone-rubber baby teats as examined by optical photothermal infrared microspectroscopy. [PubMed]. *Nature Nanotechnology*, 17(1), 76–85. Advance online publication. DOI: 10.1038/s41565-021-00998-x

Swart, B., Pihlajamäki, A., John Chew, Y. M., & Wenk, J. (2022). Microbubble-microplastic interactions in batch air flotation. *Chemical Engineering Journal*, 449, 137866. DOI: 10.1016/j.cej.2022.137866

Syberg, K., Nielsen, M. B., Oturai, N. B., Clausen, L. P. W., Ramos, T. M., & Hansen, S. F. (2022). Circular economy and reduction of micro (nano) plastics contamination. *Journal of Hazardous Materials Advances*, 5, 100044. DOI: 10.1016/j.hazadv.2022.100044

Tan, D. (2022). Assessing Indonesia's Environmental Laws Pertaining to the Abatement of Marine Plastic Pollution: A Euphemism? *Jurnal Media Hukum*, 29(1), 1–33. DOI: 10.18196/jmh.v29i1.13414

Tan, X., Yu, X., Cai, L., Wang, J., & Peng, J. (2019). Microplastics and associated PAHs in surface water from the Feilaixia Reservoir in the Beijiang River, China. [PubMed]. *Chemosphere*, 221, 834–840. DOI: 10.1016/j.chemosphere.2019.01.022

Tao, S., Li, T., Li, M., Yang, S., Shen, M., & Liu, H. (2024). Research advances on the toxicity of biodegradable plastics derived micro/nanoplastics in the environment: A review. *The Science of the Total Environment*, 916, 170299. DOI: 10.1016/j.scitotenv.2024.170299

Taran, M., Safaei, M., Karimi, N., & Almasi, A. (2021). Benefits and application of nanotechnology in environmental science: An overview. *Biointerface Research in Applied Chemistry*, 11, 7860–7870.

Taylor, S. E., Pearce, C. I., Sanguinet, K. A., Hu, D., Chrisler, W. B., Kim, Y.-M., Wang, Z., & Flury, M. (2020). Polystyrene nano- and microplastic accumulation at Arabidopsis and wheat root cap cells, but no evidence for uptake into roots. *Environmental Science. Nano*, 7(7), 1942–1953. DOI: 10.1039/D0EN00309C

Theron, J., Walker, J. A., & Cloete, T. E. (2008). Nanotechnology and water treatment: Applications and emerging opportunities. *Critical Reviews in Microbiology*, 34(1), 43–69. DOI: 10.1080/10408410701710442

Tiffin, L., Hazlehurst, A., Sumner, M., & Taylor, M. (2022). Reliable quantification of microplastic release from the domestic laundry of textile fabrics. *Journal of the Textile Institute*, 113(4), 558–566. DOI: 10.1080/00405000.2021.1892305

Timmer, J. M. K. (2001). *Properties of nanofiltration membranes: model development and industrial application.* [PhD Thesis 1 (Research TU/e / Graduation TU/e), Chemical Engineering and Chemistry]. Technische Universiteit Eindhoven. doi:DOI: 10.6100/IR545659

Tirkey, A., & Upadhyay, L. S. B. (2021). Microplastics: An overview on separation, identification and characterization of microplastics. [PubMed]. *Marine Pollution Bulletin*, 170, 112604. DOI: 10.1016/j.marpolbul.2021.112604

Tlili, I., & Alkanhal, T. A. (2019). Nanotechnology for water purification: Electrospun nanofibrous membrane in water and wastewater treatment. *Journal of Water Reuse and Desalination*, 9(3), 232–248. DOI: 10.2166/wrd.2019.057

Torabian, A., Kazemian, H., Seifi, L., Bidhendi, G. N., Azimi, A. A., & Ghadiri, S. K. (2010). Removal of Petroleum Aromatic Hydrocarbons by Surfactant-modified Natural Zeolite: The Effect of Surfactant. *Clean (Weinheim)*, 38(1), 77–83. DOI: 10.1002/clen.200900157

Treatment technologies for the removal of micro plastics from aqueous medium | AIP Conference Proceedings | AIP Publishing. (n.d.). Retrieved August 21, 2024, from https://pubs.aip.org/aip/acp/article-abstract/2463/1/030009/2825284/Treatment-technologies-for-the-removal-of-micro?redirectedFrom=fulltext

United Nation Environment Program. (2021). *From Pollution to Solution: A global assessment of marine litter and plastic pollution (unep.org)*. United Nation Environment Program.

United Nation Environment Program. (2023). Microplastics: The long legacy left behind by plastic pollution. Ocean and costs. United nations environment program. https://www.unep.org/news-and-stories/story/microplastics-long-legacy-left-behind-plastic-pollution

United Nations Sustainable Development Goals. Available online: https://sdgs.un.org/goals (accessed on 14 September 2023).

Valiyaveettil Salimkumar, A., Kurisingal Cleetus, M. C., Ehigie, J. O., Onogbosele, C. O., Nisha, P., Kumar, B. S., ... & Rejish Kumar, V. J. (2024). Adsorption Behavior and Interaction of Micro-Nanoplastics in Soils and Aquatic Environment. *Management of Micro and Nano-plastics in Soil and Biosolids: Fate, Occurrence, Monitoring, and Remedies*, 283-311.

van Schothorst, B., Beriot, N., Huerta Lwanga, E., & Geissen, V. (2021). Sources of Light Density Microplastic Related to Two Agricultural Practices: The Use of Compost and Plastic Mulch. *Environments (Basel, Switzerland)*, 8(4), 4. Advance online publication. DOI: 10.3390/environments8040036

Varghese, A. (2004). A comparative risk approach to assessing point-of-use water treatment systems in developing countries. In I. Linkov & A.B. Ramada (Eds.) *Proceedings of Comparative risk assessment and environmental decision making*. NATO Science Series IV Earth and Environment Science NAIV, *38*, 99–112.

Varshney, A. (2024). Cleaning Up the Smallest Pollutants: The Potential of Microbial Degradation in Tackling Micro-and Nano-Plastic Pollution. In Management of Micro and Nano-plastics in Soil and Biosolids: Fate, Occurrence, Monitoring, and Remedies (pp. 367–389). Springer Nature Switzerland. doi:DOI: 10.1007/978-3-031-51967-3_15

Verma, B., & Balomajumder, C. (2020). Surface modification of one-dimensional Carbon Nanotubes A review for the management of heavy metals in wastewater. *Environmental Technology & Innovation*, 17, 100596. DOI: 10.1016/j.eti.2019.100596

Verma, K. K., Song, X.-P., Xu, L., Huang, H.-R., Liang, Q., Seth, C. S., & Li, Y.-R. (2023). Nano-microplastic and agro-ecosystems: A mini-review. *Frontiers in Plant Science*, 14, 1283852. DOI: 10.3389/fpls.2023.1283852

Vouk, V. B. (1979). *Handbook on the Toxicology of Metals*. Elsevier-North-Holland Biomedical Press.

Wahl, A., Le Juge, C., Davranche, M., El Hadri, H., Grassl, B., Reynaud, S., & Gigault, J. (2021). Nanoplastic occurrence in a soil amended with plastic debris. *Chemosphere*, 262, 127784. DOI: 10.1016/j.chemosphere.2020.127784

Wakwella, A., Wenger, A., Jenkins, A., Lamb, J., Kuempel, C. D., Claar, D., Corbin, C., Falinski, K., Rivera, A., Grantham, H. S., & Jupiter, S. D. (2023). Integrated watershed management solutions for healthy coastal ecosystems and people. *Cambridge Prisms: Coastal Futures*, 1, e27. DOI: 10.1017/cft.2023.15

Walker, T. R., & Fequet, L. (2023). Current trends of unsustainable plastic production and micro (nano) plastic pollution. *Trends in Analytical Chemistry*, 160, 116984. DOI: 10.1016/j.trac.2023.116984

Walther, B. A., Kusui, T., Yen, N., Hu, C. S., & Lee, H. (2020). Plastic pollution in East Asia: Macroplastics and microplastics in the aquatic environment and mitigation efforts by various actors. *Plastics in the aquatic environment-Part I: Current status and challenges*, 353-403.

Wang, C., Tang, J., Yu, H., Wang, Y., Li, H., & Xu, S. (2022). Microplastic Pollution in the Soil Environment: Characteristics, Influencing Factors, and Risks. *Sustainability*, •••, 1–14. DOI: 10.3390/su15010001

Wang, G., Li, X., Humphries, M., Chinni, V., Uthaipan, K., & Dai, M. (2022). Coastal pollution. In *Blue Economy: An Ocean Science Perspective* (pp. 251–286). Springer Nature Singapore., DOI: 10.1007/978-981-19-5065-0_9

Wang, H., Wang, Q., Lv, M., Li, J., Zhao, X., Song, Z., Wang, X., You, J., Wang, Y., & Chen, L. (2023). Marine micro (nano) plastics determination and its environmental toxicity evaluation. *Trends in Analytical Chemistry*, 168, 117332. DOI: 10.1016/j.trac.2023.117332

Wang, H., Zhou, Y., Jiang, X., Sun, B., Zhu, Y., Wang, H., Su, Y., & He, Y. (2015). Simultaneous Capture, Detection, and Inactivation of Bacteria as Enabled by a Surface-Enhanced Raman Scattering Multifunctional Chip. *Angewandte Chemie International Edition*, 54(17), 5132–5136. DOI: 10.1002/anie.201412294

Wang, L., & Tanveer, M. (2023). Editorial to the special issue "Eco-physiological and molecular basis of stress tolerance in plants." *Biology (Basel)*, 12(3), 485. DOI: 10.3390/biology12030485

Wang, Q. W., Wang, Q., Adams, C. A., Sun, Y., & Zhang, S. (2022). Effects of microplastics on soil properties: Current knowledge and future perspectives. *Journal of Hazardous Materials*, 424, 1–36. DOI: 10.1016/j.jhazmat.2021.127531

Wang, W. X. (2023). Environmental toxicology of marine microplastic pollution. Cambridge Prisms. *Plastics*, 1, e10. DOI. 10.1017/plc.2023.9

Wang, W., Zhang, J., Qiu, Z., Cui, Z., Li, N., Li, X., & Zhao, C. (2022). Effects of polyethylene microplastics on cell membranes: A combined study of experiments and molecular dynamics simulations. *Journal of Hazardous Materials*, 429, 128323. DOI: 10.1016/j.jhazmat.2022.128323

Wang, Y., Xiang, L., Wang, F., Redmile-Gordon, M., Bian, Y., Wang, Z., Gu, C., Jiang, X., Schäffer, A., & Xing, B. (2023c). Transcriptomic and metabolomic changes in lettuce triggered by microplastics-stress. *Environmental Pollution*, 320, 121081. DOI: 10.1016/j.envpol.2023.121081

Wassénius, E., Crona, B., & Quahe, S. (2024). Essential environmental impact variables: A means for transparent corporate sustainability reporting aligned with planetary boundaries. *One Earth*, 7(2), 211–225. DOI: 10.1016/j.oneear.2024.01.014

Watkins, E., Schweitzer, J. P., Leinala, E., & Börkey, P. (2019). Policy approaches to incentivise sustainable plastic design.

Werkneh, A. A., & Rene, E. R. (2018). Applications of Nanotechnology and Biotechnology for Sustainable Water and Wastewater Treatment. In *Energy, environment, and sustainability* (pp. 405–430). https://doi.org/doi:10.1007/978-981-13-3259-3_19

WHO UNICEF. (2012). *Progress on Drinking Water and Sanitation: 2012 Update; WHO*. UNICEF.

Williams, A. T., & Rangel-Buitrago, N. (2022). The past, present, and future of plastic pollution. *Marine Pollution Bulletin*, 176, 113429. DOI: 10.1016/j.marpolbul.2022.113429

Winiarska, E., Jutel, M., & Zemelka-Wiacek, M. (2024). The potential impact of nano-and microplastics on human health: Understanding human health risks. *Environmental Research*, 251, 118535. DOI: 10.1016/j.envres.2024.118535

Wright, S. L., Ulke, J., Font, A., Chan, K. L. A., & Kelly, F. J. (2020). Atmospheric microplastic deposition in an urban environment and an evaluation of transport. *Environment International*, 136, 105411. DOI: 10.1016/j.envint.2019.105411

Wua, J. (2019). Environmental occurrences, fate, and impacts of microplastics. *Ecotoxicology and Environmental Safety*, 184, 1–16. DOI: 10.1016/j.ecoenv.2019.109612

Wu, X., Lyu, X., Li, Z., Gao, B., Zeng, X., Wu, J., & Sun, Y. (2020). Transport of polystyrene nanoplastics in natural soils: Effect of soil properties, ionic strength and cation type. *The Science of the Total Environment*, 707, 136065. DOI: 10.1016/j.scitotenv.2019.136065

Xiong, X., Wu, C., Elser, J. J., Mei, Z., & Hao, Y. (2019). Occurrence and fate of microplastic debris in middle and lower reaches of the Yangtze River–from inland to the sea. [PubMed]. *The Science of the Total Environment*, 659, 66–73. DOI: 10.1016/j.scitotenv.2018.12.313

Xiu, Z.-M., Zhang, Q.-B., Puppala, H. L., Colvin, V. L., & Alvarez, P. J. J. (2012). Negligible Particle-Specific Antibacterial Activity of Silver Nanoparticles. *Nano Letters*, 12(8), 4271–4275. DOI: 10.1021/nl301934w

Xu, Y., Ou, Q., van der Hoek, J. P., Liu, G., & Lompe, K. M. (2024). Photo-oxidation of Micro-and Nanoplastics: Physical, Chemical, and Biological Effects in Environments. *Environmental Science & Technology*, 58(2), 991–1009. DOI: 10.1021/acs.est.3c07035

Yadav, D. K., & Srivastava, S. (2017). Carbon nanotubes as adsorbent to remove heavy metal ion (Mn^{+7}) in wastewater treatment. *Materials Today: Proceedings*, 4(2), 4089–4094. DOI: 10.1016/j.matpr.2017.02.312

Yamamura, H., Hoshino, M., & Kubota, N. (2011). Membrane Filtration. *Principle and Applications in Water Treatment.*, 67, 3.

Yan, A., Wang, Y., Tan, S. N., Mohd Yusof, M. L., Ghosh, S., & Chen, Z. (2020). Phytoremediation: A Promising Approach for Revegetation of Heavy Metal-Polluted Land. *Frontiers in Plant Science*, 11, 359. DOI: 10.3389/fpls.2020.00359

Yang, C., & Gao, X. (2022). Impact of microplastics from polyethylene and biodegradable mulch films on rice (Oryza sativa L.). *The Science of the Total Environment*, 828, 154579. DOI: 10.1016/j.scitotenv.2022.154579

Yang, H., Ju, J., Wang, Y., Zhu, Z., Lu, W., & Zhang, Y. (2024). Micro-and nanoplastics induce kidney damage and suppression of innate immune function in zebrafish (Danio rerio) larvae. *The Science of the Total Environment*, 931, 172952. DOI: 10.1016/j.scitotenv.2024.172952

Yang, L., Zhang, Y., Kang, S., Wang, Z., & Wu, C. (2021). Microplastics in soil: A review on methods, occurrence, sources, and potential risk. [PubMed]. *The Science of the Total Environment*, 780, 146546. DOI: 10.1016/j.scitotenv.2021.146546

Yang, X., Zhang, S., Ju, M., & Liu, L. (2019). Preparation and Modification of Biochar Materials and their Application in Soil Remediation. *Applied Sciences (Basel, Switzerland)*, 9(7), 1365. DOI: 10.3390/app9071365

Yang, X., Zhao, Z., Yu, Y., Shimizu, K., Zhang, Z., Lei, Z., & Lee, D. J. (2020). Enhanced biosorption of Cr(VI) from synthetic wastewater using algal-bacterial aerobic granular sludge: Batch experiments, kinetics and mechanisms. *Separation and Purification Technology*, 251, 117323. DOI: 10.1016/j.seppur.2020.117323

Yang, Z., Lü, F., Zhang, H., Wang, W., Shao, L., Ye, J., & He, P. (2021). Is incineration the terminator of plastics and microplastics? *Journal of Hazardous Materials*, 401, 123429. DOI: 10.1016/j.jhazmat.2020.123429

Yan, M., Nie, H., Xu, K., He, Y., Hu, Y., Huang, Y., & Wang, J. (2019). Microplastic abundance, distribution and composition in the Pearl River along Guangzhou city and Pearl River estuary, China. [PubMed]. *Chemosphere*, 217, 879–886. DOI: 10.1016/j.chemosphere.2018.11.093

Yap, C. K., Tze, Y. A. H., Nulit, R., Syazwan, W. M., Okamura, H., Horie, Y., & Ong, M. C. (2024). Copper in Commercial Marine Fish: From Biomonitoring to the ESG (Environment, Social, and Governance). *Method. Pollutants*, 4(1), 117–135. DOI: 10.3390/pollutants4010008

Yee, M. S.-L., Hii, L.-W., Looi, C. K., Lim, W.-M., Wong, S.-F., Kok, Y.-Y., Tan, B.-K., Wong, C.-Y., & Leong, C.-O. (2021). Impact of Microplastics and Nanoplastics on Human Health. [PubMed]. *Nanomaterials (Basel, Switzerland)*, 11(2), 2. Advance online publication. DOI: 10.3390/nano11020496

Yen, N. T. H., & Dung, N. P. (2023). Marine pollution management in response to plastic waste under international law and national law–challenges and implications for developing countries: A case study of Vietnam. *Asia Pacific Journal of Environmental Law*, 26(2), 162–193. DOI: 10.4337/apjel.2023.02.02

Yi, J., Ma, Y., Ruan, J., You, S., Ma, J., Yu, H., Zhao, J., Zhang, K., Yang, Q., Jin, L., Zeng, G., & Sun, D. (2024). The invisible Threat: Assessing the reproductive and transgenerational impacts of micro-and nanoplastics on fish. *Environment International*, 183, 108432. DOI: 10.1016/j.envint.2024.108432

Yoon, J., Cao, X., Zhou, Q., & Ma, L. Q. (2006). Accumulation of Pb, Cu, and Zn in native plants growing on a contaminated Florida site. *The Science of the Total Environment*, 368(2-3), 456–464. DOI: 10.1016/j.scitotenv.2006.01.016

Yu, F., Yang, C., Zhu, Z., Bai, X., & Ma, J. (2019). Adsorption behavior of organic pollutants and metals on micro/nanoplastics in the aquatic environment. *Sci. Total Environ., 694*, 133643. https://doi.org/10.1016/j.scitotenv.2019.133643

Yuan, S., Liang, J., Zhang, Y., Han, H., Jiang, T., Liu, Y., Zhang, Y., Wang, W., & Dong, X. (2022). Evidence from Thermal Aging Indicating That the Synergistic Effect of Glyoxal and Sodium Sulfite Improved the Thermal Stability of Conformational Modified Xanthan Gum. *Polymers*, 14(2), 243. DOI: 10.3390/polym14020243

Yuan, W., Liu, X., Wang, W., Di, M., & Wang, J. (2019). Microplastic abundance, distribution and composition in water, sediments, and wild fish from Poyang Lake, China. [PubMed]. *Ecotoxicology and Environmental Safety*, 170, 180–187. DOI: 10.1016/j.ecoenv.2018.11.126

Yu, J. &.-L. (2021). Micro plastics in soil ecosystem – A review of sources, fate, and ecological impact. *Plant, Soil and Environment.*

Yu, J., Tian, J. Y., Jiang, Y., Wang, X. D., Song, X. R., Liu, L. F., & Yang, G. P. (2024). Effects of micro-and nano-plastics on growth, antioxidant system, DMS, and DMSP production in Emiliania huxleyi. *Environmental Pollution*, 351, 124084. DOI: 10.1016/j.envpol.2024.124084

Yu, Y., & Flury, M. (2021). Current understanding of subsurface transport of micro-and nanoplastics in soil. *Vadose Zone Journal*, 20(2), e20108. DOI: 10.1002/vzj2.20108

Yu, Z., Li, Q., Wang, J., Yu, Y., Wang, Y., Zhou, Q., & Li, P. (2020). Reactive Oxygen Species-Related Nanoparticle Toxicity in the Biomedical Field. *Nanoscale Research Letters*, 15(1), 115. DOI: 10.1186/s11671-020-03344-7

Zhan, M. L. (2018). Biodegradable, slow-release fertilizer prepared with superabsorbent polymer composites and controlled-release coatings. *Journal of Applied Polymer Science, 136(4), 46068*, 1-12.

Zhang, C., Lai, C., Zeng, G., Huang, D., Yang, C., Wang, Y., Zhou, Y., & Cheng, M. (2016). Efficacy of carbonaceous nanocomposites for sorbing ionizable antibiotic sulfamethazine from aqueous solution. *Water Research*, 95, 103–112. DOI: 10.1016/j.watres.2016.03.014

Zhang, G. X. (2023). Global dust cycle transports microplastics from arid and semi-arid regions to marine environments. *Nature Geoscience*, 16(3), 346–352.

Zhang, S., Wang, J., Yan, P., Hao, X., Xu, B., Wang, W., & Aurangzeib, M. (2021). Non-biodegradable microplastics in soils: A brief review and challenge. [PubMed]. *Journal of Hazardous Materials*, 409, 124525. DOI: 10.1016/j.jhazmat.2020.124525

Zhang, X., Leng, Y., Liu, X., Huang, K., & Wang, J. (2020). Microplastics' pollution and risk assessment in an urban river: A case study in the Yongjiang River, Nanning City, South China. *Exposure and Health*, 12(2), 141–151. DOI: 10.1007/s12403-018-00296-3

Zhang, Z. C. (2020). Microplastics alter soil aggregate stability and microbial community composition. *Environmental Science & Technology*, 54(10), 5905–5914.

Zhao, W., Tian, Y., Chu, X., Cui, L., Zhang, H., Li, M., & Zhao, P. (2021). Preparation and characteristics of a magnetic carbon nanotube adsorbent: Its efficient adsorption and recoverable performances. *Separation and Purification Technology*, 257, 117917. DOI: 10.1016/j.seppur.2020.117917

Zhao, Y., Zhang, B., Zhang, X., Wang, J., Liu, J., & Chen, R. (2010). Preparation of highly ordered cubic NaA zeolite from halloysite mineral for adsorption of ammonium ions. *Journal of Hazardous Materials*, 178(1-3), 658–664. DOI: 10.1016/j.jhazmat.2010.01.136

Zheng, S., Tang, B. Z., & Wang, W. X. (2024). Microplastics and nanoplastics induced differential respiratory damages in tilapia fish Oreochromis niloticus. *Journal of Hazardous Materials*, 465, 133181. DOI: 10.1016/j.jhazmat.2023.133181

Zheng, S., & Wang, W. X. (2024). Single-Cell RNA Sequencing Profiling Cellular Heterogeneity and Specific Responses of Fish Gills to Microplastics and Nanoplastics. *Environmental Science & Technology*, 58(13), 5974–5986. DOI: 10.1021/acs.est.3c10338

Zhou, B., Li, Y., Zheng, G., Dai, K., Liu, C., Ma, Y., Zhang, J., Wang, N., Shen, C., & Guo, Z. (2018). Continuously fabricated transparent conductive polycarbonate/carbon nanotube nanocomposite films for switchable thermochromic applications. *Journal of Materials Chemistry. C, Materials for Optical and Electronic Devices*, 6(31), 8360–8371. DOI: 10.1039/C8TC01779D

Ziani, K., Ioni ă-Mîndrican, C.-B., Mititelu, M., Neac u, S. M., Negrei, C., Moro an, E., Drăgănescu, D., & Preda, O.-T. (2023). Microplastics: A Real Global Threat for Environment and Food Safety: A State of the Art Review. *Nutrients*, 15(3), 3. Advance online publication. DOI: 10.3390/nu15030617

Zuverza-Mena, N., Martínez-Fernández, D., Du, W., Hernandez-Viezcas, J. A., Bonilla-Bird, N., López-Moreno, M. L., Komárek, M., Peralta-Videa, J. R., & Gardea-Torresdey, J. L. (2017). Exposure of engineered nanomaterials to plants: Insights into the physiological and biochemical responses-A review. *Plant Physiology and Biochemistry*, 110, 236–264. DOI: 10.1016/j.plaphy.2016.05.037

About the Contributors

Nisha Gaur is an Assistant Professor in the Department of Biotechnology at the Gautam Buddha University, India. She has received her Ph. D from the National Institute of Technology Warangal. She has a postdoctoral research associateship with the Defence Research Laboratory, DRDO Tezpur. Her research interests lie in water quality monitoring, water treatment, soil health maintenance, phytoremediation, organic pollutant degradation and waste management. She is an editorial board member in the International Journal of Sustainable Energy. She has served as a reviewer in reputed journals. She has published 19 studies in reputed peer reviewed journals, 20 book chapters and 1 edited book to her name. She is also a member of Association of Microbiologist of India.

Niranjan Prakash Melkania is a Professor & Dean Academics at Gautam Buddha University, Greater Noida. He obtained his Ph.D. from Kumaun University, Nainital, India. From past 30 years he has been working in the areas, viz., pollution. environment impact assessment, social impact assessment, ecology and biodiversity assessment and conservation, and eco-development. He has been listed as a noteworthy Management educator, consultant by Marquis Who's Who.

* * *

Om Akolkar is a Bachelor of Technology (B.Tech) student in Computer Science Engineering at MIT Art, Design and Technology University, Rajbaugh Loni Kalbhor, Pune, Maharashtra, India. He has a strong interest in blending technology with environmental science Om has recently authored a research paper titled "Wastewater Treatment Using Nanotechnology." This research explores innovative approaches to enhancing water purification processes through advanced nanomaterials, His involvement in this research demonstrates his dedication to advancing technology and addressing environmental challenges. Through his work, Om underscores his commitment to delivering impactful contributions to the academic community. Beyond his academic achievements, Om is deeply engaged in various tech-driven projects and participates actively in codathons and hackathons. He regularly attends workshops and seminars covering a range of topics including machine learning, artificial intelligence, data science, and software development. Om is also involved in clubs dedicated to programming, software development. Om's active learning and research, along with his passion for exploring new technologies, show his drive to make a real impact. As he grows his skills, he stays focused on pushing the limits of technology and innovation.

Manmeet Kaur Arora is a Research Scholar at the School of Law, Sharda University, Greater Noida. She is a graduate of Law and has completed her Master's in Law from Amity University, Noida. Manmeet Kaur is a qualified Company Secretary Executive Programme. She has over 8+ years of Industry Experience along with 1 year of experience as an Associate at a reputed IPR Law Firm. Manmeet Kaur has attended 20+ National & International Conferences, Webinars, Seminars, and Symposiums along with Paper Presentations, and participated in 5+ Faculty Development programs. She has over 10+ publications in various esteemed Journals and Books. She is an active member at the Family Dispute Resolution Clinic, Meditation Centre, Greater Noida. Manmeet Kaur has a keen interest in areas of Corporate Law, Intellectual Property Rights, Insolvency, Artificial Intelligence, Machine Learning, and Blockchain Technology.

Susanta Das is a seasoned educator with a proven record of teaching and mentoring diverse students, working at multiple institutions, and providing service to the various initiatives of organizations. He received his Ph.D. and M.A. degrees from Western Michigan University (WMU), USA, and M.Sc. degree from Banaras Hindu University (BHU), India all in Physics. He continued his research as a Marie-Curie post-doctoral fellow at Stockholm University, Stockholm, Sweden on beam diagnostics for the DESIREE (Double ElectroStatic Ion Ring ExpEriment) in the project DITANET (Diagnostic Techniques for particle Accelerators – a Marie-Curie initial training NETwork), at the Indian Institute of Science Education and Research-Kolkata, India on high-pressure physics as a Project Scientist-B, and at the University of Electro-Communications, Tokyo, Japan on ion-surface interactions as a Post-doctoral fellow. Throughout his career, Dr. Das worked and collaborated with many researchers, Ph.D., masters, and visiting students from different countries. He further visited many countries to discuss research and an international conference participant (UK, Italy, Germany, Greece, Belgium, Bulgaria, Romania, Brazil etc.). He co-authored several research articles in WoS/Scopus indexed international journals, conference proceedings, and scholarly book chapters published by renowned international publishing houses. Beyond his research endeavors, Dr. Das has a storied history of service to academic and administrative committees, exhibiting his commitment to the growth and development of educational institutions. His experience includes tenure at Central University South Bihar, Sri Sri University, and P.K. University before assuming his current role at Ajeenkya DY Patil University. He received the Marie-Curie post-doctoral fellowship in Sweden, Science Academies' Summer Research Fellowship in India, Gwen Frostic Doctoral Fellowship, Department Graduate Research and Creative Scholar Award by WMU, and Leo R. Parpart Doctoral Fellowship by Dept. of Physics, WMU, among many others, throughout his academic journey. At present, Dr. Das continues to delve into cutting-edge fields, with a keen interest in nanotechnology, quantum computing, and data science. His multifaceted contributions, spanning teaching, research, and administrative leadership, showcase a dedicated professional who is instrumental in advancing the vision and mission of the institutions he serves.

Harshita Jain is working at Amity University, Noida.

Christian Kaunert is Professor of International Security at Dublin City University, Ireland. He is also Professor of Policing and Security, as well as Director of the International Centre for Policing and Security at the University of South Wales. In addition, he is Jean Monnet Chair, Director of the Jean Monnet Centre of Excellence and Director of the Jean Monnet Network on EU Counter-Terrorism (www.eucter.net).

Arya Kawale is pursuing her B.Tech Computer Science and Engineering at MIT ADT Campus, Rajbaugh, Solapur - Pune Hwy, near Bharat Petrol Pump, Loni Kalbhor, Maharashtra 412201. In addition to her academic work, Arya is actively involved in tech-driven projects and hackathons. She regularly attends workshops and seminars on machine learning, blockchain, data science, or software engineering and joins clubs focused on programming, software development, artificial intelligence, or cybersecurity. Driven by a passion for applying technology to solve real-world problems, Arya Kawale aspires to continue exploring and contributing to sustainable solutions in nanotechnology while pursuing further research and academic opportunities.

Yadnyesh Khapekar, currently pursuing a B.Tech. in Computer Science and Engineering from Ajeenkya D Y Patil University, has a broad spectrum of research interests, ranging from nanoparticles, and aerodynamics to many others. His research not only focuses on the synthesis but also the characterization of nanoparticles. 'How can one minimize the air resistance for a moving object?' is his focus for studying aerodynamics. He aims to research in these fields and make them more applicable for research purposes.

Poonam Kodag has completed her graduation from Mspm's Deogiri College Aurangabad and post-graduation in botany from department of Botany, Dr. B.A.M. University Aurangabad. She is currently working as an assistant professor in Rashtriya College of Pharmacy Hatnoor. She has more than 3 year experience in field of environmental science and microbiology. She has published 2 reviewed & 1 research article in National and international journal and published 1 book chapter. Miss. Kodag has presented poster in 'Shodh A National Level Research Convention'. She has attended number of national conferences and state level seminars.

Sahil Lal is a Research Scholar at School of Law, Sharda University. He pursued his graduation from National Law University, Lucknow and completed his Master's in Law from Gujarat National Law University, Gandhinagar. Sahil has completed a certificate course on 'AI & Law' from Lund University, Sweden and on 'Cyber Law' from Indian Law Institute, New Delhi. Sahil Lal is a qualified UGC-NET & JRF candidate. He has a 10+ year Industry Experience and is an active member of Family Dispute Resolution Clinic, Mediation Center, Greater Noida. Sahil has attended 20+ International & National Conferences along with Paper Presentation and has also participated in more than 5+ Faculty Development Programme. Mr. Sahil has more than 10+ publications in various esteemed Journals & Books. Mr. Sahil has a keen interest in area of Artificial Intelligence, Machine Learning, Robotics, Block-Chain Technology and Intellectual Property Rights.

Rashmi Mittal developed a keen interest in the analysis of volatile and semi volatile organic pesticides including some of listed POPs. She pursued her academic aspirations, obtaining a Bachelor's degree in chemistry followed by a Master's and Ph.D. in environmental science, specializing in pesticide analysis. Currently working at Central Pollution Control Board.

Shashank Mittal has done his FPM in Organizational behavior and Human Resource from Indian Institute of Management Raipur. He holds B. Tech from I.E.T. Lucknow. His post FPM work experience includes almost five years of industry and academics exposures in multiple roles. Prior to joining FPM, he has over two years of industrial experience and three years of teaching experience in various organizations. He has published multiple papers in ABDC ranked and SSCI indexed journals of international repute such as Journal of Knowledge Management, Journal of Behavioral and Experimental Finance, International Journal of Conflict Management, Journal of Management and Organization and Current Psychology. His current research interest includes Social identity, Knowledge exchanges, Status,

Proactive helping, Employer branding, Organizational justice, and Humanitarian relief management. He enjoys badminton and cycling during leisure time.

Xuan-Hoa Nghiem received his PhD in Economics from RMIT university, Melbourne, Australia. His research interests include macroeconomics, monetary economics, energy economics, environmental economics, energy finance, corporate finance, sustainable development goals (SDGs). His work has been published in high-quality journals, including Technological Forecasting and Social Change, Resources Policy, Renewable Energy, Finance Research Letters, Journal of Environmental Management etc.

Ashutosh Pandey earned his doctorate in marketing from ABV-Indian Institute of Information Technology and Management, Gwalior. He holds a UGC-NET Management qualification and has more than a decade of experience in industry and academia.. He has taught postgraduate and executive students about business research methods, advanced marketing research, services marketing, marketing management, consumer behaviour, and structural equation modeling. His research has appeared in many prestigious journals, including Environment, Development, and Sustainability, Journal of Quality Assurance in Hospitality & Tourism, and International Journal of Tourism Cities. He has worked as a visiting professor at IMT Ghaziabad, Indian Institute of Foreign Trade, New Delhi and the Indian Institute of Travel and Tourism Management in Gwalior. Dr. Pandey has been a resource person for several qualitative and quantitative research workshops hosted at reputed organisations.

Sunil Sankathala is pursuing his B.Tech. in Computer Science and Engineering at School of Engineering of Ajeenkya DY Patil University. His research interests include cybersecurity, internet of things, and application of artificial intelligence in nanotechnology.

Aditya Shrivastav is currently studying B.Tech. in Computer Engineering at the School of Engineering, Ajeenkya DY Patil University. His research interests are centered around nanoparticle biosynthesis, exploring ionic liquids' use in various applications, and applying Artificial Intelligence to different fields. Aditya is particularly focused on understanding how these technologies can be utilized together to solve practical problems. He aims to advance research in these areas and contribute to innovative solutions that address real-world challenges.

Bhupinder Singh is working as Professor at Sharda University, India. Also, Honorary Professor in University of South Wales UK and Santo Tomas University Tunja, Colombia. His areas of publications as Smart Healthcare, Medicines, fuzzy logics, artificial intelligence, robotics, machine learning, deep learning, federated learning, IoT, PV Glasses, metaverse and many more. He has 3 books, 139 paper publications, 163 paper presentations in international/national conferences and seminars, participated in more than 40 workshops/FDP's/QIP's, 25 courses from international universities of repute, organized more than 59 events with international and national academicians and industry people's, editor-in-chief and co-editor in journals, developed new courses. He has given talks at international universities, resource person in international conferences such as in Nanyang Technological University Singapore, Tashkent State University of Law Uzbekistan; KIMEP University Kazakhstan, All'ah meh Tabatabi University Iran, the Iranian Association of International Criminal law, Iran and Hague Center for International Law and Investment, The Netherlands, Northumbria University Newcastle UK,

Devanshi Soni is currently pursuing a B.Tech in Computer Engineering at the School of Engineering, Ajeenkya DY Patil University. Her research interests are metal nanoparticles and their applications and the integration of Artificial Intelligence across various domains. Devanshi is particularly interested in exploring how metal nanoparticles can be combined with AI technologies to address practical challenges and develop innovative solutions. She aims to advance research in these fields and contribute to groundbreaking developments that tackle real-world issues.

Suresh Waghamare is working as an assistant Professor & HOD of Pharmaceutics at Rashtriya College of Pharmacy Hatnoor, Tq. Kannad Dist. Aurangabad (MS) India. He specializes in employing bioactivity-driven techniques for extracting and identifying phytochemicals, ensuring the consistency of drugs and herbal remedies, and analyzing medicinal plants and their tissue cultures for bioactive compounds. Additionally, he has experience with various laboratory instruments such as UV spectrophotometer, Optical Microscope, Soxhlet Apparatus, Clevenger Apparatus, Sonicator, Rota evaporator, IR and techniques like Thin Layer and Column Chromatography. She focuses on formulating and assessing herbal products for their antimicrobial properties. Also, he filed for two patents. He is credited with nine publications in reputable national and international magazines, four books, and one book chapter.

Mohit Yadav is an Associate Professor in the area of Human Resource Management at Jindal Global Business School (JGBS). He has a rich blend of work experience from both Academics as well as Industry. Prof. Mohit holds a Ph.D. from Department of Management Studies, Indian Institute of Technology Roorkee (IIT Roorkee) and has completed Master of Human Resource and Organizational Development (MHROD) from prestigious Delhi School of Economics, University of Delhi. He also holds a B.Com (Hons.) degree from University of Delhi and UGC-JRF scholarship. He has published various research papers and book chapters with reputed publishers like Springer, Sage, Emerald, Elsevier, Inderscience etc. and presented research papers in national and International conferences both in India and abroad. He has many best paper awards on his credit too. He is reviewer of various international journals like Computers in Human Behavior, Policing etc. His areas of interest are Organizational Behavior, HRM, Recruitment and Selection, Organizational Citizenship Behavior, Quality of work life and role.

Poonam Yadav is working at ICAR-Indian Institute of Maize Research Institute.

Index

N

P

S

T

W

Z

www.ingramcontent.com/pod-product-compliance
Lightning Source LLC
LaVergne TN
LVHW081340300926
849161LV00008B/2032